SETAC Special Publications

Bergman, H.L., R.A. Kimerle, and A.W. Maki, Eds. (1985). *Environmental Hazard Assessment of Effluents*. Pergamon Press, Elmsford, New York, 366 pp. ISBN 0-08-030165-7

Cairns, J., Jr., Ed. (1985). *Multispecies Toxicity Testing*. Pergamon Press, Elmsford, New York, 261 pp. ISBN 0-08-031936-X

Dickson, K.L., A.W. Maki, and W.A. Brungs. Eds. (1987). *Fate and Effects of Sediment-Bound Chemicals in Aquatic Systems*. Pergamon Press, Elmsford, New York. 449 pp. ISBN 0-317-66334-8

Bodek, I., Ed. (1988). *Environmental Inorganic Chemistry: Properties, Processes, and Estimation Methods*. Pergamon Press, Elmsford, New York, 1280 pp. ISBN 0-08-036833-6

Huggett, R.J., R.A. Kimerle, P.M. Mehrle, and H.L. Bergman, Eds. (1990). *Biomarkers: Biomedical, Physiological, and Histological Markers of Anthropogenic Stress*. Lewis Publishers, Chelsea, Michigan. 347 pp. ISBN 0-87371-505-5

Walker, C.H. and D.R. Livingstone, Eds. (1992). *Persistent Pollutants in Marine Ecosystems*. Pergamon Press, Oxford. 272 pp. ISBN 0-08-041874-0

Dallinger, R. and P.S. Rainbow, Eds. (1993). *Ecotoxicology of Metals in Invertebrates*. Lewis Publishers, Chelsea, Michigan. 461 pp. 0-87371-734-1

Soares, A.M.V.M. and P. Calow, Eds. (1993). *Progress in Standardization of Aquatic Toxicity Tests*. Lewis Publishers, Chelsea, Michigan. 208 pp. ISBN 0-87371-845-3

Graney, R.L., J.H. Kennedy, and J.H. Rodgers, Jr. Eds. (1994). *Aquatic Mesocosm Studies in Ecological Risk Assessment*. Lewis Publishers, Chelsea, Michigan. 736 pp. ISBN 0-87371-592-6

Kendall, R.J. and T.E. Lacher, Jr., Eds. (1994). *Wildlife Toxicology and Population Modeling Integrated Studies of Agroecosystems*. Lewis Publishers, Chelsea, Michigan. 592 pp. ISBN 0-87371-591-8

The SETAC Special Publications Series

The SETAC Special Publications Series was established by the Society of Environmental Toxicology and Chemistry to provide in-depth reviews and critical appraisals on scientific subjects relevant to understanding the impacts of chemicals and technology on the environment. The series consists of single- and multiple-authored/edited books on topics selected by the SETAC Board of Directors and the Council of SETAC-Europe for their importance, timeliness, and their contribution to multidisciplinary approaches to solving environmental problems. The diversity and breadth of subjects covered in the series will reflect the wide range of disciplines encompassed by environmental toxicology, environmental chemistry, and hazard/risk assessment. Despite this diversity, the goals of these volumes are similar; they are to present the reader with authoritative coverage of the literature, as well as paradigms, methodologies, controversies, research needs, and new developments specific to the featured topics. All books in the series are peer reviewed for SETAC by acknowledged experts.

The SETAC Special Publications will be useful to environmental scientists in research, research management, chemical manufacturing, regulation, and education, as well as students considering careers in these areas. The series will provide information for keeping abreast of recent developments in familiar areas and for rapid introduction to principles and approaches in new subject areas.

Thomas W. La Point
The Institute of Wildlife and Environmental Toxicology
Clemson University, U.S.A.

Peter W. Greig-Smith
Fisheries Laboratory
Ministry of Agriculture, Fisheries, and Food, U.K.

Series Editors, SETAC Special Publications

Preface

With respect to soil contamination, recently there has been a growing awareness of the complexity of the problem and the intrinsic difficulties in measuring and assessing effects on soil organisms.

Related to this, there is an increasing interest in soil ecotoxicological studies, as expressed by the interest for the soil related sessions and presentations at the SETAC conference at Sheffield, April 1991. As a consequence SETAC and the publisher asked the sessions convenors to consider the joint, coherent publication of the oral and poster presentations concerning soil ecotoxicology.

This book contains these contributions arranged according to the type of research and the organism group:

- immission of contaminants to the soil
- research with microorganisms
- research with invertebrates
- research with plants
- research with higher organisms and food chains
- ecotoxicological risk assessment procedures

Moreover, to further structurize and balance the somewhat skewed distribution of the presentations over the different organism groups, the editors invited international experts to write introductory chapters to each group of contributions.

The opening chapter characterizes the special aspects of soil ecotoxicology, or as the title states "Ecotoxicology of soil organisms: seeking the way in a pitch dark labyrinth". The other introductions give a review of the characteristics of the soil ecotoxicological research within the topic; they are not meant to give a complete overview. Within each chapter the further contributions have been arranged in a more or less logical order.

Altogether we do have the feeling that the book gives a comprehensive picture of the different aspects of soil ecotoxicology and of the complex interrelations between abiotic and biotic variables.

All contributions have been thoroughly reviewed and we gratefully acknowledge the help of the following reviewers: H. Becker, F. Bairlein, T. Crommentuijn, P. Doelman, P.J. Edwards, W.H.O. Ernst, C.A.M. van Gestel, P.W. Greig-Smith, G. Jagers op Akkerhuis, C. Kula, O. Larink, P. Leonard, L. Posthuma, R.A. Prins, G. Purvis, S. Rundgren, F. Riepert, J. Römbke, W.H.M. Stigliani, N.M. van Straalen, G. Tyler, H. Vonk, J. van Wensem and J. Wiles.

We are grateful to M. Aldham-Breary for correcting the English texts.

M. Donker, H. Eijsackers and F. Heimbach

Marianne Donker is a Post Doc fellow of the Department of Ecology and Ecotoxicology of the Vrije Universiteit in Amsterdam, The Netherlands. She studied Ecotoxicology, Plant Pathology, and Microbiology at the Vrije Universiteit of Amsterdam and the Agricultural University of Wageningen, respectively. She received her Ph.D. degree from the Vrije Universiteit in 1992, her thesis focusing on metal tolerance in terrestrial isopods. After writing her thesis she joined the Department of Ecology and Ecotoxicology as a Post Doc fellow. Dr. Donker is a member of the Society of Environmental Toxicology and Chemistry (SETAC Europe) and of the European Society for Comparative Physiology and Chemistry (ESCPB). Her present research concerns the physiological basis of metal tolerance mechanisms in terrestrial invertebrates, such as isopods and collembola, and the consequences of adaptation for their population characteristics.

Herman Eijsackers was trained as ecologist and applied entomologist at the University of Leiden, working on parasitic wasps in biological control. At the Research Institute for Nature Management he worked on side effects of herbicides (2,4,5-T) on soil fauna, and enlarged his field of interest to effects of soil contamination on soil ecosystems. Since 1986 he is program director of the Netherlands Integrated Soil Research Programme, a multidisciplinary program to support protection policies of soils and sediments. He has been active in the Netherlands' Ecology Society (chairman) and WWF-Netherlands (vice-chairman), various advisory bodies of the Dutch government and a great number of courses and workshops in the framework of soil protection and especially soil ecotoxicology.

Fred Heimbach is a well-known research scientist at the Institute of Environmental Biology in the Crop Protection Division of Bayer AG in Monheim, Germany. He obtained his Masters degree and Doctorate in Biology from the University of Cologne in 1973 and 1976, respectively. From 1976 to 1979 he conducted basic research on marine insects at the Institute of Zoology, Physiological Ecology at the University of Cologne. Since 1979 he has worked at the Institute of Environmental Biology of the Bayer AG in his main field of interest, the side-effects of pesticides on non-target organisms. In addition to his work at Bayer, he gives lectures on ecotoxicology at the University of Cologne. Dr. Heimbach has done extensive research in development of single-species toxicity tests for both terrestrial and aquatic organisms, and has worked with micro- and mesocosms on development of multispecies tests for aquatic organisms. As an active member of both European and International Working Groups, he has participated in development of suitable methods for testing pesticides and other chemicals for their potential side-effects on non-target organisms. With these groups, Dr. Heimbach has worked on interpretation and use of such tests for hazard assessment.

Contributors

A.M.M. Abdul Rida
INRA/CNRS
Laboratoire de Zooécologie du Sol
Centre Emberger
1919 route de Mende BP 5051
34000 Montpellier, France

C. Abel
Federal Biological Research Centre
  for Agriculture and Forestry
Institute for Plant Protection in
  Fieldcrops and Grassland
Messeweg 11/12
D-38104 Braunschweig, Germany

F. Andreux
Centre de Pedologie Biologique
Centre National de la Recherche
  Scientifique
17, rue Notre Dame des Pauvres
F-54501 Vandoeuvre-les-Nancy
  Cedex, France

A.L. Atkinson
The Research Centre
The University of Luton
John Matthews Building
24 Crawley Green Road
Luton Bedfordshire LU1 3 LF,
  United Kingdom

B. Berger
Institut für Zoologie, Abt.
  Ökophysiologie
Universität Innsbruck
Technikerstraße 25
A-6020 Innsbruck, Austria

H. Borén
Department of Water and
  Environmental Studies
Linköping University
S-58183 Linköping, Sweden

M.B. Bouché
INRA/CNRS
Laboratoire de Zooécologie du Sol
Centre Emberger
1919 route de Mende BP 5051
F-34033 Montpellier, France

W. Büchs
Federal Biological Research Centre
  for Agriculture and Forestry
Institute for Plant Protection in
  Fieldcrops and Grassland
Messeweg 11/12
3300 Braunschweig, Germany

A. Calderbank
CITERA (Consultancy in Toxicology,
  Environmental and Regulatory
  Affairs)
Tiberon House
Woodlands Ride
Ascot, Berks SL5 9HN, United
  Kingdom

R. Dallinger
Institut für Zoologie, Abt.
  Ökophysiologie
Universität Innsbruck
Technikerstraße 25
A-6020 Innsbruck, Austria

G. Diaz Lopez
Avda Felipe II, 10
Centro Nacional de Sanidad
  Ambiental
Instituto de Salud Carlos III
28009 Madrid, Spain

P. Doelman
IWACO B.V.
Consultants for Water and
  Environment
P.O. Box 8520
3009 AM Rotterdam, The
  Netherlands

M.H. Donker
Department of Ecology and
  Ecotoxicology
Vrije Universiteit
De Boelelaan 1087
1081 HV Amsterdam, The
  Netherlands

H. Eijsackers
Programme Director
Netherlands Integrated Soil Research
  Programme
Postbus 37
600 AA Wageningen, The
  Netherlands

B. Förster
Battelle Europe
Am Römerhof 35
60486 Frankfurt, Germany

R.O.G. Franken
National Institute of Public Health
  and Environmental Protection
P.O.B. 1
3720 BA Bilthoven, The Netherlands

D. Fraters
National Institute of Public Health
  and Environmental Protection
P.O.B. 1
3720 BA Bilthoven, The Netherlands

A. Grimvall
Department of Water and
  Environmental Studies
Linköping University
S-58183 Linköping, Sweden

A. Gruber
Institut für Zoologie, Abt.
  Ökophysiologie
Universität Innsbruck
Technikerstraße 25
A-6020 Innsbruck, Austria

M.Z. Hauschild
Laboratoriet for økologi og miljølaere
Technical University
Bygning 224
2800 Lyngby, Denmark

F. Heimbach
Bayer AG
Crop Protection
Environmental Biology
51368 Leverkusen, Germany

U. Heimbach
Federal Biological Research Centre
  for Agriculture and Forestry
Institute for Plant Protection in
  Fieldcrops and Grassland
Messeweg 11/12
D-38104 Braunschweig, Germany

P.D. Hiley
Yorkshire Water Services
Biological Services
Knostrop S.T.W.
Knowsthorpe Lane
Leeds LS9 OPJ, United Kingdom

G. Jagers op Akkerhuis
Heussensstraat 70
2023 JS Haarlem, The Netherlands

P.C. Jepson
Department of Biology
University of Southampton
Biomedical Sciences Building
Bassett Crescent East
Southampton SO9 3TU, United
  Kingdom

K.C. Jones
Institute of Environmental and
  Biology Sciences
Environmental Sciences
University of Lancaster
Lancaster LA1 4YQ, United
  Kingdom

Th. Knacker
Battelle Europe
Am Römerhof 35
60486 Frankfurt, Germany

H.H. Koehler
University of Bremen, FB2
Research Group Ecosystems and Soil
  Ecology
P.O. Box 330 440
28334 Bremen 33, Germany

L.H.M. Kohsiek
National Institute of Public Health
  and Environmental Protection
P.O.B. 1
3720 BA Bilthoven, The Netherlands

C. Kokta
Federal Biological Research Centre
  for Agriculture and Forestry
Department for Plant Protection
  Products and Application
  Techniques
Biology Division
Messeweg 11/12
3300 Braunschweig, Germany

W. Kratz
Terra Protecta
Himbeersteig 18
D-14129 Berlin, Germany

J.C. Kühle
ITEC
Forschungstelle für terrestrische
  Ökotoxicologie GmbH
Grimlingerhauser 21
DW 4000 Düsseldorf 1, Germany

H. Kula
Institute of Zoology
Technical University
Pockelstrasse 10 A
D-3300 Braunschweig, Germany

O. Larink
Zoological Institute, Technical
  University, Braunschweig
Pockelstrasse 10a
D-3300 Braunschweig, Germany

J. Latour
National Institute of Public Health
  and Environmental Protection
P.O.B. 1
3720 BA Bilthoven, The Netherlands

P. Leeuwangh
Winand Staring Centre
Marijkeweg 22
6700 AC Wageningen, The
  Netherlands

P. Leonard
DowElanco
Letcombe Laboratory, Letcombe
  Regis
Wantage Oxon OX12 9JT, United
  Kingdom

H. Løkke
Director of the Department of
  Terrestrial Ecology
National Environmental Research
  Institute
Department of Terrestrial Ecology
P.O. Box 314
Vejlsøvej 25
DK-8600, Silkeborg, Denmark

W. Ma
Institute for Forestry and Nature
  Research (IBN-DLO)
P.O. Box 23
6700 AA Wageningen, The
  Netherlands

R. Mancha
Centro Nacional de Sanidad
 Ambiental, Instituto de Salud Carlos
 III
CRTA. Najadahonda, Pozuelo km.2
28220 Madrid, Spain

M. Mansour
GSF-Research Centre for
 Environment and Health
Institute for Ecological Chemistry
Schulstraße 10
D-85356 Fresing-Attaching, Germany

A. Marcinkowski
Battelle Europe
Am Römerhof 35
60486 Frankfurt, Germany

K. Mathes
University of Bremen
Research Group Ecosystems and Soil
 Ecology
P.O. Box 330 440
D-28334 Bremen, Germany

B. Metcalfe
Yorkshire Landscape Services
Knostrop S.T.W.
Knowsthorpe Lane
Leeds LS9 OPJ United Kingdom

R. Miyakawa
DowElanco
9002 Purdue Road
3939 Building
Indianapolis, Indiana

J. Moser
Institut für Zoologie, Abt.
 Ökophysiologie
Universität Innsbruck
Technikerstraße 25
A-6020 Innsbruck, Austria

P. Nilsson
Ecological Institute
Helgonavägen 5
S-223 62 Lund, Sweden

N.E.I. Nyholm
University of Umeå
Department of Public Health and
 Environmental Studies
S-901 87 Umeå, Sweden

J.P. Obbard
Aspinwall & Company
19th Floor
Trinity House
165–171 Wanchai Road
Wanchai, Hong Kong

R. Pöhhacker
Universität Bayreuth
Lehrst Uhl für Bodenkunde
Postfach 101251
D-8580 Bayreuth, Germany

D.M. Rawson
The Research Centre
The University of Luton
John Matthews Building
24 Crawley Green Road
Luton Bedfordshire LU1 3 LF,
 United Kingdom

R. Reiling
National Institute of Public Health
 and Environmental Protection
P.O.B. 1
3720 BA Bilthoven, The Netherlands

J. Römbke
Battelle Europe
Am Römerhof 35
60486 Frankfurt, Germany

D.R. Sauerbeck
Institute of Plant Nutrition and Soil
  Science
Federal Research Centre of
  Agriculture
Braunschweig-Volkenrode (Fal)
Bundesalle 50
D-3300 Braunschweig, Germany

I. Scheunert
GSF-Research Centre for
  Environment and Health
Institüt für Bodenökologie
Ingolstädter Landstrasse I
D-8042 Neurenberg, Germany

W. Sloof
National Institute of Public Health
  and Environmental Protection
P.O.B. 1
3720 BA Bilthoven, The Netherlands

A.-B. Steen
Department of Water and
  Environmental Studies
Linköping University
S-58183 Linköping, Sweden

P.B.M. Stortelder
Institute for Inland Water
  Management and Waste Water
  Treatment
P.O. Box 17
8200 AA
Lelystad, The Netherlands

P. Van Beelen
National Institute of Public Health
  and Environmental Protection
P.O.B. 1
3720 BA Bilthoven, The Netherlands

A.M.A. Van der Linden
National Institute of Public Health
  and Environmental Protection
P.O.B. 1
3720 BA Bilthoven, The Netherlands

C.A.M. Van Gestel
Department of Ecology and
  Ecotoxicology
Vrije Universiteit
De Boelelaan 1087
1081 HV Amsterdam, The
  Netherlands

N.M. Van Straalen
Department of Ecology and
  Ecotoxicology
Vrije Universiteit
De Boelelaan 1087
1081 HV Amsterdam, The
  Netherlands

P.L.A. Van Vlaardingen
National Institute of Public Health
  and Environmental Protection
P.O.B. 1
3720 BA Bilthoven, The Netherlands

J.A.C. Verkleij
Department of Ecology and
  Ecotoxicology
Vrije Universiteit
De Boelelaan 1087
1081 HV Amsterdam, The
  Netherlands

W. Völkel
GSF Research Centre for
  Environment and Health
Institute for Ecological Chemistry
Schulstraße 10
D-8050 Fresing-Attaching, Germany

J.A. Wiles
Department of Biology
University of Southampton
Biomedical Sciences Building
Bassett Crescent East
Southampton SO9 3TU, United
  Kingdom

J.W. Vonk
TNO, Institute of Environmental
  Sciences TNO
P.O. Box 6011
2600 JA Delft, The Netherlands

W.J. Willems
National Institute of Public Health
  and Environmental Protection
P.O.B. 1
3720 BA Bilthoven, The Netherlands

Contents

SECTION IV
Bioaccumulation and Food Chain Transfer

SECTION V
Ecotoxicological Assessment Procedures

SECTION I

Introduction

---

1. Ecotoxicology of Soil Organisms:
Seeking the Way in a Pitch Dark Labyrinth

CHAPTER 1

Ecotoxicology of Soil Organisms: Seeking the Way in a Pitch-Dark Labyrinth

H. Eijsackers

TABLE OF CONTENTS

0-87371-530-6/94/$0.00 + $.50

I. GENERAL INTRODUCTION

Soil ecotoxicology is typified by the complexity of the soil matrix both with respect to its diverse components and heterogeneous spatial arrangement. As a result there is a vast complex of interactions between soil organisms and, hence, between toxic compounds and soil matrix and organisms. The first section gives a description of the characterizing elements of ecotoxicological research and the different experimental approaches used with microorganisms, invertebrates, and plants. In the second section the functioning and composition of the biological soil system are described. Special attention is given to the spatial heterogeneity and temporal variability of soil biota, to the impact of different scales, and to the complexity of the soil matrix and the ways contaminants are present in the soil. The third section deals with the resulting exposure routes in the soil, with the different ways to measure and describe dose-effect relationships, and with the different end points which comprise levels of biological organization, varying from biochemical to ecosystem characteristics.

II. ECOTOXICOLOGY

A. Description and Definitions

Ecotoxicology is a relatively recent specialism of toxicological research. The different definitions used show that ecotoxicology can be approached from

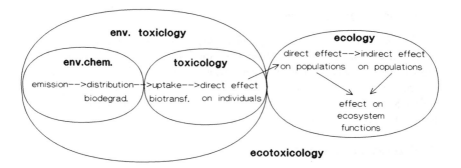

FIGURE 1. Different scientific fields within ecotoxicology arranged according to the source-path-target approach. (Murk, personal communication.)

different points of view. The main characteristics are the integration of chemical, toxicological, and ecological research, but viewed from various angles and with different emphases according to the primary background of the researchers.

Ecotoxicology started as an extension of classical toxicology towards the effects upon plants, animals, and microorganisms as formulated by Truhaut.[43] Koeman[31] defines it as a scientific approach integrating environmental chemistry, toxicology, and ecology. Butler[5] and Moriarty[35] emphasize the levels of biological organization and the interactions with the environment, whereas, according to Eijsackers,[15] the prime characteristic is the application of ecosystem-oriented functional and structural characteristics, and Van Straalen and Verkleij[48] give priority to the different aims: signalizing, analyzing, and predicting environmental problems.

Ecotoxicology comprises different scientific approaches and aims primarily at effects on (elements of) the natural environments (Figure 1). Furthermore, it combines mechanisms, processes, and responses from environmental chemistry, toxicology, and ecology. This combined approach is similar to ecological research which integrates the relation between organisms and their environment. Ecotoxicological research includes therefore the distribution and behavior of contaminants in the environment, the impact of a contaminant on the environment, and the impact on the organism and on the interrelation between organism and environment.

B. Major Elements of Ecotoxicological Research

Characterizing elements of ecotoxicological research are the difficulty in achieving a proper control situation, the complexity of field research, the combination of direct and indirect effects, and the different levels of biological organization involved. These elements will be described briefly.

1. A Proper Control Situation

To get a proper control is a problem; especially in field-oriented research, nonaffected areas will always differ from the influenced area. In experimental plots it is extremely difficult to match treated and control plots in all relevant aspects. A recent workshop on Soil Ecotoxicity Risk Assessment Systems[20] (SERAS) recommended working with matched sites, pretreatment data, regional trends, or (even) toxic-standard treatments.

2. Field Research is Complex

As the size of animal territories or home ranges can greatly surpass (and underpass) the scale of experimental fields, this will result in spatial complexation. Between-seasons and between-years variability in environmental conditions will further complicate a proper approach. Field incidents usually have to be assessed retrospectively, and consequently the data set will mostly be incomplete.

3. Direct as Well as Indirect Effects

Next to direct effects studied, indirect effects have to be taken into consideration. Relations between species may be changed, e.g., as a consequence of adverse contaminant effects on either prey or predator species. Therefore, different relationships, like predator-prey, host-parasite, and plant-mycorrhiza, but also more complex tri-trophic relations, like plant-phytophage-parasite,[6] have to be considered in this context. Changed interrelations may also result from changed environmental conditions due to the contaminant, e.g., a change in soil pH due to discharge of heavy metals in combination with acids by ore smelters or lime particles by copper and brass mills.[44]

4. Different Levels of Biological Organization

Ecotoxicology research comprises different levels of biological organization:

- The biochemical/physiological level (enzymes, hormone levels, ATP content, metallothioneins, DNA/RNA strings);
- The individual level (organ functioning, behavioral responses, growth rates);
- The population level (lethality, reproduction, genetic drift);
- The community level (species diversity, trophic and symbiotic relationships);
- The ecosystem level (Nitrogen-, Carbon, Phosphor- and Sulphur-cycles, microbial functional diversity, energy pools and fluxes).

In the SERAS system,[20] tests are grouped according to:

- Physical aspects like particle size, mineral composition, organic matter characterization
- Soil processes like nutrient and carbon cycling

- Community structure with characteristics like species diversity, abundance, and trophic structure
- Community functions including predation, mutualism, etc.

C. Different Approaches in Experimentation With Different Organism Groups

Within ecotoxicology the disciplines dealing with microorganisms, (in)vertebrates, and plants have different approaches. These are briefly described in this section.

1. Microorganisms

Soil microbiology mainly deals with natural microbial communities and to a lesser extent with individual species or strains. Microorganisms are studied at the community level with emphasis on the impact of functional responses, by what is called "microbial corsortia".

Functional responses can be general, like C-mineralization measured as soil respiration, or as more specific mineralization of a defined substrate, like glucose or glutamic acid or as enzyme activity. The last ones are selected as being part of the C-, N-, P-, or S-cycle: e.g., urease (N-cycle), sulfatase (S-cycle), and phosphatase (P-cycle). Techniques like most probable number (MPN) estimate the numbers of strains or species potentially able to break down specific substrates. The potential aspect of this capability is important as the majority of soil microorganisms in the soil are in an inactive state.

In microbial biodegradation studies, redox conditions like methanogenic, sulfate reducing, denitrifying, and aerobic conditions are used to define the electron acceptor involved in the conversion of the tested compound. If a specific electron acceptor is not involved, this may be due to the chemical degradation process or to intoxication of the organisms involved in that process.

2. Invertebrates

Soil invertebrate studies deal with the biochemical/physiological, individual, population, and community level. The great majority of laboratory studies are of individual species. More recently, species interactions are also studied in the laboratory, either in natural soil samples brought into the laboratory or in artificially composed systems with selected invertebrate species (e.g., Van Wensem).[49]

In experimental studies, lethality and reproduction are studied most extensively. During the last 5 years there has been growing attention for uptake, accumulation, and excretion mechanisms. This enables calculation of the total or effective load and to combine it with effect types and levels, as, for instance, accumulation in the reproductive organs will result in other effects than accumulation in neural tissue. Behavioral responses have been studied seldomly, although the heterogeneous distribution of contaminants in the soil supports the importance of avoidance behavior in combination with real exposure conditions.

There is growing interest in studying biochemical/physiological responses, so called "biomarkers", in invertebrates, while it comprises basic and generally valid response mechanisms and gives a good impression of the impact range of a toxicant. Walker[52] showed interesting differences as to the mixed function oxygenase (MFO), measured as monooxygenase activity in vertebrate and invertebrate groups correlated to their ability to degrade organochlorine pesticides. In field studies, species composition and population ecology characteristics of dominant invertebrate groups, like Lumbricidae, Collembola, Acari, and Carabidae, are studied.

3. Plants

Ecotoxicological research with plants mostly deals with survival, growth rate, and seed formation of individual species. Because of the large seed supply in the soil and seed dormancy, germination processes are important.

Much plant ecotoxicological research is carried out within crop protection. Pesticides are tested on different crops, weeds, and test plant species, which results in extensive ecotoxicological data sets.

Characteristic of this area of research is also the emphasis on physiognomic plasticity of plants: change of leaf size, leaf position or total leaf area index, the root to shoot ratio, or the internal transport from root to shoot. These last two reaction patterns influence uptake and internal transport of nutrients and of contaminants like heavy metals. Different root uptake systems stimulate or diminish uptake or immobilize contaminants after uptake. Interactions with different humus compounds are important in this respect.

Physiognomic responses have been classified as therophytes, geophytes, hemicryptophytes, chamaephytes, and phanerophytes by Raunkiaer.[38] Grime[23] described three stress strategies: competitive, tolerant, and opportunistic (ruderal). The first group is adaptive with a high leaf area index and root mass; the second group has an efficient use of resources, low growth rate, long lifetime, and low flowering and seed formation; and the third group has a high seed production, eventually to the cost of vegetative growth, and a short generation time.

A community-oriented approach is much experienced in plant sociological studies, but mainly deals with nutrient levels (eutrophication). Implications and potential of this plant-sociological approach for ecotoxicological purposes are not fully recognized.

Especially the root zones of plants (rhizosphere) interfere with soil processes. Here, root physiological and microbial rhizosphere processes intimately interact in decomposing organic material, thereby releasing mineral plant nutrient compounds. Specific microbe-plant root interactions, like V.A. mycorrhiza and ectomycorrhiza, play a dominant role in nutrient interactions. However, this field is only scarcely covered within ecotoxicology.

III. SOIL AND SOIL ORGANISMS

A. Definition of Soil Organisms

Soil organisms are defined as "organisms which live during an essential part of their life cycle in the soil". "Essential part" means that, e.g., surface-active beetles whose larvae complete their part of the total life cycle in the soil are part of the soil fauna, but organisms staying in the soil because of temporarily adverse environmental conditions above ground are not. "In" the soil means that surface-active insects do not belong to the soil organisms. However, in a number of cases the distinction between soil and surface (or terrestrial) fauna is still arbitrary, as can be seen above.

According to the different compartments in the soil, biota can be distinguished in pore water, mineral soil, or litter layer inhabitants. This has far-reaching consequences not only for survival strategies, but also for exposure to contaminants.

B. Activities of Soil Organisms

Soil organisms mainly contribute to litter breakdown. This is a concerted action of soil invertebrates and microorganisms. Invertebrates fragment (comminute) and partially solubilize in general organic material. Microorganisms mineralize complex organic molecules to simple molecules which can be taken up by roots, or further mineralized to CO_2 and H_2O.[18]

Comminution increases the total litter surface available for further attack. Moreover, it improves nutrient leaching from damaged plant cells. Both processes result in increased microbial growth and breakdown of organic cell constituents.

Litter consumption also results in an intimate mixing of litter fragments in the gut with the gut microflora, which further improves mineralization processes. Moreover, organic material is mixed through the soil profile by bioturbation.

Within microbiology, microorganisms are characterized according to their breakdown of specific substrates (MPN technique; see Section II.C). Microorganisms are also characterized by their functioning within specific nutrient cycles: nitrifiers, ammonifiers, sulfur-metabolizing organisms.

Similarly for soil invertebrates, functional definitions have been developed, like guild (functional similarity of taxonomically not closely related species) and league (functional similarity within the same soil layer). For earthworms are discerned: anecic, epigeic, and endogeic (Bouché)[3] with a different feeding strategy and niche. Anecic invertebrates like *Lumbricus terrestris* make permanent vertical burrows. At night they collect grass or leaf litter at the surface and draw it into their burrow so that the organic matter content alongside the burrow is elevated. Epigeics like *L. rubellus* move through the upper litter layer and comminute and consume litter. Endogeics like *Nicodrilus caliginosa* move through the upper organic-rich mineral layers and sieve out finely dispersed organic particles. Recently, Faber[21] suggested a system for euedaphic, hemiedaphic, and epiedaphic fungus-eating soil arthropods.

C. Numbers and Importance of Organism Groups

The impact of different organism groups greatly differs; the International Biological Programme quantified their roles in mineral cycles and energy flows. Figure 2 from Reichle[37] clearly emphasizes the importance of microorganisms by biomass and activities.

Recent research has revealed the importance of the promotive "grazing" by soil animals of organic matter, fungi, bacteria, and protozoa on nutrient cycling and energy flow within soil ecosystems.[18,21,51] Between different organism groups, numbers differ considerably (see Figure 3 and Table 1).

Mean numbers, moreover, give a mere indication as there are great fluctuations in numbers according to year, season, or even weather conditions, especially precipitation during prevailing days. The greatly differing sizes of soil organisms result in highly variable metabolic and growth rates and, hence, in temporal fluctuations in population numbers. Reproduction and growth rates of microfauna and microflora are very high, and large temporal fluctuations are normal. Microorganisms especially respond very quickly and massively to favorable environmental conditions, like increased moisture content after drought, as shown experimentally[36] and under natural conditions.[42] Next to these day-to-day responses there are seasonal effects related to temperature, precipitation, and availability of organic material through leaf fall. In general, the range between minimum and maximum numbers of soil microorganisms and soil fauna will be one order of magnitude irrespective of the mean population number.[10,19] However, individual situations, especially those concerning microorganisms, show ranges of three to four orders of magnitude.

These fluctuations pose very serious problems for assessment procedures: what is the reference population level? Is it the mean level, the lowest level, or something else?

As ecotoxicological effect assessment is based for a major part upon recoverability, reproduction, maturation, and growth rate, natural fluctuation rates have seriously to be taken into account.

D. Different Scales

Because of the huge differences in size of soil animals ranging from micrometers to decimeters (1:100,000), scale is an important item within soil ecotoxicology as it is in soil biology. Soil fauna is distinguished according to size in micro-, meso-, and macrofauna. In relation also habitat size will differ. Figure 4 schematizes body and habitat size together with feeding types: feeding on live plant material (phytophagous), dead organic matter (saprophagous), or live animal material (predaceous).

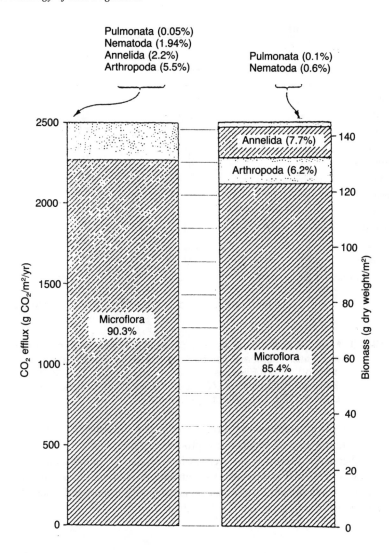

Pulmonata (0.05%)
Nematoda (1.94%)
Annelida (2.2%)
Arthropoda (5.5%)

Pulmonata (0.1%)
Nematoda (0.6%)

Annelida (7.7%)

Arthropoda (6.2%)

Microflora
90.3%

Microflora
85.4%

CO_2 efflux (g CO_2/m^2/yr)

Biomass (g dry weight/m^2)

FIGURE 2. General contribution of different soil biota groups to soil biomass and soil biological activity, expressed as CO_2 production. (From Reichle, D.E., in Lohm, V. and Persson, T., Eds., *Soil Organisms as Components of Ecosystems,* Ecol. Bull. (Stockholm) 25: 145–156, 1977. With permission.)

According to the habitat size, the environmental conditions will differ. The pH of a soil sample of 100 g is not really relevant for a soil bacteria living within a crevice of a soil particle of a few micrometers. Environmental conditions may change considerably over very short distances, partly because of the activities of the organisms themselves. Especially near roots and microorganisms, con-

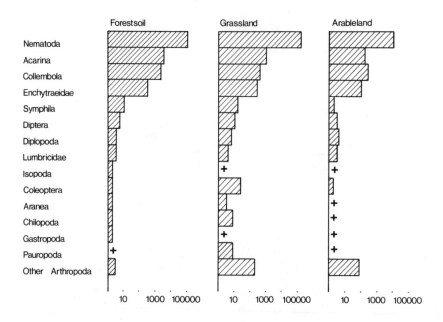

FIGURE 3. Mean composition of the soil fauna in a forest, meadow, and arable soil (mean numbers per dm² for the 0- to 30-cm layer; + means present). (From Eijsackers, H. and Van de Bund, C.F., in *Interactions Between Herbicides and the Soil*, Hance, R.J., Ed., Academic Press, London, 180, 255. With permission.)

ditions like pH change for one order over a few millimeters, which greatly influences microbial and biochemical process rates.

Another consequence of differences in size are the variable metabolic and growth rates. This will greatly influence uptake, accumulation, degradation, and excretion of contaminants. It also has consequences for temporal fluctuations in population numbers, as previously described.

Table 1. Vertical Distribution of Different Groups of Microorganisms in the Soil ($\times 10^3 \cdot g^{-1}$ soil)

Depth (cm)	Aerobic bacteria	Anaerobic bacteria	Actinomycetes	Fungi	Algae
3–8	7800	1950	2080	119	25
20–25	1800	379	245	50	5
35–40	472	98	49	14	0.5
65–75	10	1	5	6	0.1
135–145	1	0.4	—	3	—

From Eijsackers, H., van de Bund, C., Doelman, P., and Ma, W., *Fluctuating Numbers and Activities of Soil Organisms*, Rapport 88/33, Rijksinstituut voor Natuurbeheer, Arnhem, 1988, 85 pp. With permission.

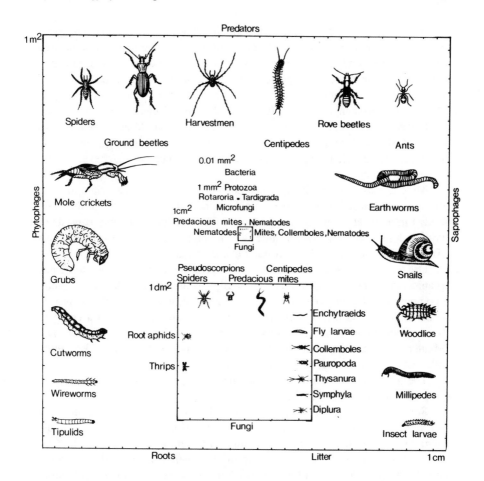

FIGURE 4. Schematic survey of the soil fauna community. Squares indicate habitat size and relevant sampling area. Species are arranged according to feeding type. Important species are drawn at scale. (From Eijsackers, H. and Van de Bund, C.F., in *Interactions Between Herbicides and the Soil,* Hance, R.J., Ed., Academic Press, London, 1980, 255. With permission.)

E. Complexity of the Soil Matrix

Soil comprises solid, liquid, and gas compartments, each with a diverse composition. The solid comprises mineral particles and organic material, the liquid water and dissolved ions nutrients and dissolved organic carbon, the gas different gases and volatile organic components. Moreover, the constituents are arranged in a certain order and according to particle size in a certain structure and texture.

In general, soils are heterogeneous, which can be observed at different levels of soil studies, ranging from intermingled soil types in river bank areas, layered soil profiles like Podzols, to differences at meso- and microscale (cm^2 to μm^2). In nature reserve plots with apparent homogeneous soil and vegetation types, considerable and discontinuous differences in standard chemical conditions (pH, OM%, cation-exchange capacity [CEC]) and chemical element contents could be observed over distances of 1 m. These differences could be more than two times the standard deviation to both sides.[12] A study of homogenized sandy grassland soils revealed impressing differences in water content over even shorter (10-cm) distances.[4]

Mineral particles are distinguished according to particle size class in clay, silt, and sand. Based on the mineral background, silt and sand are further sub-divided into silicates, micas, and feldspars. Clay (or lutum) mainly consists of clay minerals such as kaolinite, illite, and montmorillonite. The clay minerals differ in CEC, surface and electrostatic charge capacity.

Organic materials can be subdivided into dead and live organic matter. Dead organic matter consists of *H*umus, partially decomposed rohhumus, slightly decomposed (= *F*ermented) litter and *L*itter material, indicated in the soil profile by the L, F, and H layers. Organic matter, especially humus, has a complex chemical structure with very diverse binding capabilities (see Eijsackers and Zehnder).[18] Live material constitutes all organisms listed in Figure 3 together with roots and other underground plant parts.

The total composition of inorganic and organic, live, and dead material is given for a grassland soil in Table 2.

Soil water, specifically soil pore water, has a different composition from rain water and "free" soil water due to chemical exchange processes at the microscale and local changes in environmental conditions. As an example, CO_2 produced by soil respiration is taken up in the water film and will cause a pH shift. Between different soil types, water-holding capacity differs considerably.

Soil air is an overlooked, but major component of soils, differing in its mean composition slightly from atmospheric air due to production of CO_2 and other volatile organic compounds. At microsites, however, there can temporarily be considerable differences due to anaerobic conditions caused by metabolic O_2 consumption. The percentage of soil air changes according to soil type and water content. At field capacity (when a soil is saturated with water and adsorption, leaching, and evaporation are at balance) sandy soils have 30 to 40 vol% air, loamy soils 10 to 25%, and clayey soils 5 to 15%.

Because of the variable composition and spatial setting (texture) of the soil, physical and chemical characteristics will vary (Section III.C). Particle size greatly influences total surface area.

As a consequence in the habitats of different soil organisms, contaminants can be present in different ways. Attempts to simplify experimental conditions by homogenized soil samples or artificial soils underestimate the real exposure in natural soil conditions.

Table 2. Composition of a Soil of a Grassland on Dry Weight Base

Soil	mineral soil 94%			
	organic matter 6%	dead org. m. 85%		
		living org. m. 15%	living roots 8.5%	
			edaphon 6.5%	bacteria + actinomycetes 50%
				fungi 25%
				worms 14%
				macrofauna 5%
				mesofauna 2.5%
				microfauna 3.5%

From Dunger, W., *Tiere in Boden*, Ziemsen Verlag, Wittenberg Lutherstadt, 1984, 265 pp. With permission.

F. Presence of Soil Contaminants

Soil contaminants comprise organic and inorganic compounds, which can be persistent or degradable. Persistent compounds may gradually diminish from the upper soil layers by leaching, although for certain compounds (i.e., heavy metals) this takes very long periods (centuries or millennia?).

Degradation of a compound does not mean that the toxic problem has disappeared; metabolites can be quite more toxic than the parent compound. Measuring only the parent compound may result in a false positive conclusion on disappearance of the compound; therefore, degradation and leaching studies on metabolites have to be included when relevant.

Contaminants are present in the soil as distinct particles, dissolved in the soil water, vaporized in the soil air, adsorbed to mineral/organic particles, and absorbed in mineral/organic particles. Each has its own specific behavior under various environmental conditions and is subject to different physicochemical processes. Contaminants adhering to mineral particles are more subject to changes in pH and redox potential than if adsorbed to organic material. Superficially adsorbed contaminants can be more easily reached by microorganisms than when adsorbed within the very fine soil pores within soil aggregates. Contaminants absorbed within organic particles may be mobilized through breakdown of the organic matter. Contaminants adsorbed to organic colloids leach faster by facilitated transport of these colloids. As a consequence the distribution of contaminants within the soil profile will vary.

IV. DOSE-EFFECT RELATIONSHIPS

Dose-effect relationships comprise the phenomena and processes involved in the exposure of an organism to a compound, the ways to describe the relation between the actual dose and the observed effects expressed in functional target organisms and end points.

A. Exposure Routes

The complex and heterogeneous distribution of contaminants and organisms leads to different exposure routes, depending further on soil characteristics, morphological characteristics of the soil organisms, and ecological characteristics such as food choice.

Contaminants can be bound to mineral or organic particles, dissolved in soil water, and volatilized in soil air. Partitioning between mineral particles and soil air and soil water can be described according to Henry's law by K_d. For dissociating chemicals also pK_a and pH are of importance. Volatilization is described in different basic physicochemical rules, but has not been applied for uptake by soil organisms. Exchange between organic chemicals and fatty tissues (in general hydrophobic chemical characteristics) can be described by the K_{ow} partitioning coefficient.

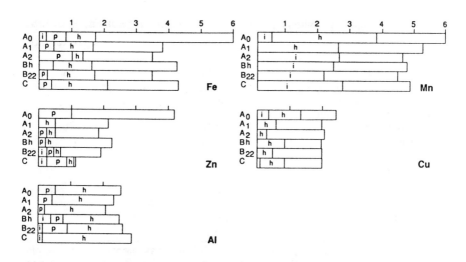

FIGURE 5. Speciation of heavy metals and metalloids in the soil pore water of a podzolic sandy soil (i = inorganic/ionic, p = phenol-bound, h = humic or fulvic acid-bound; lengths of bars represent total content). (From Van Straalen, N.M. and Verkleij, J.A.C., Eds., *Leerboek Ecotoxicologie*, VU Publ., Amsterdam, 1991, 422 pp. With permission.)

For soil microorganisms morphological characteristics are important for uptake through the outer membrane or enhanced efflux processes. Formation of exocellular enzymes or extracellular complexation processes play a role by, e.g., forming organometallic complexes. These complexes are in general more mobile than metal ions or salts. This improves mobility of these metals in the soil and, on the other hand, hampers uptake. Also, formation of low-solubility metal salts such as sulfides or oxalates can decrease uptake. Because these processes act in different directions it is difficult to properly quantify the total compound balance with respect to mobilization, uptake, and excretion.

Plant roots have effective uptake mechanisms. Exudation of organic compounds such as citric and pomic acid may enhance the mobilization of heavy metals, as formation of organometallic complexes by phenols, and fulvic and humic acids in organic matter as illustrated in Figure 5.

For soil invertebrates the distinction between hard and soft bodies is important, as the epidermis structures influence contaminant uptake. Soft-bodied organisms (e.g., Protozoa, Nematoda, Lumbricidae) greatly rely on soil pore water as niche and exposure chances are greater than as shown for earthworms with chlorophenols[45] and copper.[40]

Hard-bodied species with trachea systems may become affected by uptake of vaporized contaminants.

Intake by food is a major route in hard-bodied organisms such as Carabidae[14] and Aranea (Jagers op Akkerhuis, personal communication), as well as cleaning with mouth parts of antennae and legs which have come into contact with chemicals.

Table 3. Storage Organelles For Heavy Metals in Different Soil Fauna Groups

	Storage	Excretion
Earthworms		
Ca — gland	Pb Zn	Zn
Body wall chloragosomes	Pb Cu Cd Zn Fe	
Body wall surface	Cd	
Body wall mucoid coat	Zn	Zn
Waste nodules	Pb Cd	Pb
Intestine		Zn Pb Fe
Woodlice		
Hepatopancreas B cells	Fe Zn Pb	
S cells	Zn Cd Pb Cu	Cu
Metallothioneins	Cd	
Springtails		
Cuticula	Pb	
Gut epithelium	Pb	

From Eijsackers, H. and Doelman, P., in *Ecological Assessment of Environmental Degradation, Pollution and Recovery,* Elsevier, Amsterdam, 1989, 245. With permission.

Soil vertebrates take up chemicals through their skin, lungs, or via the food, the last being the most important routing.

For the internal metabolism of heavy metals, soil microorganisms use excretion mechanisms, oxidation processes (oxides are less toxic), trapping mechanisms with intracellular polymers such as metallothioneins, binding or precipitation to the cell wall by electrostatic binding to specific groups of the cation exchange complex (carboxyl groups to pectins and specific proteins), and biomethylation.[8]

For plants, decreased membrane permeability, increased efflux, extracellular precipitation and complexation, and intracellular detoxication are relevant.[48]

In invertebrates, uptake via organic particles is mainly associated with food uptake for which gut processes are essential. Moreover, specific soil fauna groups have specific metal-accumulating organs or organelles (Table 3) which result in different kinetic patterns (Figure 6).

Also, ecological characteristics influence exposure. Because of the heterogeneous distribution of contaminants, mobile organisms can avoid contaminated spots if they are able to observe the contaminant. Moreover, food choice and food-searching behavior are important. In different earthworm feeding groups, contaminants bound to mineral particles are taken up by endogenic species such as *N. caliginosa,* and contaminants bound to organic matter are consumed by litter feeders such as *L. rubellus* and *L. terrestris.* Contaminants in the upper soil profile layers are taken up primarily by litter species. The chances for surface-active earthworms to become affected by contaminants at the soil surface will be greater with moist than with dry weather; under dry conditions earthworms will withdraw into their burrows or become immobile.

Food chain transfer is essentially exposure via the food. For the observed magnification of contaminants in different food chains, various mechanisms have been suggested: biomass transfer, body size, metabolic rate (absolute rates and in relation to body size), specific accumulation/excretion mechanisms, and food choice. Heavy metal transfer in terrestrial and soil food chains is mainly steered by accumulation/excretion mechanisms and food choice.[29,34] The classic mechanism of bioconcentration due to biomass transfer in relation to metabolic efficiency (10% food assimilation rate results in a tenfold magnification in each step of the food chain) is therefore questioned in recent literature.

B. Dose-Effect Curves and Formula

1. Dose

Dose-effect relations can be expressed[48] as:

- The absolute dose (single or various sequentially administered applications; in the latter case dose rate is important)
- The relative dose or concentration (related to the size of the organism)
- The internal dose (the absolute or relative dose actually taken up by the organism)
- The substrate concentration (only an exposure concentration, not the actual or internal dose)
- The logarithm of the absolute or relative dose (because with log dose a zero level does not exist, an extremely low amount of the compound is taken as "zero", or reference level)

For natural compounds such as heavy metals reference levels are related to the mineral composition of the soil. Lexmond et al.[32] have derived a relation between the heavy metal and clay and humus content of a soil which enables calculation of the natural derived background level and consequently the amount added through anthropogenic activities. For organic compounds this may perhaps also be possible, but similar formulas have not been derived yet. For most organic compounds it may be assumed that they are for the greater part of anthropogenic origin, in which case the detection limit is used as policy standard. The application of sumparameters such as extractable organic chlorine compounds, assuming again the major anthropogenic origin of organochlorine, is criticized because substantial amounts of natural organochlorines may be involved as also suggested recently for dioxins.

2. Effects

Effects can be dichotomous or graded. Dichotomous responses give the frequency of a certain effect (lethality, behavioral abnormalities), whereas graded responses give a certain effort, such as the number of eggs produced.

Dose-effect curves can take different forms (Figure 7):

- With biologically necessary or useful compounds at first there is a limiting part (1a) in which the chemical has a positive effect and dose increase results in a positive increase of the response.

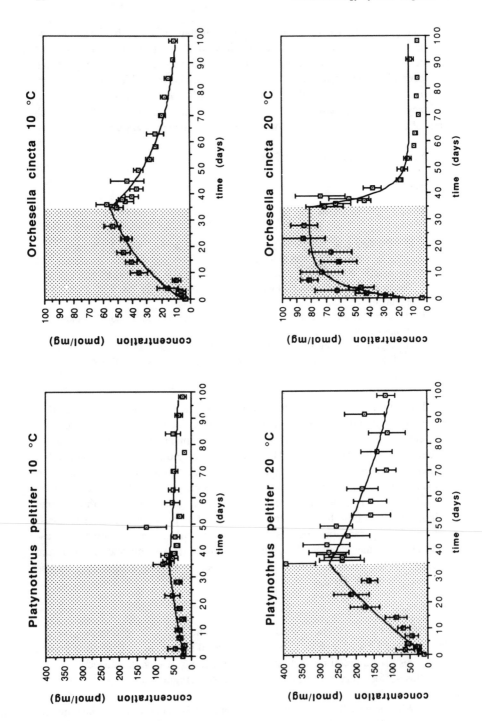

- With xenobiotic compounds (which are not necessary or useful) there will be a neutral trajectory (1b) with no effect at all.
- In the next part of the trajectory there will be an optimum phase for biotic chemicals (2a) or it may be a further extension of the neutral part (2b). This neutral part is also called the lag phase.
- The third part of the trajectory expresses the negative impact of a biotic or xenobiotic chemical. The decline in response can be either steep (3a) or gradual (3b), which has implications for the accuracy with which an effect will be statistically significant.
- Certain chemicals also cause a slight positive effect just before the negative decline (4), which is called "hormesis".

Dose-effect curves can be described as linear, a decline with a threshold value, or a gradual decrease according to a sigmoid curve. A linear relation supposes a single-hit model: each molecule has a fixed chance to cause a laesis or damage and the number of nondamaged cells/organisms decreases exponentially with concentration. The threshold relation supposes a certain regulation, buffering, or protection mechanism, and it may look almost similar to the sigmoid curve. The sigmoid curve, however, supposes an effect with a specific variance between the individual cell/organisms. The principal difference between the threshold and sigmoid relationship is that the sigmoid does not have a no-effect level, as the response decline will become smaller and smaller with decreasing doses, but will never become zero.

Sigmoid curves can be fitted by a probit (a cumulated normal distribution), a logit (a cumulated log-normal distribution), or the Weibull function. All three methods allow for the calculation of an effect level (EL) of EL_{50}, EL_{90}, or EL_{99}, but not of a no observed effect level (NOEL) as explained above.

Because the probit calculation as such is complicated, although now facilitated by computer programs, simplified methods were developed in the same period by Spearman,[39] Kärber,[30] and Thompson.[41] These are still useful to obtain quick estimates of LD_{50} in particular when its order of magnitude is unknown.

Also, sequential methods were developed[7] in which individual organisms are treated one after another, starting with the expected LD_{50} dose which decreases or increases one dose step, depending on a positive or negative response in the preceding test. These methods are time consuming, use less experimental organisms, and give the approximate median lethal dose, not the slope of the effect curve.

In medicine testing, a safety index is calculated by dividing LD_1 (1% probability of causing lethality) by ED_{99} (99% probability of causing the desired

FIGURE 6. (*Opposite*) Accumulation and elimination of cadmium in a mite (*Platynothrus peltifer*), a collembole (*Orchesella cincta*), a pseudoscorpionid (*Neobisium muscorum*), and a carabid (*Notiophilus biguttatus*). Measured data with S.E., line according to fitted model. (From Janssen, M.P.M., *Turnover of Cadmium Through Soil Arthropods*, Thesis, Free University, Amsterdam, 1991.)

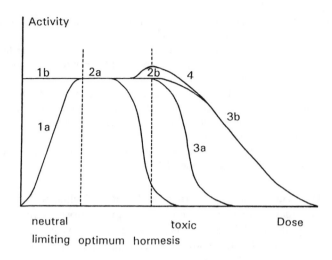

FIGURE 7. The dose-effect curve in its various forms (further explanation in the text).

therapeutic effect), thereby relating the positive and negative impacts of a compound. This is worth considering for other chemical compounds with an intended positive effect, such as pesticides.

Within the sigmoid curves the EL_{50} or ED_{50} is at the point where slope is at its maximum; small variations in dose will correspond with the largest variations in effect then. The slope of the curve affects the precision of the estimate, too. A steep slope permits a narrower dose trajectory for the ED_{50}.

The variation between the different measurements at one dose step is minimal around the ED_{50} level and maximum at the tails of the curve (ED_{99} and ED_1). The same holds for threshold curves. This is especially relevant for the NOEL, as this level is defined as the next dose step which has a statistically significant difference with the lowest observed effect level (LOEL) dose. Depending on the variance of the results at each dose step this significant interval between the LOEL and NOEL dose will vary. Small samples imply the risk that relatively large doses appear without effect, and that the use of greater samples makes it more likely that the NOEL will decrease.

As an alternative, Babich et al.[1] suggest the ecological dose range (the range at which effect changes from ED_{90} to ED_{10}), together with the ED_{50} and its coefficient of variance. Glowa[22] determines the dose with an expected measurable effect in a defined small proportion of the population (i.e., 10% effect in 1 out of 1000 individuals). Hoekstra & Van Ewijk[25] suggest a two-step procedure: first the 25% effect concentration to be established with reasonable certainty and precision, followed by linear extrapolation to a concentration where the effect is acceptably small. Extrapolating the slope in the middle part to 100% is questionable, as dose-effect curves over a wide dose trajectory show variable slopes in different parts of the curve.[44]

Another way to overcome variance is by using larger test groups close to the expected NOEL/LOEL level. Normally, a probit or logit test will comprise an equidistant dose trajectory and treatment groups of constant size. Alternatively, a range finding test should provide preliminary sensitivity data, followed by a second test with variable distances between the doses and group sizes.

In air pollution research in which dose fluctuation in time can be extremely large, an effect-limit curve is constructed. For exposure by long-lasting fluctuating air pollutants the percentile value is used. This expresses the percentage of observations encompassing the given value: perc98 comprises 98% of all observations, hence, 2% of the observations exceed the percentile level.

This may also be useful in aquatic, especially stream water ecotoxicology with fluctuating contamination, but perhaps also in terrestrial ecotoxicology with pesticides applied very intensively at a limited scale. Under these conditions, that specific area will have a certain diffuse background level due to the frequent application of the pesticide which is exceeded temporarily in that specific lot shortly after application.

With nonpersistent chemicals, exposure is related to gradual disappearance by degradation or leaching. When the application or emission rate is higher than the degradation or leaching rate the total dose or load will increase (accumulation). The relation between a (standard) dose-response curve and exposure duration must therefore be taken into account, like the effect-limit curve in air pollution research.

C. Functional Target Groups and End Points

The ultimate species protection from an ecological point of view should be prevention from extinction. Given the time scale, it is difficult to measure extinction. Intrinsic survival capacity (or intrinsic population growth capacity) and colonization capacity could provide approximations. And ecotoxicity testing should focus on the potential (local) disappearance of a particular species. At the ecosystem level, gradual extinction of interconnecting species will result in malfunctioning of the system. Disturbance of species equilibria have been shown after frequent application of pesticides which eliminate not only specific pathogens, but also their natural antagonists. This stimulates new pathogens which have to be treated next.

Selection of ultimate aims or end points depends on the purpose of the assessment. Test systems can be preventive or auditing, respectively assessing potential compound toxicity before release or actual toxicity of combinations of toxic compounds already present in contaminated areas.

Another distinction can be made for the duration of the impact and the nature of the area being undisturbed, temporarily disturbed (e.g., pesticide-treated arable land), or permanently disturbed (some mining and industrial areas). Specific sets of responses can be defined for each area related to natural and anthropogenic (= naturally derived) functions of soils (Table 4).

Table 4. Relevance of Soil Tests to Different Types of Habitat and Different Reasons For Protecting and Monitoring Soil

	Natural (protected)	Temporarily stressed (managed)	Permanently stressed (contaminated)
Use for roads and buildings	PHYSICAL	PHYSICAL	PHYSICAL
Soil fertility			
Nutrient cycles		PROCESS	
Primary production		STRUCT/FUNCT	
Groundwater retention and quality	PHYS/FUNCT/PROC	PHYS/FUNCT/PROC	PHYS/FUNCT/PROC
Cycling of essential elements	PROCESS	PROCESS	PROCESS
Biological conservation	STRUCT/FUNCT	STRUCT/FUNCT	?
Monitoring environmental change	PROC/FUNCT/STRUCT	PROC/FUNCT/STRUCT	
Biodegradation	PROCESS	PROCESS	PROCESS

Note: The terms of PHYSICAL, PROCESS, FUNCTIONAL, and STRUCTURAL test characteristics are described below:

- Physical aspects (particle sizes, mineral composition, organic matter)
- Soil processes (nutrient and carbon cycling)
- Community structure (species diversity, abundance, and trophic structure)
- Community functions (predation, mutualism)

From Eijsackers, H. and Løkke, H., SERAS Soil Ecotoxicity Risk Assessment System, report from a European workshop, January 1992, NERI Silkeborg, 1992, 52 pp. With permission.

These tests can be grouped as physicochemical characteristics, system processes, community structures (e.g., species diversity), and community functions (interacting species relations). From a biological point of view, tests could be classified according to the level of biological organization: cell and tissue, organ, individual, population, community, and ecosystem.

Testing at the cell and tissue level, primarily aiming at biochemical and physiological processes, has already been described in Section II.C under microorganisms and invertebrates. These types of end points, sometimes called biomarkers, have received growing interest recently. Our knowledge of the extent to which these processes have, on the one hand, a general meaning for various types of organisms or, alternatively, can discriminate between different organism groups is still limited. The study of Walker[52] on MFO indicates the potentials of this approach. Distinct similarities and differences in MFO activity within and between different vertebrate and invertebrate groups, respectively, markedly coincide with the abilities of these groups to break down organochlorine pesticides and, hence, become affected.

Individual responses can be easily measured experimentally, mostly in the laboratory. Besides, it is possible to link external and internal exposure as well as kinetical behavior to the type of effects. According to Bengtsson et al.,[2] distribution of heavy metals over specific organs of earthworms coincides with the sensitivity of the responses: high gonad contents will primarily influence reproduction capacity, while high contents in nerves and ganglia affect mobility and avoidance responses, and presumably colonization capacity.

Individual responses, however, give partial answers not to be extrapolated to the population level. The mite *P. peltifer* and the springtail *O. cincta,* for instance, have cadmium EC_{50} levels of 7.3 and 1.6 $\mu mol \cdot g^{-1}$. Moreover, the springtail shows a decreased egg production with increasing Cd content of the food and the mite a reduced growth rate[47] with distinct consequences for the intrinsic population growth rate, contrary to the EC_{50}. The growth rate response coincides with field observations around heavy metal smelters where *O. cincta* can maintain itself, whereas *P. peltifer* declines.

Another population response is the intrinsic biomass turnover rate. This is also relevant for food chain transfer (given the compound transfer rates are mainly dominated by biomass transfer) and can be used for modeling of ecosystem responses on energy flow and nutrient cycling.

A special application of population parameters could be r and K selection. As an ecological concept it is criticized, but it may relate survival strategy and stress responses. Table 5 summarizes behavioral, survival strategy, and exposure differences of earthworms related to r and K selection. More r-selected species are more sensitive for reproduction and K-selected species for survival. In a field study by Brown and Edwards (1982), the r-selected *L. festivus* recovered well after a pesticide application, whereas the K-selected *L. terrestris* did not.

Colonization and succession responses of soil organism populations have hardly been studied, although these are directly related to extinction. Colonization

Table 5. Survival Strategies of Different Earthworm Species Expressed in Cu EC$_{50}$ Values For Lethality, Growth, and Reproduction in Relation to Differences in Behavior and Exposure to Soil Contaminants

	Type	Strategy	Substrate	Behavior	Cocoons · yr^{-1}	Maturity time (weeks)	Offspring · yr^{-1}	LC$_{50}$ ng · g	EC$_{50}$ growth	EC$_{50}$ cocoon prod
N. caliginosa	endo	K	Mineral soil, soil org. matter	Siever	35	50	30	>300	405	27
A. chlorotica	endo	K/r	Surface soil	Siever	31	29	27	127	121	28
L. terrestris	anec	K/r	Burrows, litter/soil org. matter	Collector	41	50	35	161	139	—
L. rubellus	epig	K/r	Litter/dung	Crawler	94	40	80	205	240	80
Eisenia fetida	epig	r	Litter/dung heaps	Crawler	140	9	350	231	285	191

has been investigated with surface-active beetle and spider species and with earthworms.[24,26,33] Behavioral observations in the laboratory, combined with data on field dispersal and recolonization could assess the rehabilitation of temporarily stressed (e.g., after pesticide application) remediated (cleaned) soils.

Early succession processes may provide ecological insight in recoverability potentials. Succession of soil arthropods has been followed in mine spoil heaps[27] and reclaimed industrial sites.[28] Litter bag studies[16] provide good possibilities to quantitatively study early succession of arthropods in litter in the field in relation to litter breakdown. As such it connects population (succession) and ecosystem (litter breakdown) responses.

Ecosystem responses can be grouped in:

- Soil processes such as energy flow and nutrient cycling
- Community structure measured as species diversity, abundance, and trophic structure
- Community functions such as predation, mutualism, and commensalism

Soil processes are mainly studied as metabolic characteristics: biomass turnover; C- and N-cycles measured per organism group, as general soil respiration, or as induced substrate respiration; specific enzyme levels representative for certain nutrient cycles, etc. Structural responses can be expressed as the total number or abundance of species present. In various diversity measures the number of species is combined with the number of individuals per species. In recent literature a more functional grouping in leagues and guilds has been developed (see Section III.B).

Community functions have been studied within food chain research. However, other types of interaction, such as mycorrhiza, microbial antagonism, host-parasite relations as used in biological control, and prey-predator processes have hardly received any attention within an ecotoxicological context.

An example of such impacts on prey-predator processes is the differential influence of pesticides on springtails and predatory mites resulting in a distinct decrease of the last group and a resulting increase of the first.[13] This kind of indirect consequence should be properly confirmed experimentally. Only few experimental studies on predator-prey interactions are reported in the literature.[14,29]

Functional grouping as suggested for arthropods by Van Straalen et al.[46] and Faber[21] could combine soil processes and community structure. Doelman et al. (in press) investigated functional diversity by measuring the capacity of communities/consortia of microorganisms which had been cultured from clean or contaminated soils, to degrade organic substrates of different complexity. In clean soil, more strains were capable of degrading these substrates than in contaminated soil, which indicates that the functional diversity was reduced. Moreover, it suggested that adaptation to contamination may have a certain price: a loss in physiological potential.

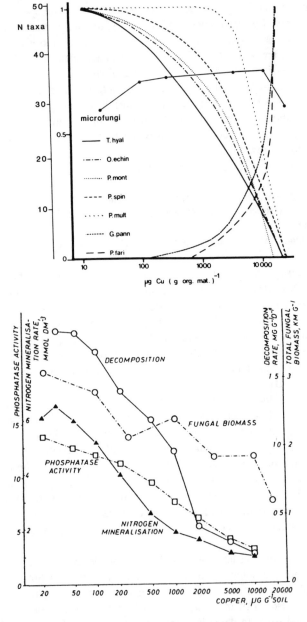

FIGURE 8. Changes in the composition (total number of microfungi taxa and relative appearance of seven dominant fungal species) and various microbial activities along a copper gradient in the soil around a copper brass mill. (After Tyler, G., *Ambio*, 13, 18, 1984. With permission.)

The application of reference or key species enables the combination of laboratory experiments, field studies, and biomonitoring by using a limited number of key species. Selection of key species is very sensible, as final assessment otherwise may become based on species which are more relevant from a political than an ecological point of view.

To study species interactions, microcosms may be promising tools. Experiments on the use of litter microcosms with microorganisms and isopods for assessment procedures[49] provided variable results, however, and did not indicate an overall greater sensitivity than single-species experiments.

The need for an integrated approach combining the aforementioned three groups of responses is illustrated by the results of a study in the surroundings of a copper and brass mill (Figure 8).

Structural aspects such as the total number of fungal fruiting appear to stay unaffected by very high copper contents. This is partly explained by the considerable differences in sensitivity of individual fungi species; decline and disappearance of one fungal species leaves room for a more resistant species which takes over. This shift in community composition has its consequences for system functions as shown by the instant decline of various microbial process parameters. Moreover, these various functions show different response rates, suggesting different survival strategies: fungal biomass decreases slower than specific enzyme levels, suggesting that maintaining biomass is a primary aim. Therefore, in general the best approach seems to be to combine functional and structural responses.

Within the ecotoxicological risk assessment procedure used by the Dutch and Danish authorities it is assumed that protecting the structure will implicitly protect ecosystem functions. Therefore, the primary starting point is to protect the structure of the system by protecting as much individual species as possible. The Gusum example illustrates that it seems wise to combine all three approaches and study soil processes, system structure, and system function in an integrated way.

To provide a more practical framework, standardized tests for effects measured at the individual and population levels could provide data for first-tier testing. Moreover, these tests may provide background data for more complicated measurements, such as intrinsic population growth rate and biomass turnover rate, to be used in further tier testing, aiming at a more integrated approach.

REFERENCES

1. Babich, H., Bewley, R.J.E., and Stotzky, G., Application of "Ecological Dose" concept to the impact of heavy metals in some microbe-mediated ecological processes in soil, *Arch. Environ. Contam. Toxicol.*, 12, 421, 1983.
2. Bengtsson, G., Nordström, S., and Rundgren, S., Population density and tissue metal concentration of Lumbricids in forest soils near a brass mill, *Environ. Pollut. Ser. A*, 30, 87, 1983.
3. Bouché, M.B., Stratégies lombriciennes, in Lohm, U. and Persson, T., Eds., *Soil Organisms as Components of Ecosystems*, Biol. Bull. (Stockholm) 25: 122, 1977.
4. Bronswijk, J.J.B., Dekker, L.W., and Ritsema, C.J., Preferent transport van water en opgeloste stoffen in de Nederlandse bodem: meer regel dan uitzondering?, *H2O*, 23, 594, 1990.
5. Butler, G.C., *Principles of Ecotoxicology*, Scope 12. John Wiley & Sons, Chichester, England, 1978.
6. Dicke, M., Infochemicals in Tritophic Interactions, Thesis, Agricultural University, Wageningen, 1991, 235 pp.
7. Dixon, W.J. and Mood, A.M., A method for obtaining and analyzing sensitivity data, *J. Am. Stat. Assoc.*, 43, 109, 1948.
8. Doelman, P., Resistance of soil microbial communities to heavy metals, in *Microbial Communities in Soil*, Jensen, V., Kjoller, A., and Sorensen, L.H., Eds., Elsevier, London, 1986, 369.
9. Doelman, P., Hannstra, L., Janssen, E., Michels, M., and van Til, M., Effects of heavy metals in soil on microbial diversity and activity, *Biol. Fertil. Soils*, in press.
10. Domsch, K.H., Jagnow, G., and Anderson, T.H., An ecological concept for the assessment of side-effects of agrochemicals on soil micro-organisms, *Residue Rev.*, 86, 65, 1983.
11. Dunger, W., *Tiere im Boden*, Ziemsen Verlag, Wittenberg Lutherstadt, 1984, 265 pp.
12. Edelman, Th., Achtergrondgehalten van een aantal anorganische en organische stoffen in de bodem van Nederland. RIN-rapport 83/8, Bodembeschermingsreeks nr. 38, SDU, Den Haag, 1983.
13. Edwards, C.A. and Thompson, A.R., Pesticides and the soil fauna, *Residue Rev.*, 45, 1, 1973.
14. Eijsackers, H., Side effects of the herbicide 2,4,5-T affecting the carabid Notiophilus biguttatus Fabr., a predator of springtails, *Z. Angew. Entomol.*, 86, 113, 1978.
15. Eijsackers, H., De betekenis van ecotoxicologisch onderzoek, een nabeschouwing, *Vakbl. Biol.*, 65(13/14), 122, 1985.
16. Eijsackers, H. and van de Bund, C.F., Effects on the soil fauna, in *Interactions Between Herbicides and the Soil*, Hance, R.J., Ed., Academic Press, London, 1980, 255.
17. Eijsackers, H. and Doelman, P., The impact of heavy metals on terrestrial ecosystems: biological adaptation through behavioural and physiological avoidance, in *Ecological Assessment of Environmental Degradation, Pollution and Recovery*, Ravera, O., Ed., Elsevier, Amsterdam, 1989, 245.

18. Eijsackers, H. and Zehnder, A.J.B., Litter decomposition: a Russian matriochka doll, *Biogeochemistry*, 11, 153, 1990.
19. Eijsackers, H., van de Bund, C., Doelman, P., and Ma, W., Fluctuerende aantallen en activiteiten van bodemorganismen (Fluctuating numbers and activities of soil organisms), Rapport 88/33, Rijksinstituut voor Natuurbeheer, Arnhem, 1988, 85 pp.
20. Eijsackers, H. and Løkke, H., SERAS Soil Ecotoxicity Risk Assessment System, report from a European workshop, January 1992, NERI Silkeborg, 1992, 52 pp.
21. Faber, J.H., Functional classification of soil fauna: a new approach, *Oikos*, 62, 110, 1991.
22. Glowa, J.R., Dose-effect approaches in risk assessment, *Neurosci. Biobehav. Rev.*, 15, 153, 1991.
23. Grime, J.P., *Plant Strategies and Vegetation Processes*, John Wiley & Sons, Chichester, England, 1979.
24. Hengeveld, R., *The Dynamics of Biological Invasions*, Chapman and Hall, London, 1989, 176 pp.
25. Hoekstra, J.A. and van Ewijk, P.H., Alternatives for no observed effect level, *Environ. Toxic. Chem.*, Submitted.
26. Hoogerkamp, M., Rogaar, H., and Eijsackers, H.J.P., Effect of earthworms on grassland in recently reclaimed polder soils in The Netherlands, in *Earthworm Ecology; From Darwin to Vermiculture*, Satchell, J.E., Ed., Chapman and Hall, London, 1983, 85.
27. Hutson, B.R., The invertebrate fauna of a reclaimed pit heap, in University of Newcastle upon Tyne, Landscape reclamation; a report on research into problems reclaiming derelict land, IPC Buss. Press, Guilford.
28. Hutson, B.R., Colonization of industrial reclamation sites by Acari, Collembola and other invertebrates, *J. Appl. Ecol.*, 17, 255, 1980.
29. Janssen, M.P.M., Turnover of Cadmium Through Soil Arthropods, Thesis, Free University, Amsterdam, 1991, 136 pp.
30. Kärber, G., Beitrag zur kollektiven Behandlung pharmakologischer Reihenversuche, *Arch. Exp. Pathol. Pharmacol.*, 162, 480, 1931.
31. Koeman, J.H., General introduction to ecotoxicology, in *Ecological Indicators for Quality Assessment of Air, Water, Soil and Ecosystems*, Environmental monitoring and assessment 3, Best, E.P.H. and Haeck, J., Eds.
32. Lexmond, Th., Edelman, Th., and van Driel, W., Voorlopige referentiewaarden en huidige achtergrondgehalten van een aantal zware metalen en arseen in de bovengrond van natuurterreinen en landbouwgronden (Reference values of heavy metals and arsenic in nature areas and arable land), VCTB-advies A 80/02, VTCB, Leidschendam, 1986.
33. Ma, W.C. and Eijsackers, H., 1990. The influence of substrate toxicity on soil fauna return in reclaimed land, in *Animals in Primary Succession of Fauna in Reclaimed Land*, Majer, J.D., Ed., Cambridge University Press, Cambridge, 1990, 1.
34. Ma, W.C., Denneman, W., and Faber, J., Hazardous exposure of ground-living small mammals to cadmium and lead in contaminated terrestrial ecosystems, *Arch. Environ. Contam. Toxicol.*, 20, 266, 1991.
35. Moriarty, F., *Ecotoxicology. The Study of Pollutants in Ecosystems*, Academic Press, London, 1983.

36. Orchard, V.A. and Cooke, F.J., Relationship between soil respiration and soil moisture, *Soil Biol. Biochem.*, 15, 447, 1983.
37. Reichle, D.E., The role of soil invertebrates in nutrient cycling, in Lohm, U. and Persson, T., Eds., *Soil Organisms as Components of Ecosystems,* Ecol. Bull. (Stockholm) 25, 145, 1977.
38. Raunkiaer, C., The life forms of plants and statistical plant geography, Clarendon Press, Oxford, 1934.
39. Spearman, C., The method of 'right and wrong cases' ('constant stimuli') without Gauss's formulae, *Br. J. Psychol.*, 2, 227, 1908.
40. Streit, B., Effects of high copper concentrations on soil invertebrates (earthworms and oribatid mites): experimental results and a model, *Oecologia,* 64, 381, 1984.
41. Thompson, W.R., Use of moving averages and interpolation to estimate median effective dose, *Bacteriol. Rev.*, 11, 115, 1947.
42. Tietema, A., Nitrogen Cycling and Soil Acidification in Forest Ecosystems, Thesis, University of Amsterdam, 1947, 139 pp.
43. Truhaut, R., Ecotoxicology: objectives, principles and perspectives, *Ecotoxicol. Environ. Saf.,* 1, 151, 1977.
44. Tyler, G., The impact of heavy metal pollution on forests. A case study of Gusum, Sweden, *Ambio,* 13, 18, 1984.
45. Van Gestel, K., Earthworms in Ecotoxicology, Thesis, University of Utrecht, 1991, 197 pp.
46. Van Straalen, N.M., Verhoef, H.A., and Joosse, E.N.G., Functionele classificatie van bodemdieren en de ecologische functie van de bodem, *Vakbl. Biol.,* 65, 131, 1985.
47. Van Straalen, N.M., Schobben, J.H.M., and de Goede, R.G.M., Population consequences of cadmium toxicity in soil microarthropods, *Ecotoxicol. Environ. Saf.,* 17, 190, 1989.
48. Van Straalen, N.M. and Verkleij, J.A.C., Eds., *Leerboek Ecotoxicologie,* VU Publ., Amsterdam, 1991, 422 pp.
49. Van Wensem, J., Isopods and Pollutants in Decomposing Leaf Litter, Thesis, Free University, Amsterdam, 1992, 134 pp.
50. Van der Werff, M., Ecotoxicity of Heavy Metals in Aquatic and Terrestrial Higher Plants, Thesis, Free University, Amsterdam, 1981.
51. Verhoef, H.A. and Brussaard, L., Decomposition and nitrogen mineralization in natural and agroecosystems: the contribution of soil animals, *Biogeochemistry,* 11, 175, 1990.
52. Walker, C.H. Species variations in some hepatic microsomal enzymes that metabolize xenobiotics, *Prog. Drug Metab.,* 5, 113, 1980.

SECTION II

Occurrence and Fate of Contaminants in Soil

CHAPTER 2

The Pollution of Soils and Groundwater in the European Community

L.H.M. Kohsiek, R. Franken, D. Fraters, J. Latour, A.M.A. van der Linden, R. Reiling, and W.J. Willems

TABLE OF CONTENTS

0-87371-530-6/94/$0.00 + $.50

ABSTRACT

In all member states of the European Community, current practices lead to nonsustainable use of soil and groundwater systems. The main problem areas are found in the agricultural and industrial core regions of the EC. The scale and complexity of the problems vary per region. The most serious problems are pollution from pesticides and nitrate and pollution from industrial and urban areas. Drinking water reservoirs and natural ecosystems are seriously threatened. Provided that in the coming years proper measures are implemented the recovery time will last centuries.

I. GENERAL INTRODUCTION

Intensive land use in Europe results in continual changes to land and soil. These changes are mainly physical (homogenization, soil erosion, etc.) and chemical (pesticides, over-manuring, etc.). As a consequence, soil properties are heavily disturbed in a great number of places and groundwater is being threatened on a large scale.

These types of problems are well known and extremely visible throughout Europe. In 65% of the EC, for example, concentrations of pesticides exceed drinking water standards. However, the underlying physical phenomena are regional in nature, and transboundary fluxes, in these cases, are the exception rather than the rule. Also, with respect to policy process and legislation (directives), the problems might be dealt with on the European, rather than the regional, scale.

In order to identify the extent of these problems the total pollution load originating from such sources as *industrial activities and urbanization* (waste deposits, landfills, mining waste, etc.) and *agriculture* (pesticides, including herbicides and fungicides; manure, and fertilizers) have been calculated, as well as pollution by heavy metals origination from both sources.

Atmospheric deposition has an indirect impact. Where possible or necessary, the direct and indirect loads are estimated. Provided that in the coming years proper measures are implemented, the recovery time will be centuries. As a result, the multifunctional use of soil and groundwater systems, now and in the future, and the diversity of groundwater-related ecosystems are in danger all over the EC.

Data originate from the OECD,[73-75] FAO,[40] Eurostat, and the EC (mainly Corine[30]), and from reports from most of the EC countries and obtained bilateral contacts. It should be noted that the information on most subjects is not entirely representative for the entire EC. The collected data are inconsistent. In order to establish a comparable EC overview of the problems, assumptions had to be made. These assumptions have, as far as possible, been checked, for instance, by comparison with monitoring data. However, it should be stressed that the purpose of computing EC overviews is to give an impression of the magnitude of the problems and an indication of the areas where these problems may exist, rather than giving precise information on concentrations and locations.

In order to assess the extent of the pollution problem, the total pollution load from diffuse sources such as agriculture and atmospheric deposition, as well as the load from point and line sources such as industrial plants, waste disposal sites, etc., have been calculated. As far as possible these calculations have been checked with monitoring data and/or illustrated with case studies.

For this study much literature has been used which is incorporated in the reference list but is not always referred to in the text.

II. DIFFUSE POLLUTION OF SOIL AND GROUNDWATER

A. Agricultural Threats

1. Introduction

The areal use of land for agriculture in EC-Europe is on the order of 57% of the total land mass of approximately 226 million ha; 52% of the cultivated area is arable and 37% is used for grass and green fodder production. Approximately 10% of the area is used for the production of fruit, wine grapes, and vegetables. Although the contribution of agricultural production to the gross national product (GNP) of several countries has fallen in relative terms, the

Table 1. Total Sales of Pesticides (tonnes active ingredients) in Ten EC Countries[a]

	Disinfectants	Fungicides	Herbicides	Insecticides
Belgium	900	2,160	4,770	360
Denmark	76	1,555	4,506	289
France	4,807	49,775	36,075	6,657
Germany	— [b]	10,151	14,756	4,558
Greece	640	10,384	3,411	2,818
Italy	— [b]	27,934	9,234	3,941
Netherlands	9,830	4,063	3,271	1,554
Portugal	— [b]	21,288	1,055	587
Spain	4,518	33,496	6,360	2,643
U.K.	76	5,522	19,625	690

[a] Amounts calculated from confidential information obtained from experts in the different countries; for Belgium data were obtained from the Dutch policy document MJPG (long-term crop protection plan) and for Portugal data were obtained from the FAO statistical yearbook.
[b] No data available.

production in absolute value has increased enormously owing to technological developments, i.e., increased mechanization and use of fertilizers and pesticides. However, most forms of agricultural land use provide an important source for diffusive contamination of soils and groundwater.

2. Pesticides

a. Their Use   The EC is the largest user of agrochemicals in the world, and five of the individual countries are within the top ten.

The amount of pesticides used may vary from year to year and is primarily dependent on weather conditions. This is especially true for fungicides and insecticides. During the past 5 years the amounts used (expressed as kilograms of active ingredient) seem to have stabilized. In Table 1 the total sales of pesticides for ten EC countries are given. In general, fungicides comprise the largest use. In France, Italy, and Spain, fungicides are very important in the viniculture; sulfur and copper compounds are still used in large quantities. In the Netherlands, soil disinfectants are used in large quantities in the culture of potatoes, vegetables, and flower bulbs.

In Spain, pesticide use is low except in the culture of vegetables, fruit, and viniculture. For each EC member state an analysis has been made of the average use of soil disinfectants, fungicides, herbicides, and insecticides on the major crops. For France, this is illustrated in Table 2 where figures were obtained by assigning major crops to the different pesticides (information was gathered from specialists in the different countries) and dividing total sales over the different crop areas (data from Eurostat). Such an analysis has been made for each of the EC member states.

Table 2. Average Pesticide Use on Different Crops (active ingredients in kg/ha) For France

Crop	Disinfectants	Fungicides	Herbicides	Insecticides
Fruits	—[a]	—[a]	—[a]	1.12
Vegetables	—[a]	4.50	1.19	0.42
Cereals	0	0.36	1.07	0.17
Oil-plant crops	0	0.52	2.51	0.67
Potatoes	0	1.97	0.94	0.14
Sugar beet	0.14	2.20	6.26	0.37
Maize (fodder and corn)	0	—[a]	2.69	0.53
Vine	2.67	39.10	2.81	0.48

[a] No data available.

In general, pesticide loads are low on grass and green fodder, except fodder maize; loads are up to approximately 1 kg/ha. Loads between 1 and 5 kg/ha are encountered in maize, cereals, and oil-plant crops. The average load per hectare is high in viniculture, vegetable and fruit cultures, flower bulbs, potatoes, and sugar beets.

The average load of total pesticides per hectare on agricultural land (including arable land, for example, cereals, maize, sugar beets, potatoes and vegetables and permanent crops including viniculture, orchards, olive and citrus culture) in the different regions of Europe are given (Figure 1).

b. Atmospheric Deposition   Pesticides may reach the atmosphere by drift and volatilization during and after application. Quantitative data on the amounts going into the atmosphere, however, are rather scarce (mainly due to experimental and interpretation problems). Also, data on occurrence in the atmosphere and rainwater are sparse. Based on rather crude assumptions (for instance a volatilization rate of 1.50% of the application rate), Baart and Van Diederen[4] calculated deposition rates for 12 pesticides to be between 8.5×10^{-5} and 1.1×10^{-2} kg/ha/year (atrazine: 8.1×10^{-3}; lindane 1.9×10^{-3}). Given the uncertainties of the calculations, these figures compare reasonably well with the observed values.[11] One study gives an impression of the deposition rates for the insecticide lindane and the herbicide atrazine[1] (Braun et al.).[11] In comparison to normal application rates these figures are negligible (approximately 1000 times smaller), even when the sum of all pesticides is considered.

c. Computed Pesticide Leaching   Once released into the environment pesticides can undergo a large number of processes which determine their ultimate fate. After entrance into the soil the most important processes are degradation, sorption, plant uptake, and transport. These processes determine, to a very large extent, the leaching of pesticides.

In simulation models the interaction between the different processes is described conceptually, making these models useful for extrapolation when relevant information is available. A simulation model, PESTLA (Boesten and Van der

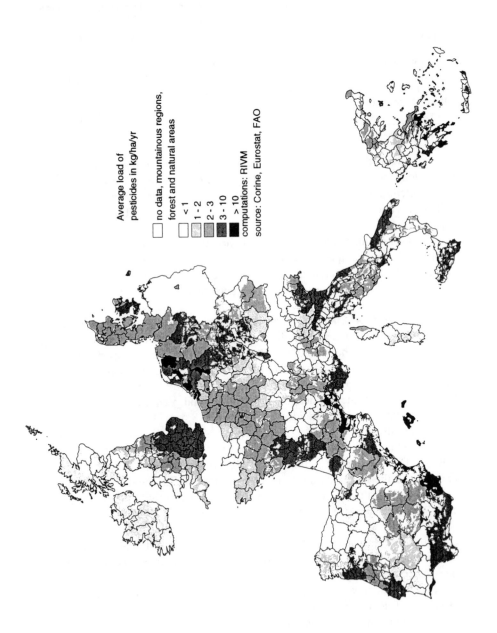

Average load of
pesticides in kg/ha/yr

no data, mountainous regions,
forest and natural areas

< 1
1 - 2
2 - 3
3 - 10
> 10

computations: RIVM
source: Corine, Eurostat, FAO

Table 3. Leachability of Selected Pesticides in Relation to Atrazine

1 Low-leaching pesticides	2 Leachability in the same order of magnitude as atrazine	3 Leachability greater than atrazine (factor 1 to 20)
1,3-dichloropropene	Atrazine	Aldicarb
Metam-sodium	2,4-D	Bentazon
Copper compounds	Metribuzin	Dicamba
Pyrethrines	Metsulfuron-methyl	Gluphosinate
Glyphosate	Thiameturon	TCA
Paraquat	Triclopyr	Zineb
Parathion	Linuron	Maneb
Sulfur compounds	Isoproturon	Pyridate
Mineral oils	Pyrifenox	Metalaxyl

Note: For some of the pesticides mentioned in the table a metabolite may leach instead of the active ingredient.

Linden),[9] was used to calculate the amount of pesticide accumulated to soil and leached into the groundwater. For this paper, all calculations were carried out for atrazine and the behavior of other pesticides is described relative to the result of these calculations. In general, other triazines are close to atrazine, pyrethroids have much less leaching potential, and bentazon and the acetanilides have a higher leaching potential. At the moment the model is not able to simulate the behavior of ionizable compounds of the phenoxycarbonic acids (e.g., 2,4-D, MCPA, and mecoprop), inorganic sulfur compounds, and the heavy metals contained in pesticides. Table 3 illustrates the leachability of a number of pesticides in relation to atrazine.

Case

Germany, Bavaria

Twenty-six rainwater monitoring points (dry and wet deposition) were regularly sampled in the period of 1987 to 1989. Lindane was found in every sample in concentrations from 1 to 420 ng/l; atrazine was found in approximately half the samples in concentrations up to 2200 ng/l. Most of the positive data and the highest values were found in the periods May to July (the main application period). From these measurements it was calculated that the average load per hectare in Bavaria is approximately 850 mg for lindane and approximately 700 mg for atrazine. For other pesticides it was not possible to calculate the load as a result of too little (positive) data.[11]

Combining this apparent load with the leaching of atrazine results in a calculated expectation of the total pesticide concentration in the uppermost layer of the groundwater (Figure 2).

FIGURE 1. (*Opposite*) The average load (kg/ha) of pesticides (soil disinfectants, fungicides, herbicides, and insecticides) on agricultural soils. The use of pesticides is widespread with relatively high loads in parts of almost all EC countries.

Computed leaching of
pesticides in the topsoil
(0-1m) in ug/l

no data, mountainous regions,
forest and natural areas

< 0.05

0.05 - 0.50

-- drinking water standard

0.50 - 5.00

> 5.00

computations: RIVM

source: Corine, Eurostat, FAO

It is clear that the EC standard for the total concentration of pesticides in drinking water (0.5 μg/l) will be exceeded in a very large part of Europe in the groundwater. Under the assumptions mentioned, the EC standard will be exceeded in 65% of the area and even more than ten times in 25% of the area.

d. Monitoring Data   For a large number of pesticides allowed in EC member states analytical methods appropriate for measurement at the 0.1 μg/l level are not available. For other pesticides the cost of analysis is too high. For these reasons, data on the occurrence of pesticides in groundwater are available in only six countries. Data are most abundant on triazine compounds (principally atrazine) and phenoxycarbonic acids. Atrazine is found in groundwater throughout Europe. The EC drinking water directive (0.1 μg/l) is often exceeded by a factor of 10 to 100. Exact locations of the findings are often not reported, so that comparison with the sensitivity on the local scale is not possible. In general, however, fair agreement is found; findings are reported in areas where sensitive soils are abundant. Comparison between the computed leaching and monitoring data on total concentrations of pesticides is not possible; no data were given in literature.

e. Conclusions

- Atmospheric deposition of pesticides is negligible compared to the total pesticide load in agricultural areas. As in contrast for natural areas, this source can account for up to 100% of the total load; the absolute loads for these areas, however, are small (presumably <10 g/ha).
- Organic pesticides have been in use now for 45 years (since World War II). Monitoring data have been available only for about 7 years. For a large number of pesticides an analytical detection method at the 0.1 μg/l level is lacking. For these reasons and because monitoring programs, in general, are small or even absent in the different EC member states, it is expected that the actual situation is more severe than may be deduced directly from the monitoring data.
- Groundwater is threatened by pesticides in all EC member states. This is obvious from both the available monitoring data and calculations for pesticide load, soil sensitivity, and leaching. Very high pesticide loads (10 kg/ha and higher) are often used on highly sensitive soils. It has been calculated that in 65% of all agricultural land the EC standard for the sum of pesticides (0.5 μg/l) will be exceeded. In approximately 25% of the area this standard will be exceeded by more than ten times.

FIGURE 2.   (*Opposite*) Leaching of pesticides at agricultural soils. The expected sum of pesticide concentrations calculated in groundwater at about 1 m below surface level. It is assumed that these areas receive an average load as given in Figure 7. The load on the area is divided into a nonleaching fraction (Table 3), a fraction leaching like atrazine, and a fraction leaching five times as much as atrazine (Table 3). The EC standard for the concentration of pesticides in drinking water (0.5 μg/l) will be exceeded in large parts (about 65%) of the agricultural soils in the EC.

3. Nitrates

Nitrogen (N) has become an important pollutant due to three types of non-point sources:

- Application of (organic) manure to arable land and grassland
- Use of synthetic fertilizers in agriculture
- Atmospheric deposition of NO_x and NH_y

On a local scale, point sources may exert a negative influence on the quality of groundwater and thus may not be disregarded as a source of pollution.

a. Total Nitrogen Load   The total input of nitrogen from manure and fertilizer in agricultural soils in the EC is depicted in Figure 3. The general picture shows a high input of N in the northern part of the EC, notably in northwestern France, and in Belgium, England, the Netherlands, Germany, and Denmark, but also in the northern and western part of Italy, and some small parts of Portugal, Spain, and Greece.

b. Atmospheric Deposition   The atmospheric deposition of NO_x and NH_y is partly induced by agriculture itself and therefore partly autorelated to the pattern of manure production and to a lesser degree to the fertilizer supply pattern (ammonia volatilization). Atmospheric deposition also originates from all combustion processes; i.e., traffic, electric power production, home heating, industrial activities.

In central England, the Benelux countries, Germany, and part of northern Italy the N supply from the atmosphere amounts to more than 24 kg/ha/year. In the northwestern and southern part of the EC the N deposition is below 8 kg/ha/year. Generally speaking, the amount of nitrogen compounds deposited on the soil is about 10% of the total N supply from direct application of manure and fertilizer. In forests and nature reserves, atmosphere-derived nitrogen is the only external supply. This constitutes, however, a significant threat to oligotrophic ecosystems.

c. Estimation of Actual Nitrate Leaching   The effects of a high N load on agricultural soils mainly threatens the quality of groundwater. Therefore, the total N supply as shown in Figure 3 is combined with the sensitivity characteristics of soils in the EC to estimate the nitrate concentration in the leachate of agricultural soils (Figure 4). In order to avoid double counting the N, atmospheric deposition on agricultural soils was not considered. Computations with a stationary leaching model in which both load and sensitivity are taken into account[36] lead to an estimation of the annual nitrate leaching. Net precipitation is assumed to be on average 300 mm/year.

It can be concluded that in Denmark, Germany, the Netherlands, and Belgium a combination of high N input and sensitivity of the soil leads to nitrate concentrations in soil leachate above the drinking water standard of 50 mg/l for most parts of these countries. In France, Italy, and to a lesser degree in England, the areas with high nitrate concentration in the leachate show a more localized distribution.

In a number of EC countries, data on groundwater quality indicating nitrate pollution (e.g., in Spain, France, England, Denmark, FR Germany, and the Netherlands) have been published in recent years. No nationwide data were available for Belgium, Italy, Greece, and Portugal. The available information is either based on individual groundwater samples or on regional groundwater quality surveys. Reported nitrate pollution is in most cases confined to phreatic aquifers, of interest from the water supply point of view. Moreover, the depth at which the groundwater samples originate is not specified in most reports. Based on the available data in only six countries, a generalized map (Figure 5) has been drawn indicating nitrate levels in groundwater above the 50 mg/l drinking water standard.

Comparison with the computed leaching (Figure 4) shows a more localized picture, but the location of the areas where nitrate pollution has been observed agrees reasonably well with the sensitive areas on the leaching map. The observed pollution areas may be an underestimation because no extensive national groundwater quality surveys have been conducted.

d. Conclusions

- The total N load on agricultural soils is relatively high (>170 kg/ha) in the Netherlands, Belgium, Germany, Denmark, England, and parts of France, Italy, and Greece.
- The leaching map indicates that present agricultural practices may lead to nitrate pollution of groundwater in at least 9 of the 12 EC member states. In about 43% of the area, nitrate concentrations between 25 and 50 mg/l have been computed. In 25% of the area, nitrate levels exceeding 50 mg/l may be expected. The greatest problems in terms of surface area occur in northwestern Europe (France, England, Belgium, The Netherlands, Northern Germany, and Denmark). This is either a result of a large surplus in the regional N balance (input greater than crop uptake), a high sensitivity of the soil, or a combination of these two factors.
- Special attention needs to be paid to the ecological effects of nitrate leaching from agricultural soils. In particular, surface runoff and leaching of nitrate into surface waters leads to eutrophication and damage to aquatic ecosystems. The impact of nitrate, especially on nutrient-limited ecosystems, is considerable. This effect is predominantly caused by direct nitrate input in the upper soil and surface water. Many groundwater-related ecosystems are known to be affected severely by nutrient-enriched groundwater.

B. Heavy Metals

Pollution by heavy metals from diffuse sources in rural areas is partly caused by atmospheric deposition. Substantial sources of heavy metals on agricultural

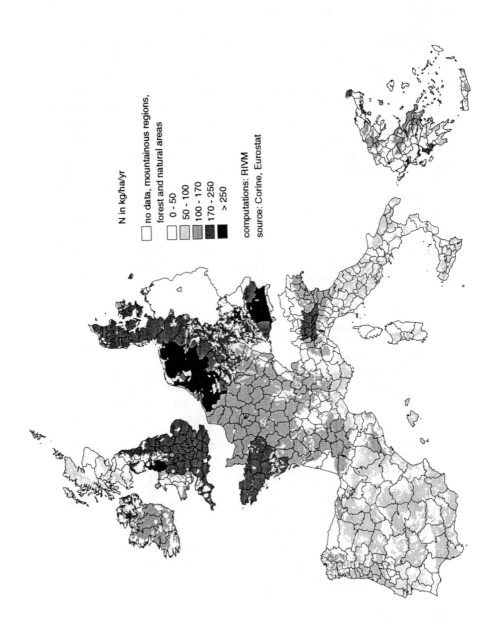

land are fertilizers, manure, sewage sludge, and metal-containing pesticides. In rural areas used for hunting, shot can increase the lead and arsenic loads locally.

After uptake by plants, metals may enter the primary consumers in the ecosystem and accumulate in the food chain. Even a small metal surplus may cause malfunctioning in the organs of both plants and animals. Because different species react differently to the pollution of their environment with heavy metals, the ecosystem may become instable and eventually change completely. An excess of heavy metals may either cause reduced crop production or products with unacceptable levels of heavy metals.

Where the soil has a high binding capacity for heavy metals, metal concentrations in the soil solution will remain low for decades or even centuries as long as the binding capacity is not saturated. An enormous reservoir of metals is built up, which may turn out to be a so called ''chemical time bomb.''[82] As long as heavy metals are immobilized by being bound to soil particles, they will not interfere with the system because they are relatively unavailable. A change in land use may decrease the binding capacity of the soil, and metals sorbed to the soil will be released. A small decrease in the binding capacity will only have a minor effect on the amount of metal absorbed into the soil, but may have an enormous effect on the metal concentrations in the soil solution and thus on the availability of the metals for biota.

Cadmium has been selected to illustrate the heavy-metal problem. The total accumulation of cadmium and the mobile concentration in the upper 30 cm of agricultural soils in the EC are calculated as a function of atmospheric deposition, manure, and fertilizer application.[43]

1. Cadmium Load

The total cadmium load on agricultural soils is the sum of the cadmium load due to atmospheric deposition and the load due to use of fertilizer, manure, sewage sludge, and cadmium-containing pesticides. The net cadmium load can be estimated by subtracting the cadmium uptake by crops or vegetation from the total load, i.e., as far as crops are removed by harvest or an above-ground reservoir is built up in the vegetation.

The yearly atmospheric cadmium deposition varies from less than 0.1 g/ha in remote areas (northern Scandinavia and southern Spain, Italy, and Greece) to over 2.4 g ha^{-1} in industrialized regions (Ruhr area, central Poland, and the Cantabria region in northern Spain). Atmospheric deposition may be higher in natural areas than in agricultural areas. The presence of heather and especially forest increase surface roughness and thereby dry deposition in the order of 10

FIGURE 3. (*Opposite*) The total N supply from manure and fertilizer on agricultural soils. The map presents an average picture at the regional level. Specific areas with high application rates within regions are not visible on the map. The total N supply is relatively high in the Netherlands, Belgium, Germany, Denmark, England, and parts of France, Italy, and Greece.

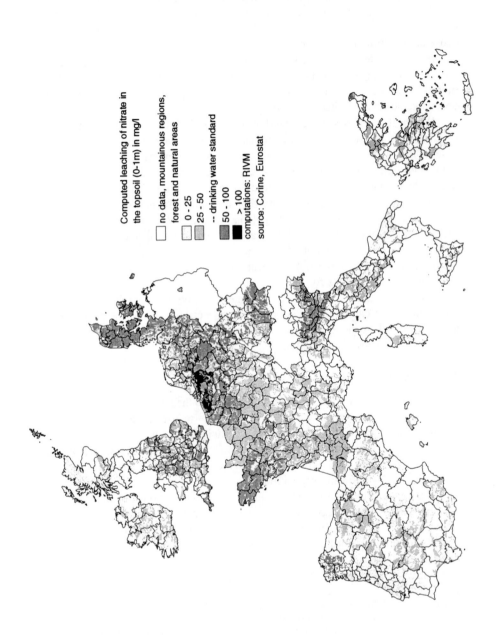

to 25%.[53a] The main source of cadmium for agricultural land is fertilizer and manure. Due to higher fertilizer use the cadmium load on arable land is higher than on grassland. The mean cadmium load per year for EC countries due to fertilizer and manure use varies from 0.6 to 0.8 g/ha for grassland and from 4.4 to 8.5 g/ha for arable land. The regional cadmium load due to fertilizer application may be as high as 7.5 to 8.5 g/ha in the south of the Netherlands and the north of Belgium.[43]

2. Cadmium Accumulation and Concentrations in the Soil Solution

To establish a qualitative and quantitative risk estimation, both soil groups and cadmium loads were classified (see Table 4). Three cadmium accumulation classes for European soils are differentiated: soils with a low or very low binding capacity for cadmium, soils with a very high binding capacity, and an intermediate class of soils. The first group of soils are characterized by a high acidity (pH below 5) and a low organic matter content (less than 5%). The intermediate class soils are either slightly acid (pH 5 to 6.5) or are acid soils with a pH above 4 and a high organic matter content (more than 5%). The soils with a very high binding capacity for cadmium are the neutral to alkalinic soils (pH more than 6.5) and slightly acid, organic soils (eutric peat soils).

In Figure 6 the resulting cadmium accumulation rate in the topsoil and the concentration in the soil solution are presented. The cadmium accumulation rate is given as the cadmium accumulation in the topsoil in 100 years as a percentage of the current cadmium content of the soil. The computed cadmium concentration in the soil water of the top soil is given in μg/l.

In acid soils, cadmium accumulation will be negligible (<1%), but if the cadmium load is high (>2.5 g ha^{-1} a^{-1}) an elevated concentration of cadmium in the soil water of the topsoil occurs and forms an ecological risk. For about 20% of the agricultural area in the EC it is computed that the (Dutch) reference value of 1.5 μg/l is exceeded. For 15% of the soils, the cadmium concentration in the topsoil probably also exceeds the (German) test value of 2.5 μg/l. For 40% of the soils the accumulation in 100 years is more than 10% of the current cadmium content and may be as high as 50 to 60%. Reference and test values in the soil solution are exceeded, and high accumulation rates of cadmium in

---

FIGURE 4. (*Opposite*) An estimation of actual leaching of nitrate in agricultural soils, based on model computations using the total N supply data. The computed nitrate concentrations refer to the depth of 1 m below surface level. In areas with shallow groundwater levels the data refer to nitrate concentrations in the upper groundwater. The map can be conceived as a potential groundwater pollution map with respect to nitrate. Denitrification and dispersion between 1 m below surface level and the groundwater table have not been taken into account. The computed nitrate concentration at 1 m below surface level exceeds the standard for drinking water of 50 mg/l in large parts of northern EC as well as in Brittany and the Po delta. This includes about 25% of the agricultural soils in the EC.

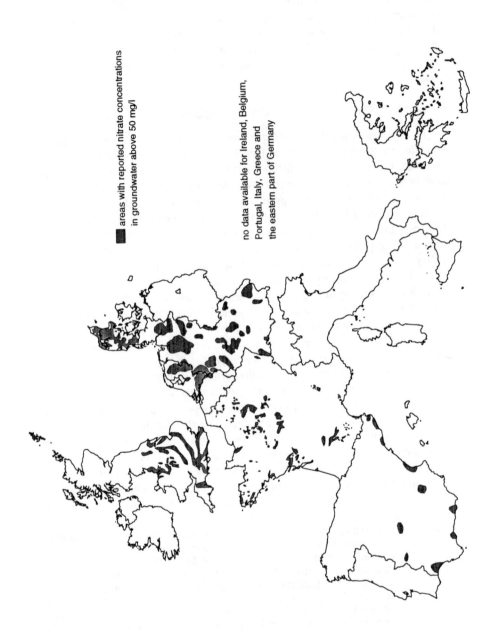

areas with reported nitrate concentrations in groundwater above 50 mg/l

no data available for Ireland, Belgium, Portugal, Italy, Greece and the eastern part of Germany

Table 4. Qualitative and Quantitative Risk Assessment For Soil Water Cadmium Concentration and Accumulation of Cadmium in the Topsoil

	Soil type		
Cadmium load (g ha)	Acid soils humus poor	Slightly acid soils and humus-rich acid soils	Neutral and alkalinic soils, organic slightly acid soils
	Groundwater ($\mu g \cdot l^{-1}$)	Accumulation in a hundred years (as percent of current Cd content)	
Low <2.5	No risk <1	No risk <1	No risk <1
Medium 2.5–5	Available[a] 1–2	Slight accumulation 1–10	Medium accumulation 10–25
High >5	Available >2	Medium accumulation 10–25	High accumulation >25

[a] Cadmium concentration in groundwater exceeds standard value, and availability for plant and soil organisms is relatively high.

the topsoil occur in arable soils; in particular, of course, liming of agricultural soils is an important factor in this discussion.

For a humus-poor sandy soil with a crop rotation of 50% potatoes, 25% sugar beets, and 25% other crops, it is advised to lime up to pH 5.2. In case once every 8 to 10 years maintenance liming is given, the pH may drop the pH 4.4 to 4.6. If sugar beets are the most important crop, liming may be done every 4 years, and the pH will drop to around 4.9. For sandy soils with a higher organic matter content, optimal pH is lower as well as in case the crop rotation is 50% potatoes and 50% cereals. This kind of action has not yet been taken into account.

III. POINT/LINE SOURCES AND THE SOIL/GROUNDWATER

A. Categories of Contamination

Man has lived in urbanized areas of Europe for centuries and in some cases for more than two millennia. Concentrations of population have created a slowly

FIGURE 5. (*Opposite*) Concentration of nitrate in the groundwater based on monitoring data. An impression of the groundwater quality with respect to nitrate from available data. No national data were available from Belgium, Italy, Greece, and Portugal. Comparison with the computed leaching gives a reasonable fit for the remaining countries. One should remember that monitoring data lead to an underestimation because extensive groundwater surveys have been scarce, and often the groundwater samples have been taken at a greater depth. This means that a large part of the nitrate load did not yet reach the groundwater at a greater depth. In other words: one may speak of a time lag between the computed results and the measurements.

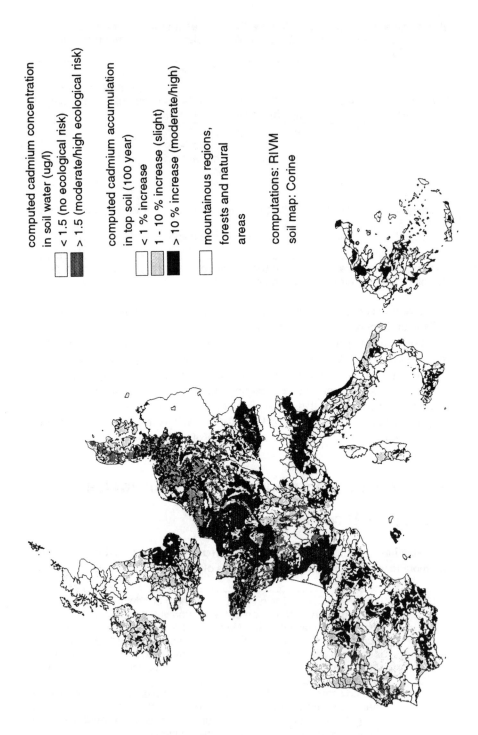

computed cadmium concentration
in soil water (ug/l)

< 1.5 (no ecological risk)

> 1.5 (moderate/high ecological risk)

computed cadmium accumulation
in top soil (100 year)

< 1 % increase

1 - 10 % increase (slight)

> 10 % increase (moderate/high)

mountainous regions,
forests and natural
areas

computations: RIVM
soil map: Corine

Table 5. Point Sources and the Expected Contamination of Groundwater

Source	Inorganic contaminant	Organic contaminants
Urban areas	Heavy metals (Pb, Zn) and salt, also Cd	Oil products, biodegradable organics
Industrial sites	Heavy metals	PAH, chlorinated hydrocarbons (tri- and tetra-chloroethylene), hydrocarbons (benzene, toluene, xylene), oil products
Landfills	Salts (Cl^-, NH_4^+) and heavy metals	Biodegradable organics and xenobiotics
Mining disposal sites	Heavy metals, salts, and arsenics	Xenobiotics
Dredged sediment and sludge disposal	Heavy metals	Xenobiotics
Hazardous waste sites	Heavy metals (concentrated)	Concentrated xenobiotics
Leaking storage tanks		Oil products (petrol)
Line sources (motorways, railways, sewerage systems, etc.)	Heavy metals (Pb), salts	PAH, oil products, pesticides

increasing contamination of the soil beneath the cities. Wastewater infiltrating from primitive sewage systems, canals, and wells leads to increased concentrations in groundwater of substances such as chloride, sulfate, nitrate, etc. Since the 18th century industrialization has proceeded without disposal of the generated waste in a proper way. Often fluids, heavily contaminated with metals, organic micropollutants, and similar substances, were disposed of via wells into groundwater. This has led to an extensive contamination of soil and groundwater in and around the industrialized towns of Europe. At the same time, the need for coal as a source of energy and the intensified exploitation of metal ore via mining has produced contamination in originally rural areas. Since the last century, traffic and transport via roads, railways, and pipelines have created a "line" pattern of pollution. The sources of pollution creating a threat to groundwater quality can be divided into the following categories (Table 5):

- Urban areas
- Industrial sites
- Waste disposal sites: landfills, surface impoundments, injection and underground storage, hazardous waste sites, dredged habor sediments and sewage sludge disposal, mining spoils
- Fuel storage tanks
- Line sources: motorways, railways, sewage systems

FIGURE 6. (*Opposite*) Accumulation or leaching of cadmium to groundwater.

The main categories of contamination will be discussed in brief. A general estimation of the area in EC countries where the groundwater system is possibly contaminated by these types of pollution will be given.

It should be noted that the variety of climatic, geohydrological, and geological conditions create a great variability in soil and groundwater vulnerability and of relevant processes.

1. Urban Areas

In urban areas there is a permanent load on the soil from traffic emissions and combustion processes (exhaust gases), fuel storage, disposal of wastewater, leaking of sewage, etc. Particular contaminants are chloride, sulfate, nitrate, oil products, volatile halogenated hydrocarbons, and heavy metals (Pb and Zn). In urban areas there are often a large number of small industries which are potential sources of soil and groundwater contamination.

2. Industrial Sites

More than 50 large industrial regions (with chemical and metallurgical industries, iron and steel production) are found all over Europe. In general, hardly any quantitative data were found on the state of pollution at industrial sites. Nevertheless, there are indications that many industrial sites are contaminated by (liquid and solid) waste disposal, spoiled fuel, leaking pipe systems, and accidental spills.

3. Landfills

Landfill leachate consists of a wide range of heavy metals and organic micropollutants. In general, landfills are created by dumping domestic and clean industrial wastes into excavations. In most European countries, especially in Eastern Europe, landfilling is still the most popular way of removing municipal and industrial waste, though the amount of incinerated waste is increasing.

Because of their operation, magnitude, and large number, landfills represent a significant threat to groundwater resources. Several landfills were (are) used improperly; illegal dumps of hazardous waste or insufficient disposal facilities are not unusual. Only during the last decade and in a few European countries new landfills have been equipped with liners and other facilities to prevent soil and groundwater contamination. However, the total number of landfills (municipal and industrial) in Europe is estimated at about 120,000 to 240,000 (one landfill to every 2500 to 5000 inhabitants).

4. Surface Impoundments

Domestic and industrial surface impoundments are often referred to as water-filled pits, ponds, lagoons, and basins, lined or unlined and range these impoundments from a meter in diameter to hundreds of acres in size. Surface impoundments are often used for the treatment, storage, or disposal of wastewater from all kind of industries. The liquids stored in industrial ponds may contain heavy metals, gasoline products, radioactive substances, and an extended list of miscellaneous xenobiotics.

5. Injection and Underground Storage

Deep-well disposal is practiced for liquid wastes from oil and gas exploration and specific liquid chemical wastes. Chemicals are also injected. Several cases of failure, leading to groundwater contamination, have been reported. Mining operations using wells contaminate large areas in the long run. Abandoned coal mines and salt cavities are sometimes used for liquid waste disposal.

6. Sediment Sludge

Dredging is necessary, especially in the delta areas of the rivers Rhine, Meuse, Scheldt, Thames, Po, Rhône, and Garonne, to keep harbors, canals, and other facilities in repair and waterways navigable. The sediment surplus is often contaminated by heavy metals and organic pollutants; in the Rotterdam harbor area the annual amount of heavily contaminated dredged sediment is between 10 million to 20 million m^3. In the past the dredged sediment was used in the surrounding areas for elevating land. A large area of land around these harbors will be contaminated (several sites have been identified) and are indirectly threatening the groundwater quality.

7. Mining Waste

The waste arising from mining practices can be divided into three categories: hard coal and lignite, metal, and salt. Dewatering of deep coal mines often results in a large production of saline water, which contaminates surface water (for example, in Poland, 7000 t of salt per day is produced in this way). Mining activities and spoils are concentrated in several large regions all over Europe, especially in England, France, Germany, Czechoslovakia, Poland, the Ukraine, and Russia. The exploitation of these mining regions started often more than 100 years ago. The acid leachate from coal mining waste is characterized by high concentrations of total dissolved solids and may be saline. Metal mining wastes often contain high heavy metal concentrations with sulfides and sulfates in particular. The spoil of salt mining contains several salts, such as chloride and potassium, in high concentrations. Large mining areas in Europe are Alsace-Lorraine (France): salt; Niederlausitz, Halle-Leipzig, Rhineland (Germany): lignite; Ruhr (Germany): hard coal; Yorkshire-Humberside-East Midlands (UK): hard coal; Upper Bohemia (Czechoslovakia): lignite; Upper Silesia (Poland): hard coal, salts; Donbass (Ukraine): hard coal; Moscow (Russia): lignite.

8. Fuel Storage and Distribution

A source of potential groundwater contamination is the leakage of fuel storage tanks and leakage by fuel distribution. One major category of underground storage tanks are those containing gasoline. The number of fuel storage tanks in Europe is not known exactly, but there must be several millions. Important leakage factors are spill and corrosion. Gasoline that has seeped down into an aquifer will tend to float on top of the water table. However, the more soluble components will intrude into the groundwater and, due to the deterioration of taste, each liter of gasoline could contaminate up to 1 million l of groundwater.

9. Line Sources

Important line sources are motorways, railways, and sewerage systems. Motorways (via runoff and deposition of air pollution) and railways are significant sources of soil pollution and probably also of groundwater pollution. The discharge of sewage into the soil and the leakage of sewage systems are significant sources of soil pollution and probably also of groundwater pollution. Extrapolating from regional studies, it is expected that the extent of this problem might be considerable.

B. Synthesis of Urban and Industrial Threats

Although there are only a few available, it seems clear from country-level surveys that the above mentioned sources constitute a severe threat to sustainable use of groundwater.

There is, for example, a definite threat of groundwater contamination by landfills within the EC (Figure 7). This figure shows the groundwater area "potentially" contaminated by landfills as a percentage of the total area per region. Only in a part of the EC the landfills are properly installed and even then they are only safe for 30 years.

As expected, the waste load in urban areas is relatively high. The figure also shows differences between (urban) areas in several EC countries. Reasons for these differences include the varying amounts and treatment of waste in individual countries.

A combination of data on numbers and dimensions of the aforementioned sources and estimations on dispersion of contaminants have resulted in a (very rough) estimation of the total groundwater area which is threatened in Europe (see Table 6).

The area which is potentially contaminated is estimated using the total (estimated) area of the most important sources (industrial estates and landfills) and/or assumptions concerning dispersion (mining waste and injection of fluids, line sources, fuel storage tanks). The lower and upper range is estimated assuming that the dispersion of contaminants in groundwater over a period of 50 years will be a factor 0.5 to 5 times the source area. Since there is a great deal of uncertainty about the area of most sources, the range of the total contaminated groundwater area is large. It is concluded that:

- Point sources of pollution induce a threat of groundwater contamination in a large area of Europe, especially in urban and industrial areas.

- The point sources of primary importance threatening the groundwater in Europe are industrial sites and waste disposal sites (including disposal of industrial and municipal waste, mining waste, and injection of fluids).
- Illegal or improper disposal of industrial, municipal, mining, or hazardous waste is a severe threat for groundwater. Rapid implementation of the EC directive concerning the landfill of waste is recommended and priority should be given to vulnerable sites. An inventory of abandoned and existing sites in Europe has to be made, including a list of risks and sanitation plans.
- If no remedial action and further precautions for new situations are taken, the potential polluted area from the groundwater system in Europe can be estimated to be in the order of 20,000 to 150,000 km² within a period of 50 years.

IV. SUSTAINABILITY

A. Introduction

Soil and groundwater are being severely threatened by pollution. As a result, their use for a large number of natural and anthropogenic functions is in danger. Given the slow, but definite accumulation in soils and the long-term response times of groundwater systems, the current abusive practices may not mean immediate problems in all cases, although these practices will certainly lead to problems in the future.

When confronting the threats to groundwater resources with the fundamental right of all human beings to live in an environment adequate for their health and well-being, as formulated in the Brundtland report[14] and widely accepted as a leading objective for development, we must conclude that some of the current developments are in conflict with this objective.

Whether or not the use of soil and groundwater is sustainable is not easy to access. Direct and straightforward conditions have not yet been formulated and it is questionable if, once they are, sufficient data are available to evaluate the situation. On the other hand, the assessment of these conditions is essential for the proper use of both soil and groundwater resources by both present and future generations.

The sustainable use of soil and groundwater basically requires two conditions:

1. No loss of the potential functions of groundwater and
2. Preservation of the diversity of the ecosystem and maintenance of the richness of species.

1. Criteria for Sustainability

If current pollution loads exceed the EC quality standards or if the abstractions exceed the natural recharge (critical levels), the function of drinking water

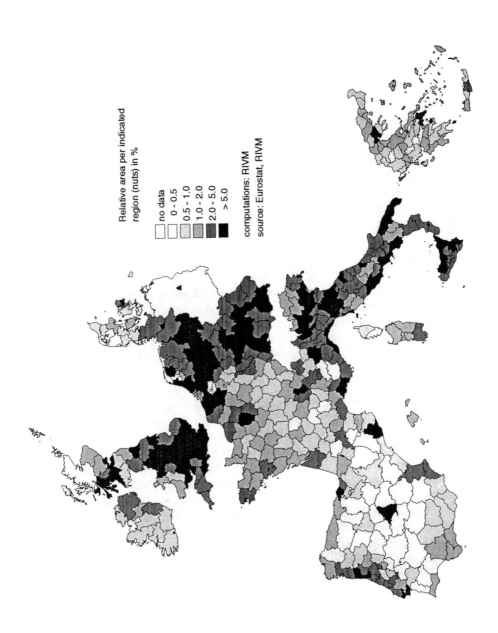

Relative area per indicated region (nuts) in %

no data
0 - 0.5
0.5 - 1.0
1.0 - 2.0
2.0 - 5.0
> 5.0

computations: RIVM
source: Eurostat, RIVM

is not sustainable as the first condition (1) is violated. The second condition (2) would be violated if due to polluted soil or groundwater flows the diversity of the existing ecosystems in the region is threatened. Criteria that can be used for the most important environmental threats to the sustainability of the drinking water function are described as follows:

a. Criteria for Condition 1: Prevent Loss of Functionality

Threats	Critical levels
Leaching of nitrate	EC standard for drinking water quality (50 mg/l)
Leaching of pesticides	EC drinking-water standard for the sum of all pesticides (0.5 μg/l)
Pollution by waste disposal sites	From the perspective of sustainability there is no acceptable level of contamination by landfill leachate. In this analysis, however, 2% of the area per region contaminated by landfills is used as a maximum tolerable level. This operationalization has been chosen on practical grounds and distinguishes the regions that are known to be seriously affected.
Overexploitation	Extraction volumes exceed natural recharge. Other critical elements of overexploitation are substantial drawdowns, salt water intrusion, mineral-rich water upcoming, and reduction or even stoppage of groundwater discharge to surface water.

Drinking water standards are, in general, stricter than standards for other functions such as agriculture (irrigation) and industry. Therefore, these standards can be used to indicate sustainability of the other groundwater uses as well.

FIGURE 7. (*Opposite*) The estimated groundwater area potentially contaminated by landfills as a percentage of the total area per region. Data have been obtained from Eurostat (population density per region), OECD (landfilled generated waste per country), and the Commission of the EC, 1991 (draft). The percentages are estimated by using the population total per region, the area, and the (estimated) average amount of waste per capita per country which is disposed to landfills (both municipal and industrial) during a period of 50 years. The amount of landfill is corrected for the amount of waste which is incinerated, composted, recycled, or exported. Furthermore, it is assumed that the leachate of landfills (in the long term) will reach the groundwater. As expected, the waste load in urban areas is relatively high. The figure also shows differences between (urban) areas in several EC countries. These differences are, for example, due to different amounts and treatment of waste in the various countries. In France, for example, the percentage of the municipal waste disposed to landfills is about 48%; on the other hand, in Greece and Ireland this is almost 100%.

Table 6. An Estimation of the Magnitude of the Contaminated Urban and Industrialized Areas in the EC

Source	Number	Amount of waste (tonnes)	Area (km²)	Potentially contaminated area (km²)
Industrial estates	12 × 10⁶		10,000	16–40 thousand
Landfills/ impoundments	60–120 thousand	3–6 thousand	600–1,200	900–7,200
Fuel storage tanks	3–6 × 10⁶			250–4,000
Mining waste dump sites	A few thousand	17,000 × 10⁶	250–500	350–5,000
Line sources			10–25 thousand	1,500–7,500
Dredged sediment dump sites	Hundreds			Hundreds
Hazardous waste sites	Hundreds			Hundreds
Estimated total contaminated area				20–60 thousand

Source: Franken and Glasbergen + (42)

b. Criteria for Condition 2: Maintain Diversity of Ecosystems   Ecological critical levels are, in general, stricter than the critical levels to avoid loss of functionality. This will be illustrated with some examples.

Nitrate
Nitrate concentrations in some major unpolluted rivers range between 0 and 1 mg/l,[66] which is considerably lower than the EC drinking water standard for nitrate (50 mg/l).

Acidification
The critical value for aluminum in natural freshwater ecosystems is 0.08 mg/l (CCE Technical report No. 19), whereas the EC drinking water standard is 0.2 mg/l.

Pesticides
On the base of observed no-effect concentrations of various test species, a logistic probability density function of the no-effect concentrations of all species in an ecosystem can be assessed. This function is used to estimate the concentration below which the long-term protection of 95% of all species is guaranteed. These concentrations can be used as ecological standards.[65] For pesticides these standards in surface water are, in general, considerably lower than EC drinking

water standards. For instance, the ecological standard for malathion is 0.004 μg/l, whereas the EC drinking water standard is 0.1 μg/l.

Dehydration
Groundwater-related ecosystems are very sensitive to changes in groundwater level. A lowering of the groundwater table of less than 0.1 m may have severe impact on these ecosystems. There are, however, no general ecological standards or critical levels for systematic change in the groundwater regime.

It is not possible to formulate general criteria for the sustainability of groundwater-related ecosystems due to variability in geophysical properties and to variability in sensitivity of ecosystems. These criteria will have to be assessed for each specific situation. Many valuable ecosystems with critical species are located in geographically restricted areas (nature reserves).

2. Threats to the Diversity of Ecosystems

In principle, the overview of threats to ecosystems should be obtained using a similar analysis which summarizes the extent to which current loads exceed criteria for sustainability of ecosystems. However, actual analysis is quite complex since:

- Various stress factors have cumulative effects on ecosystems.
- The sensitivity of ecosystems varies regionally due to variation in physical and chemical properties of the geosphere and variation in the sensitivity of species.
- The distribution of valuable ecosystems is geographically restricted in relatively small areas.

Ecosystems are subject to several environmental stress factors simultaneously. Generally, the cumulative effects are more severe than might be expected from each stress factor separately. There are many reports that ecosystems have declined on a large scale in the EC countries.[26] Often, the causes for decline are related to combinations of various environmental problems such as acidification, eutrophication, dehydration, pollution, disturbance, habitat destruction, and land-use management. A case study illustrates this.

Case
Germany
Table 7 illustrates various factors responsible for the decline of 581 plant species in the former Federal Republic of Germany.[6] General causes of decline are attributed for 68% of the species to agriculture (high nutrient input, use of herbicides and pesticides, drainage, and habitat destruction) and for 16% of the species to water management. The most important reason for decline is habitat destruction (36% of the species). Drainage is mentioned as a reason for decline for 29% of the plant species.

Table 7. Causes for the Decline in 581 Plant Species in the Former Federal Republic of Germany[a]

Specific reasons for the decline	Number of species
Destruction of special habitats	210
Drainage	173
Abandoning management	172
Landfill and construction	155
Changes in land use	123
Mineral exploitation	112
Erosion and other mechanical impacts	99
Use of herbicides	89
Deliberate removal by clearance, weeding, fire, etc.	81
River and lake management works	69
Collection	67
Eutrophication of water	56
Discontinuation of occasional soil disturbance	42
Water pollution	31
Urbanization of rural areas	20

General causes of loss of plant species	Number of species
Agriculture	397
Tourism	112
Raw material exploitation	106
Urban and industrial use	99
Water management	92
Forestry and hunting	84
Solid and liquid waste disposal	67
Pond management	37
Military activities	32
Transport	19
Science	7

[a] Adapted from Sukopp.[84]

Note: In general, the decline is caused by more than one factor.

Sensitivity of ecosystems varies regionally. For instance, critical sulfur and nitrogen deposition loads for various types of forest ecosystems depend on the buffering capacity of the soil and may differ up to tenfold in Europe among the various ecosystems (CCE Technical Report No. 1, 1991).[19] The sensitivity of species also varies. The distribution of ecosystems is geographically restricted to comparatively small areas. Exceeding of critical loads is only relevant on actual locations of ecosystems. Analysis of exceeding of critical limits needs an extensive analysis of the geophysical and ecological system, including regional differences in sensitivity of ecosystems and the precise locations of the relevant ecosystems.

In view of the previous findings it is not yet possible to give scientifically supported criteria for diversity of ecosystems (condition 2).

However, the severe decline of many ecosystems, both outside and inside nature reserves, and the decline of populations of many species clearly indicate that the environment is nonsustainable in many parts of the EC.

This analysis of threats is based on current environmental stresses only. If loads of the past decennia had been taken into consideration, the conclusions would be probably more pronounced.

Strategies aimed at realizing sustainability should start by defining a general protection level that ensures a multifunctional use. This level of sustainability can be described as follows:

1. Nitrogen load: If groundwater is to be suitable for drinking water the EC drinking water standard of 50 mg/l should not be exceeded. Model computations for highly sensitive soils indicate that the maximum nitrogen application rate is to be about 100 kg N/ha/year on arable land and 300 kg N/ha/year on grassland in order to achieve this drinking water standard.
2. Pesticides: Several hundreds of active ingredients are registered for use in agriculture. There are large differences in the physical and chemical behavior of these ingredients. Setting a maximum allowable load for total pesticides (per hectare) will not necessarily reduce the leaching to the groundwater. Effective control of pests may require a certain minimum amount of active ingredients. From the point of view of pest control, reductions of pesticide below this amount are not always sensible. The only solution to the environmental problem caused by pesticides is to ban those persistent pesticides having a high potential of leaching to the groundwater or accumulation in the soil.
3. Landfills: Isolation of new dump sites and sanitation of old dump sites.

The resulting measures aimed at achieving the general protection level may differ per region, depending on the kind of environmental problems in the particular region and the magnitude by which critical levels for sustainability are exceeded. Therefore, different approaches are suggested in relatively undisturbed and disturbed areas.

- Undisturbed areas: In relative undisturbed areas the environmental stresses do not exceed critical limits for sustainability. Measures should be taken to prevent deterioration (standstill). In other words: keep the last "good" parts of the EC in their current state. The "green areas" of the integration map give a first impression of their location.
- Sanitation (or improvement) areas: In relatively disturbed areas the environmental stresses exceed critical limits, indicating nonsustainability. Considerable reductions of environmental stresses (loads/abstractions) are needed in order to realize an improved situation meeting the critical levels of sustainability. Priorities have to be defined, both in time and space, as the central part of a functional groundwater management plan.

The highest priority should be given to improvement in areas located around drinking water resources and groundwater-dependent ecosystems. Possible measures are reductions of environmental stresses in the vicinity of these vulnerable areas and reallocation of polluting "source" functions.

Secondly, priority should be given to all other regions, where generic measures must lead to improvements that meet criteria for sustainability. A relevant time path must be agreed upon, within which this goal can be achieved without allowing irreversible deterioration.

These generic measures must be implemented differently in various regions because of such regional differences as:

- Environmental loads
- Abiotic sensitivity due to the natural geohydrological and geochemical properties of the soil and groundwater system. Comparable loads may, for instance, have regionally different impacts in terms of leaching and/or lowering of groundwater tables.
- Critical levels for sustainability of the use due to differences in sensitivity of ecosystems. In Section IV, some examples have been given to show that critical levels for ecosystems are in general stricter than those for the use of groundwater as drinking water.

If the above mentioned strategies are not pursued in the near future, some areas will be irreversibly affected. In these "abandoned" areas, improvement towards sustainability of the use may lose its feasibility, even with extensive (but too late) reduction measures and physical planning. Recovery of such "abandoned" areas by purification is unlikely due to the enormous volumes, unreachable position, and slow flow of groundwater.

These "abandoned" areas may necessitate an isolation strategy in order to prevent further dispersion.

REFERENCES

1. Aldwell, C.R., Burdon, D.J., and Sherwood, M., Impact of Agriculture on Groundwater in Ireland, 1983.
2. Aldwell, C.R., Burdon, D.J., and Sherwood, M., Impact of Agriculture on Groundwater in Ireland, 1983.
3. Aust, H. and Kreysing, K., Geologische und geotechnischer Grundlagen zur Tiefversenkung von flüssigen Abfälle und Abwässern, *Geol. Jahrb.*, Reihe C, Heft 20, 1978.
4. Baart, A.C. and Diederen, H.S.M.A., Calculation of the atmospheric deposition of 29 contaminants to the Rhine catchment area, TNO Environmental and Energy Research, Report No. R91/219, Delft, the Netherlands, 1991.
5. Balades, J.D., Cathelain, M., Marchandise, P., Peybernard, J., and Pilloy, J.C., Chronic pollution of intercity motorway runoff waters, *Water Sci. Tech.*, 17, 1165, 1984.

6. Baldock, D., Agriculture and habitat loss in Europe, WWF International CAP Discussion Paper No. 3, 1990.

7. Barber, C., Young, C.P., Blakey, N.C., Ross, C.A.M., and Williams, G.M., Groundwater contamination by landfill leachate: distribution of contaminants and factors affecting pollution plume development at three sites, UK, in *Quality of Groundwater,* van Duijvenbooden, W., Glasbergen, P., and van Lelyveld, H., Eds., Proc. Int. Symp., 23–27 March 1981, Noordwijkerhout, Elsevier Scientific Publishing, Amsterdam, 1981, 239.

8. Bartolami, G.C., Molfetta, A. di, Fenoglio, T., and Lachello, A., Interaction between human activities and groundwater resources in the province of Turin (Italy), in *Hydrogeology in the Service of Man. Memoires of the 18th Congress,* International Association of Hydrogeologists, Cambridge, 1985, 169.

9. Boesten, J.J.T.I. and Linden, A.M.A., van der, Modelling the influence of sorption and transformation on pesticide leaching and persistence, *J. Environ. Qual.,* 20, 425, 1991.

10. Brammer, H. and Brinkman, R., Changes in soil resources in response to a gradually rising sea-level, presented at int. workshop: effects of expected climate change on soil processes in the tropics and subtropics, Nairobi, 12–14 February 1990.

11. Braun, F., Schüssler, W., Wanzinger, M., and Wehrle-von Borzyskowski, R., Neue Untersuchungen zur Analytik un Vertbreitung von Polychlorbiphenylen (PCB) and Pflanzenbehandlungsmitteln. Bayerische Landesanstalt für Wasserforschung, 1990.

12. Breukel, R.M.A. et al., Das Hochwasser 1988 im Rheingebiet. CHR/KHR Bericht No. I-9, CHR, Lelystad, the Netherlands, 1990.

13. British Geological Survey, Trace element occurrence in British groundwaters, Natural Environment Research Council, Swindon, Wiltshire, England, 1989.

14. Brundtland, G.H. et al., *Our Common Future,* Commission on Environment and Development (WCED), Oxford University Press, Oxford, 1987.

15. Bundesminister für Umwelt, Naturschutz und Reaktorsicherheit (1987) Schwer punkte des Grundwasserschutzes.

16. Bureau de recherches geologiques et minières: Ministère l'environment (1986) Teneurs en nitrate des nappes phreatiques de la France, 2nd ed., map on scale 1:1,500,000.

17. Burkl, G., Hagenguth, H., Schlaffer, T., and Scholz, A., Survey of the condition of underground water in the Munich area, with special reference to the content of volatile halogenated hydrocarbons, *Vom Wasser,* 68, 43, 1987.

18. Canter, L.W., Knox, R.C., and Fairchild, D.M., *Ground Water Quality Protection,* Leis, Chelsea, MI, 1987.

19. CCE Technical Report No. 1, Mapping Critical Loads for Europe, Hettelingh, J.P., Ed., 1991.

20. Commission of the European Communities, Directive on waste substances, No. L 194/47–49, 1975.

21. Commission of the European Communities, Die Ausbringung tierischer Excremente auf landwirdtschaftlich genütze Flächen in der Gemeinschaft, EC Report No. 48, August 1978.

22. Commission of the European Communities, Directive on hazardous waste substances, No. L 84/43–48, 1978.

23. Commission of the European Communities, Directive on the protection of ground-water against contamination by discharges of hazardous substances, No. L 20/43–48, 1979.

24. Commission of the European Communities, Directive on the quality of drinking water for human consumption, No. L 299/11–29, 1980.

25. Commission of the European Communities, The state of the environment in the European Community 1986, 1991.

26. Commission of the European Communities, The state of the environment in the European Community, 1986. Office for official publications of the EC, Luxembourg, 1987.

27. Commission of the European Communities, Resolution on waste management and old disposal sites, No. C 190/154–161, 1987.

28. Commission of the European Communities, The state of the environment in the European Community, 1991 (draft).

29. Commission of the European Communities, Directive on the protection of fresh water, coastal water and sea water against pollution caused by nitrates from diffuse sources of pollution (draft), 1991.

30. Corine, Examples of the use of the results of the programme 1985–1990. Directorate-General Environment, Civil Protection and Nuclear Safety, Commission of the European Communities, 1991.

31. Custodio, E., Characterisation of aquifer over-exploitation: comments on hydrological and hydrochemical aspects: the situation in Spain, presented at XXIII Int. Congress of Int. Assoc. Hydrogeologists, Puerto de la Cruz, 1990.

32. Delgado Rodrigues, J., Lobo Ferreira, J.P., Braga dos Santos, J., and Nélida Miguéns, Caracterização Sumária dos recursos hídricos subterrâneos de Portugal, Laboratório Nacional de Egenharia Civil, Lisbon, 1989.

33. Department of the Environment, Assessment of Groundwater Quality in England and Wales, Her Majesty's Stationery Office, London, 1988.

34. Dorhofer, G. and Fritz, J., Hydrogeological experiences with hazardous waste disposal sites in near-surface claystone formations and conclusions for the planning of future sites, *Hydrogeology and Safety of Radioactive and Industrial Waste Disposal*, International Association of Hydrogeologists, Documents du BRGM No. 160, Vol. 1, Orleans, 1988, 283.

35. Dornier System GMBH, *Contaminated Land in the European Community*, Umweltbundesamt, Berlin, 1987.

36. Drecht, G. van, Modelling of nitrate leaching from agricultural soils on a regional scale, Proc. of 2nd Int. Symp. Environmental Geochemistry, September, 1991.

37. Duijvenbooden, W. van, Ed., De kwaliteit van het grondwater in Nederland. RIVM, Bilthoven, the Netherlands. Report No. 728820001, 1989.

38. EFMA, Application rates of fertilizer nitrogen to different crops within the European Fertilizer Manufacturers Association, 1991.

39. Egmond, van N.D. and Minderhoud, A., Eds., The Environment in Europe: a Global Perspective, RIVM Report No. 481505001, 1992.

40. FAO, *Yearbook on Fertilizers*, 1989, 39.

41. Flemming, G., Ed., Landfills and Contaminated Land: Time Delayed Responses, based on the workshop organised within the framework of the Chemical Time Bombs Project of VROM, IIASA, and Mondiaal Alternatief by the University of Strathclude, Glasgow, Scotland, February, 1992.

42. Franken, R.O.G. and Glasbergen, P., Groundwater contamination in the European Community; non diffusive sources, urban and industrial areas, RIVM Report, in preparation.
43. Fraters, D. and van Beurden, A.U.J.C., Cadmium mobility and accumulation in soils of the European Communities, RIVM Report, No. 481505005, 1993.
44. Fried, J.J. et al., Groundwater resources in EC countries, investigation by the Commission of the European Communities, 1982.
45. Fried, J.J., IEE, IEA, EIW, European Institute for Water, 1991.
46. Grassi, D. and Tadolini, T., The effects of karstic groundwater overdrafting in Apulia (Southern Italy), IAH, Proc. Aquifer Overexploitation, April 1991, Canary Islands, Spain, 1991.
47. Headworth, H.G., The influence of human development on groundwater quality, *Groundwater Water Resourc. Plan.*, 1, 233, 1983.
48. Heij, G.J. and Schneider, T., Eds., Dutch Priority Programme on Acidification: final report, second phase Dutch priority programme on acidification, Report No. 200-09, 1991.
49. Hekstra, G.P., Global warming and rising sea levels: the policy implications, *Ecologist*, 19, 4, 1989.
50. Houghton, J.T., Jenkins, G.J., and Ephraums, J.J., Eds., *Climate Change: the IPCC Scientific Assessment*, Cambridge University Press, Cambridge, 1990, 365 pp.
51. *Hydrological Processes and Water Management in Urban Areas*, Proc. Duisberg Symp., April 1988, IAHS Publ. No. 198.
52. Institute for Environmental Studies, Verdroging van Natuur en Landschap in Nederland, Vrije Universiteit, Amsterdam, 1989.
53. Izrael, Y.A., Hashimoto, M., and Tegart, W.J.M., Eds., *Climate Change: the IPCC Impact Assessment*, Australian Government Publishing Services, Canberra, 1990, 358 pp.
53a. Jaarsveld, Van, personal communication.
54. Kohsiek, L.H.M. and R., v.d. Ven FHM, Sustainable Use of Groundwater Problems and Threats in the European Communities, RIVM/RIZA Report No. 600025001, 1991.
55. Krebs, G., Contrôle et surveillance de la salinité de la nappe phréatique dans le Département du Haut-Rhin. Rapport de synthèse des mesures effectuées en 1990, BRGM, Alsace, 1991.
56. Lahl, U. et al., Groundwater pollution by nitrate. (IAH/IAHS Proc. Symp. on Groundwater Resources Planning), 1983, 1159.
57. Landini, F. and Pranzini, G., Hydrogeologic balance of an overexploited aquifer (Prato Fan, Italy) and corrective measures for the water resources management, IAH, Proc. Aquifer Overexploitation, April 1991, Canary Islands, Spain, 1991.
58. Latour, J.B. and Groen, C.L.G., The state of the environment of the lowland peat area and the calcareous dunes, Report No. 711901001, RIVM, Bilthoven, the Netherlands, 1991.
59. Linden, A.M.A. and van der Boesten, J.J.T.I., Berekening van de mate van uitspoeling en accumulatie van bestrijdingsmiddelen als functie van hun sorptie-coëfficiënt en omzettingssnelheid in bouwvoormateriaal, Report No. 728800003, RIVM, Bilthoven, the Netherlands, 1989.
60. Llamas, M.R., National Parks Tablas de Daimel and Doñana.

61. Margat, J., La surexploitation des aquifèrs. Bureau de Recherches Géologi ques et Minières, Orleans, 1991.
62. Margat, J. and Roux, J.C., L'impact des activités humaines sur les eaux continentales, Bureau de Recherches Géologiques et Minières, Orleans, 1986.
63. Margat, J., La Gestion des Eaux Souterraines en France, Bureau de Recher ches Géologiques et Minières, Orleans, 1986.
64. Margat, J., La Gestion des Eaux Souterraines, Seminaire Evaluation des Politiques Publiques de l'Environnement, 15 February 1990.
65. Marsland, P.A. and Hall, D.H., Gravel extraction and water resources management of the Denge gravel aquifer, Kent, England, 1989.
66. Meybeck, M., Carbon, nitrogen, and phosphorus transport by world rivers, *Am. J. Sci.,* 282, 401, 1982.
67. Milbowa, Milieukwaliteitsdoelstellingen bodem en water, Tweede Kamer, vergaderjaar 1990–1991, 21990, No. 1, 1991.
68. Miljøministeriet, Danmarks Geologiske Undersøgelse, Status for grundvand og drikkevand i Danmark, 1990.
69. Minister für Umwelt Saarland, Sauberes Wasser sichern, 1989.
70. Ministry of the Environment, Environmental Impacts of Nutrient Emissions in Denmark, 1991.
71. Ministry of Transport, Public Works and Water Management, Water in the Netherlands, National Policy Document on Water Management, 1989.
72. Norsk Hydro, Agriculture and fertilizers, Norsk Hydro, Oslo, Norway, 1990.
73. OECD, OECD Environmental Data, Compendium 1987.
74. OECD, OECD Environmental Data, Compendium 1989.
75. OECD, *The State of the Environment,* OECD Publications, Paris, 1991.
76. Outcrop of principal aquifers and the vulnerability of groundwater to nitrate pollution, map presented to: Seminaire European Eaux-Nitrates, Toulouse, 1989.
77. Owen, M., Groundwater abstraction and river flows, presented at the IWEM Conference, 1991.
78. Razack, M., Baitelem, M., and Drogue, C., Impact of an urban area on the hydrochemistry of a shallow groundwater (alluvial) reservoir, town of Narbonne, France, 1990.
79. Redegørelse fra Miljøstyrelsen, Environmental Impacts of Nutrient Emissions in Denmark, Ministry of the Environment, Denmark, 1991.
80. RIVM (National Institute of Public Health and Environmental Protection), Pollution of groundwater in the EC caused by the use of nitrogen, phosphorus and potassium in agriculture: background document to the EC ministers seminar on groundwater, in preparation.
81. Sahuquillo, A., Andreu, J., and O'Donnell, T., *Groundwater Management, Quantity and Quality,* IAHS Publ. No. 188, 1989.
82. Stigliani, W.M., Doelman, P., Salomons, W., Schulin, R., Smidt, G.R.B., and van der Zee, S.E.A.T.M., Chemical time bombs: predicting the unpredictable, *Environmental,* 33, 4, 26, 1991.
83. Solomon, A.M. and Leemans, R., Climatic change and landscape-ecological response: issues and analyses, in *Landscape Ecological Impact of Climatic Change,* Boer, M.M. and de Groot, R.S., Eds., IOS Press, Amsterdam, 1990, 293.
84. Sukopp, H., 'Veränderungen von Flora und Vegetation', Beachtung ökologi scher Grenzen bei der Landbewirtschaftung, Sonderheft 197, 255, 1981.

85. UN, Groundwater in Eastern and Northern Europe, Natural Resources/Water Series No. 24, Department of Technical Co-operation for Development, New York, 1990.
86. UN, Groundwater in Western and Central Europe, Natural Resources/Water Series No. 27, Department of Technical Co-operation for Development, New York, 1991.
87. UN, Mapping critical loads for Europe, Report No. 259101001, Coordination Centre for Effects, RIVM, Bilthoven, the Netherlands, 1991.
88. Verbaan, A.A., Perspectives in Europe: a survey of options for a European spatial policy, Netherlands National Physical Planning Agency, 1991.
89. VROM, European aspects of soil protection, Reeks Bodembescherming No. 47, SDU, The Hague, 1985.
90. VROM (Ministry of Public Housing, Physical Planning and Environment), Risico- en effectanalyse van diverse opslagsystemen voor vloeibare aard-olieproducten. Reeks bodembescherming No. 76, SDU, The Hague, 1987.
91. VROM, Aanzet 10 jaren-scenario bodemsanering. Reeks bodembescherming No. 77, SDU Uitgeverij, The Hague, 1989.
92. Young, C.P., The impact of point source pollution on groundwater quality, in *Quality of Groundwater,* Van Duijvenbooden, W., Glasbergen, P., and van Lelyveld, H., Eds., Proc. Int. Symp. 23–27 March 1981, Elsevier Scientific Publishing, Amsterdam, 1981, 207.
93. Zonas con concentration significativas de Nitratos en Espana peninsular, map presented to: Seminaire European Eaux-Nitrates, Toulouse, 1989.

CHAPTER 3

The Consequences of Bound Pesticide Residues in Soil

A. Calderbank

TABLE OF CONTENTS

0-87371-530-6/94/$0.00 + $.50

ABSTRACT

Depending on their nature, pesticides and/or their degradation products bind to a greater or lesser extent with soil colloids, particularly as the residues "age" in soil. When residues become firmly bound in this way they lose their biological activity but are nevertheless subject to slow microbiological degradation. Most bound pesticide residues cannot be monitored satisfactorily in soil. They should not be of concern, since binding to soil colloids may be considered as an effective and safe method of deactivation of residues. The bound residues become a tiny part of the organic matter of the soil. The binding process also restricts leaching and allows slow degradation to products which pose no short term or long term problems.

I. INTRODUCTION

A significant proportion, which typically ranges from 20 to 70%, of all pesticides applied in agriculture, remains in the soil as a persistent residue bound to the soil colloids. The bound residue, which generally increases in quantity and firmness of binding with increasing time in the soil (aging), may be parent chemical or degradation product(s) or combinations of both. In only a few instances can the soil-bound residue be extracted from the soil and fully characterized or monitored.

Bound pesticide residues in soil tend to become more resistant to degradation and show little evidence of biological activity. Binding occurs mainly to the organic matter and clay minerals of the soil. The exact nature of the binding is in most cases uncertain, but is probably initiated by physical adsorptive processes and in some cases, such as the chloroanilines, by actual covalent binding to the humic acids.

The discovery of pesticide-bound residues in soil almost 30 years ago, which often constitute the major part of the soil residue, created a regulatory problem which has still not been adequately resolved by many authorities.

The subject has been reviewed in some detail previously.[2-4] The purpose of this paper is to highlight the salient points and draw attention to the anomalies in approach which currently exist.

II. NATURE AND EXTENT OF BINDING

There are, broadly, two mechanisms by which organic chemicals, or pesticides in particular, interact with the soil colloids and become bound, viz., by physical processes, involving adsorption or by chemical reaction. In the latter case a stable chemical linkage is formed between the parent or degradation product and the natural organic matter of the soil. Thus the chloroanilines liberated by partial degradation of the urea herbicides, acylanilides, and phenyl-carbamates form bound residues by chemical incorporation into soil organic matter.[5] These bound residues can constitute up to 90% of the "pesticide" residue entering the soil from such products.

The second broad mechanism which results in binding between organic chemicals and soil colloids is by adsorption. Mechanisms of adsorption, discussed by Hamaker and Thompson,[6] vary and can include ion exchange, van der Waals interaction, charge transfer, hydrogen bonding, and others. In many cases more than one type may be involved for particular compounds. The amount of chemical which is adsorbed to different soils can be determined by shaking solutions of chemical with a slurry of the soil for several hours until equilibrium is established. The adsorption constants Kd or Koc (related to organic carbon content) derived from such studies give a measure of short-term adsorption (adsorption occurring over a few hours or days) and take no account of further adsorption or binding which occurs through "aging". There is clearly a further slow adsorption, or binding to soil, occurring over weeks, months, or even years. This is of immense practical importance since it is usually irreversible and results in deactivation or removal of biological activity just as effectively as metabolic or chemical degradation. Virtually all our knowledge relates to short-term adsorption, and no progress has been made to unravel the more complex problems involved in longer-term binding of pesticide residues in soil.

III. BIOLOGICAL CONSEQUENCES OF BINDING

When pesticides become bound in soil they lose their biological activity, especially with longer residence time in soil.[7,8] There have been several reports, summarized by Fuhr,[9] of uptake of small amounts of radioactivity by plants, earthworms, and even release by microorganisms from bound pesticide residues. Such amounts are generally below 1% of the total bound residue in the soil and may have been derived from small amounts of unbound residue or from $^{14}CO_2$.

Such experimental uptake studies have little relevance to the real life situation in field soils.[9] Even if traces of pesticide or metabolite could be released from the bound state by reversible desorption or by microbial action, such processes would be expected to result in slow release of very small quantities. These supposed "free" residues are largely hypothetical, but, if present, would be subject to the normal, more rapid degradative processes affecting "free" pesticides in soil. It is entirely predictable that the very slow release of bound pesticide residues into soil solution should be of no toxicological or ecological concern.

IV. DEGRADATION OF BOUND RESIDUES

There is little doubt that the binding of pesticides to soil colloids slows down their rate of decomposition. There is nothing surprising about this; the phenomenon has been recognized to occur with other organic chemicals, both synthetic and natural in soil.[4] Although the rates of degradation of organic compounds are considerably retarded when bound to soil, degradation always continues in the bound state, albeit at a slower rate.[10-12] Nevertheless, if bound residues persist for many years, the question arises: could there be an indefinite accumulation of pesticide-bound residues in soil — the so-called "burdening" of soils? This seems improbable when considering the substantial evidence for the slow microbial decomposition of bound residues and the analogy with natural organic compounds which become part of the organic matter of the soil with half-lives of many years. Bound residues of pesticides will not accumulate indefinitely, but will plateau when the amount being degraded will be equal to the quantity of new pesticide reaching the soil each year.[4,9] We have to realize we are considering the turnover of a few kilograms per hectare of organic carbon derived from pesticides in a total carbon pool which is about 20,000 Kg ha^{-1} in an average depth of an agricultural soil.[13]

The bound pesticide residue will become a tiny part of the slowly degrading organic complex of the soil, which contains a vast array of organic substances, largely of unknown structure and toxicity. From the organic matter turnover point of view, the amounts of pesticide-bound residues are hardly significant.

V. REGULATORY CONSEQUENCES

Clearly, the most important question concerning a pesticide residue in soil is the persistence of any biological activity and/or its leachability to groundwater. Pesticide residues should not be of concern if they are firmly bound in the soil and show no significant biological effects. Thus, from a regulatory, as well as scientific point of view, it is important to differentiate between the free pesticide residue and the bound, or unavailable residue and likewise between persistence of a residue, as characterized by biological persistence or chemical persistence. The two do not necessarily go hand in hand. In some instances a pesticide with a short half-life in soil, such as an organophosphate insecticide, may exert quite harmful effects for a short period. Another pesticide which forms biologically inactive bound residues with a long half-life may be absolutely free of harmful effects. Thus persistence of pesticide residues in soil, per se, should not be regarded as harmful, or indeed as an undesirable property.

Mostly, bound pesticide residues cannot be detected in soil, except by using radiolabeling techniques. In such cases, monitoring techniques for bound pesticide residues in the field are not usually available and half-life values, based on loss of extractable (unbound) parent, are usually accepted by regulatory authorities. This makes sense, since it is the biological half-life which is, in effect, being measured.

A few pesticides, notably paraquat, diquat, and glyphosate, can withstand vigorous extraction with strong acid or alkali, which destroys the soil structure and releases the bound residue from the soil adsorption sites without decomposing the parent chemical.[11,14] Since these bound residues can be monitored in the field, long half-lives can be assigned to these chemicals, and on this account they are regarded adversely by some regulatory authorities. To differentiate between bound residues which cannot be adequately monitored and those which can, hardly makes scientific sense and responsible government scientists need to guard against applying such double standards.

It can be argued that the binding of pesticide residues to soil colloids represents the most effective and safest method of decontamination by rendering the molecule innocuous and allowing slow degradation in the bound state to products that pose no short-term or long-term problems. After all, the formation of "bound organic residues" is a natural and essential process — in essence it is the genesis of the stable fractions of soil organic matter.

REFERENCES

1. Bailey, G.W. and White, J.L., Review of adsorption and desorption of organic pesticides by soil colloids, with implications concerning pesticide bioactivity, *J. Agric. Food Chem.*, 12, 324, 1964.
2. Kaufman, D.D., Bound and conjugated pesticide residues, in *Bound and Conjugated Pesticide Residues*, Kaufman, D.D., Still, G.G., Paulson, G.D., and Bandal, S.K., Eds., ACS Symp Series 29: 1, 1976.
3. Roberts, T.R., Non-extractable pesticide residues in soils and plants, IUPAC Reports on pesticides (17), *Pure Appl. Chem.*, 56, 945, 1984.
4. Calderbank, A., The occurrence and significance of bound pesticide residues in soil, *Rev. Environ. Contam. Toxicol.*, 108, 71, 1989.
5. Mansour, M. et al., Formation and binding of chlorinated anilines to soil organic matter and their breakdown products, this volume, Chapter 4.
6. Hamaker, J.W. and Thompson, J.M., Adsorption, in *Organic Chemicals in the Soil Environment*, Vol. 1, Goring, C.A.I. and Hamaker, J.W., Eds., Marcel Dekker, New York, 1972, 49.
7. Klein, W. and Scheunert, I., Bound pesticide residues in soil, plants and food with particular emphasis on the application of nuclear techniques, in *Agrochemicals: Fate In Food And Environment*, Proc. Intern. at Symp. IAEA, Vienna, 1982, 177 (1982).
8. Khan, S.U. and Dupont, S., Bound pesticide residues and their bioavailability, in *Pesticide Science and Biotechnology*, Greenhalgh, R. and Roberts, T.R., Eds., Proc. 6th Int. Congr. Pesticide Chemistry, IUPAC. Blackwell Scientific, London, 1987, 417.
9. Fuhr, F., Non-extractable pesticide residues in soil, in *Pesticide Science and Biotechnology*, Greenhalgh, R. and Roberts, T.R., Eds., Proc. 6th Int. Congr. Pesticide Chemistry, IUPAC, Blackwell Scientific, London, 1987, 381.

10. Zunino, H., Borie, F., Afuilera, S., Martin, J.P., and Haider, K., Decomposition of ^{14}C-labelled glucose, plant and microbial products and phenols in volcanic ash-derived soils of Chile, *Soil Biol. Biochem.*, 14, 347, 1982.
11. Nomura, N.S. and Hilton, H.W., The adsorption and degradation of glyphosate in five Hawaiian sugarcane soils, *Weed Res.*, 17, 113, 1977.
12. Jenkinson, D.S. and Rayner, J.H., The turnover of soil organic matter in some of the Rothamsted classical experiments, *Soil Sci.*, 123, 298, 1977.
13. Sauerbeck, D.R., Influence of crop rotation, manurial treatment and soil tillage on the organic matter content of German soils, in *Soil Degradation*, Boels, D., Davies, D.B., and Johnston, A.E., Eds., Balkema, Rotterdam, 1980, 163.
14. Tucker, B.V., Pack, D.E., and Ospenson, J.N., Adsorption of the bipyridylium herbicides in soil, *J. Agric. Food Chem.*, 15, 1005, 1967.

CHAPTER 4

Formation and Binding of Chlorinated Anilines to Soil Organic Matter

M. Mansour, W. Völkel, I. Scheunert, and F. Andreux

TABLE OF CONTENTS

0-87371-530-6/94/$0.00 + $.50

ABSTRACT

Within the pesticides used in agricultural practice, one of the most important chemical classes are substituted anilines (phenylurea derivatives, alkylanilides, phenylcarbamic acid esters, dinitroanilines). Due to the high percentage of the formation of bound residues in the case of anilines (up to 95% in one vegetation period), covalent binding of pesticide residues to soil humus is receiving increasing attention. Faced with the difficulty of characterizing the pesticide residue-humus bond, model approaches have been employed. It is known that various anilines react strongly in the presence of enzyme with phenolic humus constituents and the reaction rates are dependent on the substituent groups. Due to the nature of soil humus, the study of incorporation of irreversible fixation of pesticides or their transformation products is difficult. Thus model reactions are preferred and have contributed to the understanding of interactions of pesticides or other organic pollutants with soil organic matter and of the chemical reactions during polycondensation. The objective of this work was to identify the reaction products between a humic acid monomer (catechol) previously used for polycondensation studies and anilines during oxidation in the laboratory under controlled conditions, and to study the influence of catalysts on the formation of additional products.

I. INTRODUCTION

Multiple binding mechanisms are effective between organic xenobiotics and soil organic matter, physical, physicochemical, and chemical, most of them being widely unexplored. Organic xenobiotics may be bound to low-molecular, water-soluble humic substances, as well as to high-molecular insoluble ones. In the first case, mobility in soil and leaching from soil are promoted, resulting in the presence of pesticides in leachate, even if the nonbound substance is water insoluble. In contrast, in the second case mobility in soil and leaching from soil are strongly reduced; the same applies to degradability and bioavailability. Even the extractability of the pesticide residues by organic solvents is strongly inhibited; therefore, these residues are called "unextractable" soil-bound residues. Due to the steady molecular changes and rearrangements within humic macromolecules, there is a steady transition between the residues bound to lower- and those bound to higher-molecular humic substances.

In this paper, chlorinated anilines are taken as examples for binding processes between xenobiotics and soil organic matter. This group of compounds shows a high tendency to form bound residues, both to water-soluble humic substances in leachate and to high-molecular humic substances in so-called "unextractable pesticide residues".[1,2]

Chlorinated anilines are widespread soil pollutants originating in most cases from the use of pesticides containing aniline moieties (Table 1), such as phenylurea herbicides (monuron, monolinuron, buturon, metobromuron, diuron, linuron, neburon, metoxuron, maloran, chlortoluron, and isoproturon), acylanilides (monalide, propanil, swep, dicryl, solan, chlorpropham, and fenfuram), acylureas, and others.[3] After the release of chlorinated anilines from these pesticides, in many cases they are bound immediately to soil constituents and thus are inaccessible to normal residue analysis.

II. BINDING OF CHLORINATED ANILINES TO WATER-SOLUBLE HUMIC SUBSTANCES

In Figure 1 an example for the long-term leaching of 4-chloroaniline from soil after binding to low-molecular water-soluble humic substances is given. ^{14}C-labeled 4-chloroaniline was applied to the topsoil of an outdoor lysimeter at a dose of 0.75 kg/ha. The percolate water was collected at a depth of 60 cm for 4 years and assayed for radioactivity. The radioactivity increased up to about 0.015 mg/l ^{14}C, equivalent to 4-chloroaniline, after about 1.5 years, then decreased again. The radioactivity was fully water soluble and was not extractable with organic solvents. However, after strong alkali treatment, 16% of this radioactivity was hydrolyzable and was identified as 4-chloroaniline.[4]

III. BINDING OF CHLORINATED ANILINES TO UNEXTRACTABLE RESIDUES IN SOIL

The binding of chlorinated anilines to unextractable residues in soil is considerable. Table 2 presents the formation rates of various chlorinated anilines in soil, 20 to 25 weeks after application of the ^{14}C-labeled compounds. Unextractable residues are defined here as those left in soil after soxhlet extraction with methanol for 48 h. The absolute amount of bound residues, expressed as % of ^{14}C applied, is highest for 3,4-dichloroaniline. However, if we consider the relative amount as compared to radioactivity recovered, bound residues are high-

Table 1. Pesticides Containing Aniline Moieties

Product	X	R
Fenuron	—	–CH$_3$
Monuron	4-Cl	–CH$_3$
Monolinuron	4-Cl	–OCH$_3$

Buturon	4-Cl	–CH–C≡CH
		  |
		  CH$_3$

Metobromuron	4-Br	–OCH$_3$
Diuron	3,4-Cl,Cl	–CH$_3$
Linuron	3,4-Cl,Cl	–OCH$_3$
Neburon	3,4-Cl,Cl	–C$_4$H$_9$
Metoxuron	3-Cl,4-OCH$_3$	–CH$_3$
Maloran	3-Cl,4-Br	–OCH$_3$
Chlortoluron	3-Cl,4-CH$_3$–	–CH$_3$
Isoproturon	4-CH-(CH$_3$)$_2$	–CH$_3$

Monalide	4-Cl	CH$_3$
		  |
		  –C–CH$_3$
		  |
		  CH$_3$

Propanil	3,4-Cl,Cl	–CH$_2$CH$_3$
Swep	3,4-Cl,Cl	–OCH$_3$

Dicryl	3,4-Cl,Cl	–C=CH$_2$
		  |
		  CH$_3$

Solan	3-Cl,4-CH$_3$	–CH–C$_3$H$_7$
		  |
		  CH$_3$

Chlorpropham	3-Cl	CH$_3$
		  |
		  –OCH
		  |
		  CH$_3$

Barban	3-Cl	–OCH$_2$–C≡C–CH$_2$Cl
Fenfuram	—	2-Methyl-furan

est for 4-chloroaniline and lowest for 2,4,6-trichloroaniline.[5] This agrees well with findings from other chemical groups where the formation of bound residues decreased with increasing chlorine content of the molecules.

IV. PLANT UPTAKE OF UNEXTRACTABLE RESIDUES IN SOIL

The question of bioavailability of unextractable soil-bound residues is decisive for the evaluation of their toxicological and ecotoxicological significance. Possible uptake by higher plants is of outstanding importance. It has to be considered that the uptake is different from the various soil fractions. Figure 2 shows the uptake of bound ^{14}C from various soil fractions by barley seedlings

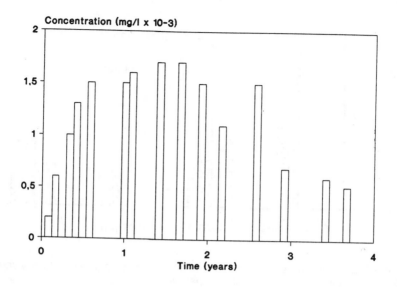

FIGURE 1. Concentration of radioactive compounds in leachate (depth 60 cm) after application of 4-chloroaniline-^{14}C in the upper soil layer (0 to 10 cm depth), (mg equivalent to 4-chloroaniline), (dose: 0.75 kg/ha).

2 weeks after application of ^{14}C-labeled chlorinated anilines. It is evident that the uptake is generally low. It is highest from the fulvic acid fraction and lowest from the humic acids.[6] Furthermore, chlorinated anilines are bound also in inorganic fractions.

V. METHODS TO DISSOLVE UNEXTRACTABLE RESIDUES

The formation of soil-bound pesticide residues takes place by multiple mechanisms that are poorly understood. In many cases, the xenobiotic may be en-

Table 2. Recovery of ^{14}C in Soil, One Growing Season (20 to 25 Weeks) After Application of ^{14}C-Labeled Chlorinated Anilines in Soil (Outdoor Experiments)

Chlorinated aniline (experimental conditions)	Total recovery [14]C (% of ^{14}C applied)	Bound residues[a] (% of ^{14}C applied)	Bound residues[a] (% of ^{14}C recovered)
4-Chloroaniline-[14]C (lysimeter)	32.4	30.8	95.1
3,4-Dichloroaniline-[14]C (lysimeter)	69.4	60.7	87.5
2,4,6-Trichloroaniline-[14]C (1.5 kg-pot)	25.2	4.3	17.1

[a] After soxhlet extraction with methanol for 48 h.

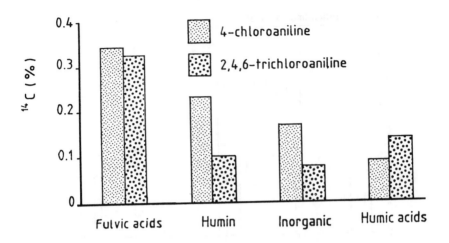

FIGURE 2. Uptake of residues of ^{14}C-labeled chlorinated anilines by barley seedlings (in % of ^{14}C in the respective fractions).

trapped in cavities of the organic macromolecule and becomes released when the humic acids are dissolved in dilute alkali, as shown for chemical groups different from chlorinated anilines.[7] Further methods to dissolve unextractable xenobiotic residues from soil are high-temperature distillation and supercritical fluid extraction. The latter method was applied to aged residues (12 years after application) of the phenylurea herbicide buturon applied to an outdoor lysimeter in a ^{14}C-labeled form.[8]

After 16 years, only 12% of the total ^{14}C recovered could be extracted by normal soxhlet extraction with methanol (Figure 3); the remaining radioactivity was bound in soil. However, by supercritical methanol, half of this radioactivity could be brought into solution. This radioactivity was not due to the parent compound, but to its metabolite 4-chloroaniline (15% of ^{14}C in soil).

It is assumed that the bound residues released by the methods discussed are not fixed by covalent bonds. However, between xenobiotics and soil organic matter covalent bonds may be formed. From these bounds the xenobiotic can be liberated in the laboratory only by chemical methods such as hydrolysis or degradative oxidation, or they may be inaccessible even to hydrolytic or oxidative attack. 4-Chloroaniline is an example also for bonds of this kind. Unextractable ^{14}C residues in soil after application of ^{14}C-4-chloroaniline were partly released after alkaline hydrolysis followed by steam distillation.

VI. MODEL EXPERIMENTS FOR COVALENT BONDS

It is assumed[9] that humic substances are formed by the polymerization of phenols liberated by the decomposition of plant material, especially lignin, or

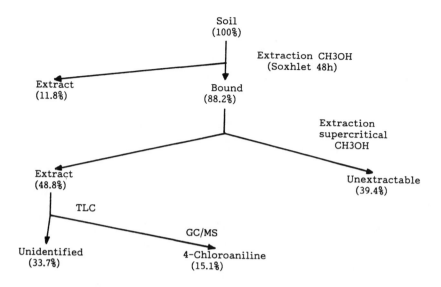

FIGURE 3. Supercritical fluid extraction of soil-bound residues of the phenylurea herbicide buturon-^{14}C, 16 years after application.

synthesized by microorganisms. In particular, the polymerization of *o*- and *p*-diphenols is accompanied by oxidation, to which those phenols are highly sensitive. Amino acids can be involved in these reactions in two ways: one is the addition, and the other is deamination and decarboxylation of the amino acids. These oxidative reactions and the resulting polymerization occur spontaneously, but generally a neutral or alkaline pH is necessary, and it seems that a catalyst is needed to achieve a high yield of polymers from phenolic units. The main catalysts found in soils are, besides microbial enzymes, inorganic compounds such as various minerals or clays, amorphous or crystallized oxides, hydroxides, or metal salts.[9,10]

In these reactions, the participating natural amino compounds may be replaced by amines of xenobiotic origin such as chlorinated anilines. These reactions have been demonstrated by various model experiments using enzymes as catalysts.[11] In our laboratory, various catalysts, including inorganic ones, have been used for such reactions.

4-Chloroaniline was reacted — under conditions similar to those in soil — with the humid acid monomer catechol.[8,12] The oligomer formed (Figure 4) was identified by NMR, IR, and HPLC/MS and found to be an anilinoquinone which may be regarded as a building block for the polymerization of humic macromolecules.[13]

The reaction occurs spontaneously with a low yield of oligomer only (0.32%) (Figure 5). Various enzymatic and inorganic catalysts, as present in normal soils, increase the yield, the most effective being the mineral pyrolusit and the oxidizing

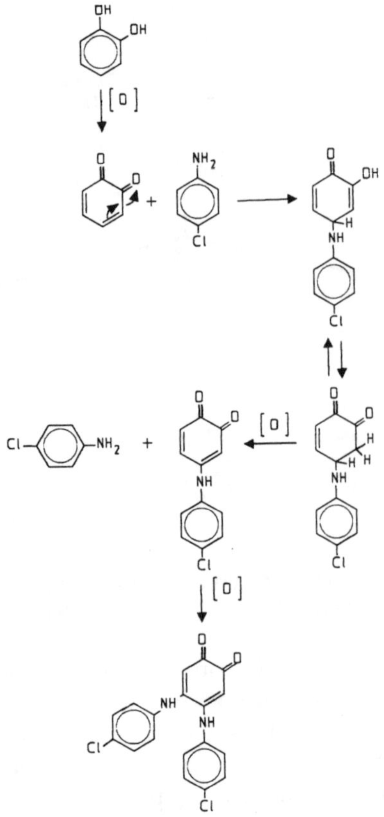

FIGURE 4. Identified structure of the main addition product of 4-chloroaniline and catechol.

enzyme tyrosinase. Further model reactions were performed with pyrolusit as a catalyst in order to obtain high yields, keeping in mind that the reaction occurs also in the absence of this substance.[14]

The humid acid monomers catechol, 5-methyl-catechol, and 5-isobutyl-catechol react with 3,4-dichloroaniline to form oligomers (Figure 6) with similar chemical structures as shown in Figure 4.

In the same conditions, Metribuzin, an asymmetric triazinone herbicide containing an amino group also, react with the humic acid monomer 5-methyl-catechol to the product shown in Figure 7.

VII. CONCLUSION AND FURTHER PROBLEMS

This study of a synthetic approach for the preparation of model substances of pesticide-bound residues has confirmed the high reactivity of NH_2- and NH-

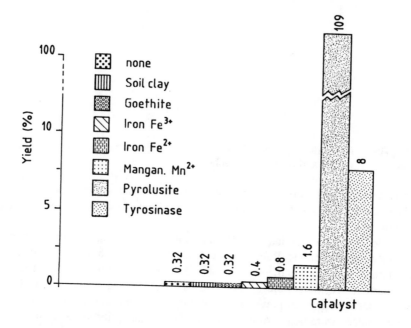

FIGURE 5. Yields of anilinoquinone in the presence of different catalysts.

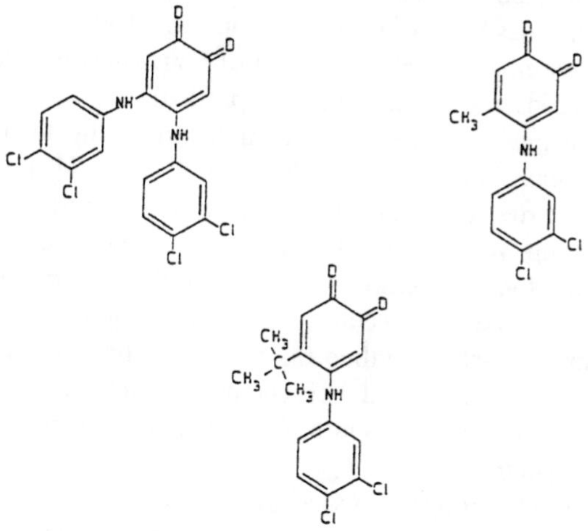

FIGURE 6. Identified structures of the main addition products of 3,4-dichloroaniline, catechol, 5-methyl-catechol, and 5-isobutyl-catechol.

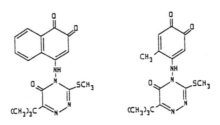

FIGURE 7. Reaction products of metribuzin with the humic acid monomers catechol and 5-methyl-catechol.

bearing molecules to form covalent addition products with oxidizing polyphenols. The evidence and possible predominance of this kind of reaction have suggested several remarks.

The first remark is that bound residues are formed mainly with soil humic and fulvic acid fractions. Rather than well-established structures, these fractions are defined according to their extraction procedure in alkaline medium. The fact that pesticide residues were extracted in the alkaline reagent together with the humic material, and not previously in an organic solvent, clearly confirms that these residues were bound to the alkali-soluble humus prior to the extraction.

The second remark is related to the nature of the bonds between humic material and pesticides. It is probable that noncovalent adsorption could occur, especially through coulombic and hydrogen bonds. However, the model approach also establishes the possibility of irreversible and covalent bonds. Further studies would have to determine the relative extent of both kinds of interactions under determined conditions and for each kind of pesticide.

For the same reason, it has seemed more promising to prepare addition products between catechol and the xenobiotic molecules. This means that, after the reaction had started and yielded an intense red or purple pigment, it was stopped by lowering the pH, and the pigment was extracted with an organic solvent. This was done successfully in case of the enzymic oxidation of 4-chloroaniline with phenolic molecules,[2] and in the case of the chemical autoxidation of a mixture of 4-chloroaniline and catechol.[14] Intermediates and products of the reactions were isolated, and their chemical structures were studied on the basis of elementar analysis, UV and infrared absorption spectroscopies, proton NMR, and mass spectroscopy.

A mechanism was proposed, and a structure of di-substituted iminoquinone was established for the main, red-colored reaction product. More recently, a similar structure was proposed in the case of 3,4-dichloroaniline.[15] Similar reactions are currently studied in the case of the dealkylated metabolites of atrazine, and only hypothetical structures have been proposed for the addition products.[16]

Finally, what is to be established now is whether or not the red addition products are the main intermediates in the incorporation process of the pesticides into the polymers. For instance, in the case of chloranilines, the red pigments are insoluble in water, but can finally be converted into water-soluble polymers under oxidative conditions. There are few works about the biodegradability of the polymer-bound pesticide, but it is probably much lower than that of the free molecule, as it occurs with biomolecules.

ACKNOWLEDGMENTS

This paper is part of the studies which have been carried out by GSF-Forschungszentrum für Umwelt und Gesundheit, Institut für Ökologische Chemie and CNRS-France, Centre de pédologie Biologique de CRNS Nancy-France since 1989, in the framework agreement AGF/CNRS. We are thankful to Prof. Dr. E. Heintz, representative of the CNRS in Bonn, and Dr. U. Deffner (GSF München) for providing the necessary facilities. The authors thank Mrs. T. Choné, Mr. J.P. Portal (Nancy), and Dr. M. Schiavon (E.N.S.A.I.A.) for their contributions to this study. Thanks are due to the CNRS/Paris France for our research project and financial assistance in the form of a post-doctoral fellowship (1992 to 1993) for W. Völkel.

REFERENCES

1. Klein, W. and Scheunert, I., in *Agrochemicals: Fate in Food and the Environment*, International Atomic Energy Agency, IAEA-SM-263/38, Vienna, 1982, 177.
2. Bollag, J.M. and Loll, M.J., Incorporation of xenobiotics into soil humus, *Experientia*, 39, 1221, 1983.
3. Mansour, M., Metabolically Resistant Chemicals, paper presented at XXIème Congres des Groupes Français des Pesticides: Reactions et Biotransformations des Xenobiotiques dans l'environement, May 22–23, 1991, Nancy Brabois, France.
4. Scheunert, I., Transport of Chemicals from Soil to Plants, paper presented at the AGF-Meeting, Chemicals in Food Chains, November 4–5, 1982, Bonn, Germany.
5. Scheunert, I., Mansour, M., and Adrian, P., Formation of conversion products and of bound residues of chlorinated anilines in soils, *Toxicol. Environ. Chem.*, 31–32, 107, 1991.
6. ter Meer-Bekk, C., Bildung, Charakterisierung und Bedeutung sogenannter "gebundener Rückstände" im Boden, Doctoral thesis, Technical University of Munich, 1986.
7. Völkel, W., Chonet, T., Mansour, M., and Andreux, F., Degradation dans un Sol Brun de Culture, de la 3,4-Dichloroaniline Libre ou Liée a un Monomère de Type Humique, paper presented at XXIème Congrés des Groupe Français des Pesticides, Reactions et Biotransformations des Xenobiotiques dans l'environment, May 22–23, 1991, Nancy Brabois, France.
8. Scheunert, I., Mansour, M., and Andreux, F., Binding of organic pollutants to soil organic matter, *Int. J. Environ. Anal. Chem.*, 46, 189, 1992.

9. Ladd, J.N. and Buttler, J.H.A., Comparison of some properties of soil humic acids and synthetic phenolic polymers incorporating amino derivatives, *Austr. J. Soil Res.*, 4, 41, 1966.

10. Lehmann, R.G., Cheng, H.H., and Harsch, J.B., Oxidation of phenolic acids by soil iron and manganese oxides, *Soil Sci. Soc. Am. J.*, 51, 352, 1987.

11. Bollag, J.M., Cross-coupling of humus constituents and xenobiotic substances, in *Aquatic and Terrestrial Humic Materials*, Christman, R.F. and Gjessing, E.T., Eds., 1983, 127.

12. Saxena, A. and Bartha, R., Modeling of the covalent attachment of chloroaniline residues to quinoidal sites of soil humus, *Bull. Environ. Contam. Toxicol.*, 30, 485, 1983.

13. You, I.-S. and Bartha, R., Evaluation of chemically defined model for the attachment of 3,4-dichloroaniline to humus, *Bull. Environ. Contam. Toxicol.*, 29, 476, 1982.

14. Adrian, P., Andreux, F., Mansour, M., Scheunert, I., and Korte, F., Reaction of the soil pollutant 4-chloroaniline with the humic acid monomer catechol, *Chemosphere*, 18, 1599, 1989.

15. Mansour, M., The influence of Soil Properties on the Rate of Degradation of Pesticides, paper presented in Rome, University La Sapienza, October 1992.

16. Völkel, W., Doctoral theses, Technical University, Munich, 1993.

SECTION III

Effects on Organisms

Effects on Microorganisms
5. Soil Microorganisms of Global Importance to Consider Ecotoxicology in an Economical and Ecotoxicological Way

6. A Method For the Ecotoxicological Risk Analysis of Polluted Sediments by the Measurement of Microbial Activities

7. Biosensors For Pollution Monitoring and Toxicity Assessment

8. The Effects of Heavy Metal Contaminated Sewage Sludges on the Rhizobial Soil Population of an Agricultural Field Trial

Effects on Plants
9. Effects of Heavy Metals, Organic Substances, and Pesticides on Plants

10. Phytotoxic Organic Compounds in Spruce Forest Soil: Chemical Analyses Combined With Seedling Bioassays

11. The Yorkshire Water Standard Plant Growth Trial for Toxicity Testing of Soils, Sludge and Sediments

12. The Use of Barley Root Elongation in the Toxicity Testing of Sediments, Sludges, and Sewages

13. Accumulation of Putrescine in Chromium Exposed Barley and Rape: A Potential Biomarker in Higher Plants

Effects on Invertebrates
14. Ecotoxicological Test Systems For Terrestrial Invertebrates

CHAPTER 5

Soil Microorganisms of Global Importance to Consider Ecotoxicology in an Economical and Ecological Way

P. Doelman and J.W. Vonk

TABLE OF CONTENTS

I. INTRODUCTION

To diagnose and to prognose the impact of environmental pollutants on soil ecosystems is hardly possible. Sheenan et al. (1984) have provided theoretical time table schedules with respect to changes on a biochemical level as well as on an ecosystem level. However, applicable data based on valuable research are hardly available, since it needs time periods of at least 10 years to collect these data.

The prediction of side effects of the application of pesticides on nontarget microorganisms has already been a point of discussion for almost 20 years and is still considered to be difficult nowadays (Domsch, 1991).

Nevertheless, it is our opinion that microorganisms are especially suitable to act as a sharp mirror of environmental pollution, i.e., to function as first warning systems due to their ubiquity, size, and versatility and their important role in foodwebs and recycling of elements. They are ubiquitous with population densities as high as 10^6 to 10^8 microorganisms per gram of soil. Moreover, due to their size they are relatively vulnerable: the ratio surface area over volume is extremely large. Unfortunately, their caress factor is extremely low in comparison to deer. Due to their physical framework, their occurrence as single cells, the growth of microorganisms on specific substrates in relation to their specific environment (e.g., pH, redox potential, moisture content) can easily be studied. However, for special chemicals with a specific biological action, e.g., inhibitors of choline esterase and photosynthesis inhibitors, microorganisms are not very sensitive.

It is our aim to discuss various conceptual approaches which are already applied or potentially to be applied. It concerns determination of inhibitory or other toxic effects of commonly present diffuse soil compounds such as heavy metals, pesticides, and chlorinated organic compounds towards the soil microflora. Special attention will be given to the ability of microorganisms to transform compounds which may partly bioremediate soils, but may also cause more serious ecotoxicological effects due to formation of higher toxic metabolites. In the following sections the following aspects will be discussed:

- Criteria for ecotoxicological tests (2)
- Heterogeneity of soil and sediment and its consequences (3)
- The soil microbial vs. the soil faunal approach (4)
- Dose-effect relations (5)
- Transformation of environmental pollutants (6)
- Awareness of limited biotransformation (7)
- Recommendations for overall measurements (8)

II. CRITERIA FOR ECOTOXICOLOGICAL TESTS

The purpose of considering ecotoxicology in its broadest sense is to avoid reduction of biological diversity of ecosystems.

It is basically assumed that for the degradation of each organic compound, a suitable microbe and biochemical pathway exist irrespective of its molecular complexity. This phenomenon is called "the principle of biological infallibility". In this context, strong irreversible encroachments on what has been established during microbial evolution cannot be tolerated, because it will hamper microbial mediated organic matter decomposition. In principle, applied ecotoxicological tests should reflect this array of degradation pathways. However, the suitability of those tests to properly assess toxic effects is limited by our incapability to determine real no-effect levels and to define the time period for reversibility.

The intensive use of pesticides demanded a practical approach which was carried out by Domsch et al. (1983). They indirectly defined irreversible as an agent which caused still 100% inhibition 30 d after treatment or 70% inhibition 90 d after treatment (Figure 1). These criteria were based on natural fluctuations occurring in populations of soil microorganisms. As a consequence, approximately 12% of all pesticides used around 1975 were classified as having persistent effects and considered to be critical or harmful for the functioning of the soil. Eijsackers and Doelman (1981) elaborated on this concept and concluded that heavy metals had to be classified as harmful when their effect was still measurable after 200 d.

Although in principle tests on soil microorganisms should be carried out with the most sensitive processes and organisms (Domsch, 1991), from a practical point of view there are some other requirements. Tests have to be

- Practically applicable
- Easily to be interpreted by regulatory bodies
- Ecologically relevant for many ecosystems
- Discriminating from natural fluctuations
- Giving the causal relation between compound and effect
- Carried out fast against low cost
- Standardizable

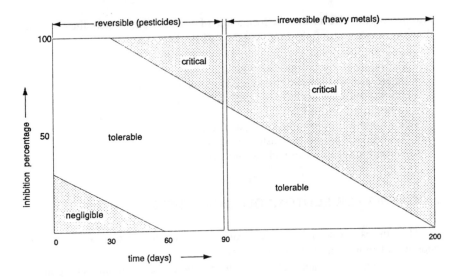

FIGURE 1. The ecotoxicological profile of pesticides and heavy metals.

Due to the latter requirements indicators generally used are (Domsch, 1991; Genber et al., 1991)

- Microbial biomass
- Soil respiration
- Ammonification
- Nitrogen mineralization, after addition of lucerne meal
- Nitrification
- Enzyme activities
- Counting various bacteria, actinomycetes, and fungi

However, many very sensitive parameters, such as symbiotic nitrogen fixation and vesicular arbuscular mycorrhiza, have been omitted so far (Domsch, 1991). Obviously, the generally used tests are not the most sensitive and we should be aware of the above mentioned testing compromises.

III. HETEROGENEITY OF SOIL AND ITS CONSEQUENCES

Soils are very heterogeneous, especially on the microscale. Consequently, many environmental conditions exist with specific microbial activities and interactions, which are influenced by diurnally fluctuating temperatures and other fluctuations.

By defining and recording the specific testing conditions, such as pH, redox potential, moisture content, and temperature, the heterogeneity problem can largely be avoided. This may be called the stratification or the niche approach and reflects the heterogeneity in defined ways. Particularly the microbial approach demands more research on a micro-site scale.

IV. THE SOIL-MICROBIAL VS. THE SOIL-FAUNAL APPROACH

Again, due to their size, ubiquity, versatility, and numbers, research on the functioning of soil microorganisms differs in general from investigating larger representatives of soil inhabitants, such as soil fauna species. Lethality, behavior, food choice, survival, hatching, growth, reproduction, or development of larval stages are key parameters mostly studied in ecotoxicology with soil fauna populations such as earthworms, springtails, and beetles.

Moreover, effect studies should be combined with biodegradation of the pollutant compound and the rehabilitation of other microbial processes. Those studies provide integrated insight and therefore are more appropriate in ecotoxicological research. An example of such a combined approach is the study of Loonen et al. (1987). They simultaneously measured the degradation of alpha hexachlorocyclohexane and microbial features as respiration glutamic acid decomposition, enzyme levels, and numbers of organisms (Figure 2).

Another difference is that the soil microbiologist tends to carry out his investigations in a more defined way; the approach usually is exact, more physiological/biochemical, and the laboratory experiments are reproducible, since carried out under defined conditions. Work on soil fauna is often done in the field where conditions are not exactly manageable. Soil microbial parameters investigated so far can be classified in activity (functional) and biomass diversity (structural) oriented. Of the former, soil respiration or rate of C decomposition, nitrification, enzyme activity in various macroelement cycles, and kinetics of substrate use are important examples. Functional diversity, adaptation and selection, and shifts towards other dominant consortia are important examples of the biomass-oriented structural parameters.

Research work in terrestrial environments so far has mainly been carried out under aerobic conditions, i.e., restricted to oxygen as the electron acceptor. Since diffusion of oxygen in soil easily becomes limited, anaerobic processes also need attention. These processes might be far more sensitive for heavy metals and chlorinated compounds, as has been shown for aquatic microbes (Walker, 1988). In sediments as well as in deeper soil layers anaerobic conditions are prevailing. In that case denitrifying, sulfate-reducing, but also methanogenic conditions occur where other types of microorganisms emerge and other degradation pathways may function. The difference between the presence and absence of oxygen is gradual and small, both on spatial and temporal scales. As soon as a strictly aerobic microorganism has used all oxygen it creates favorable conditions for anaerobic microorganisms. The difference in redox potentials (Eh) can be a few millivolts, although the Eh difference between extreme aerobic

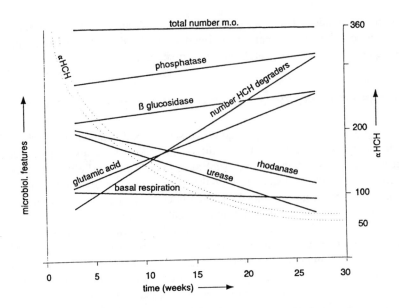

FIGURE 2. α-HCH degradation in relation to total number of microorganisms, number of α-HCH degraders, basal respiration, glutamic acid decomposition, phosphatase, β-glucosidase, rhodanase, and urease.

and extreme anaerobic can be 800 mV (Stigliani, 1988). Thus the use of the words "aerobic" and "anaerobic" preferably can better be expressed as redox potential. Under various redox potentials different electron acceptors (NO_3^-, SO_4^{2-} function by the oxidation of organic compounds (Figure 3). Especially methanogenic processes turned out to be very sensitive for heavy metals and therefore can act as good ecotoxicological parameters.

V. DOSE-EFFECT RELATIONS

Dose-effect relations in soil can be described as linear, logarithmic, and other forms, depending on the physicochemical properties of the soil involved. A classical example of this dependency is given for the heavy metal Pb in Figure 4 (Doelman and Haanstra, 1979). The effect of lead is strongly influenced by the cation exchange capacity (CEC) of the soils, being 80, 30, 10, and, respectively, 5 milli-equivalent/100 g soil for peat, clay, loam, and sand.

For effects of various heavy metals on microbial activities in soil a logistic dose-response curve (Figure 4) has been observed for respiration, urease, phosphatase, and arylsulfatase (Haanstra and Doelman, 1991). A similar dose-response curve has been found for effects of ethylene-ethiourea and nitrapyrin on ammonium oxidation in soil (Vonk, 1991). Preceding retardation or inhibition,

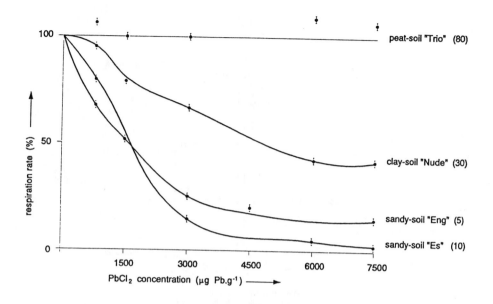

FIGURE 3. The influence of Pb on microbial soil respiration.

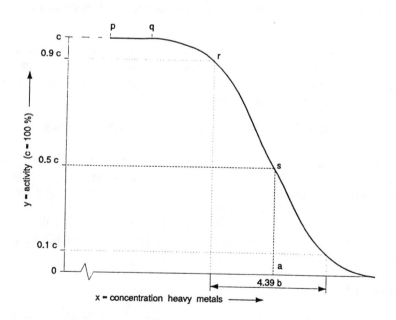

FIGURE 4. The logistic dose-response curve of heavy metals and respiration, urease, phosphatase, or arylsulfatase.

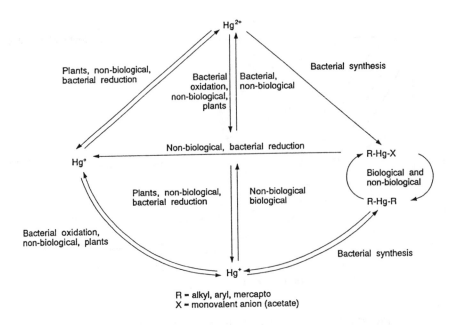

FIGURE 5. The methylation of mercury, as an example of heavy metal ecotoxicology.

the population may endure a qualitative shift, which takes place in the concentration range p-q (see Figure 5). As shown by Eijsackers (1993, by quoting Ruhling et al., 1984) earlier in this book for microfungi, an analog dose-response curve relation may exist in the p-r range.

VI. TRANSFORMATION OF ENVIRONMENTAL POLLUTIONS

The degradative or transformative potential of microbes is an intriguing phenomenon in nature. It is not only restricted to natural organic matter, but is also working with xenobiotic organic chemicals and even with heavy metals, provided the organisms benefit from the reaction or the reaction occurs due to cometabolism. Therefore, in microbe-oriented ecotoxicology both the microflora, toxicity of the chemical under study, and its biodegradation have to be studied.

A. Heavy Metals

Microbial transformation of metal compounds, especially those containing iron, have been known for a long time. Table 1 lists the metals for which documentation of such conversions exist (Summers and Silver, 1978).

From an ecotoxicological point of view, there is considerable interest in the role of bacteria in methylation of metals because of the enhanced toxicity of the

Table 1. Bacterial Transformation of Metals and Metalloids

Transformation	Metal
Reduction	As^{5+}, Cr^{6+}, Fe^{3+}, Hg^{+}, Hg^{2+}, Mn^{4+}, Se^{4+}, Te^{4+}
Oxidation	As^{3+}, Cr^{3+}, Fe^{0}, Fe^{2+}, Mn^{2+}, Sb^{3+}
Methylation	As^{5+}, Cd^{2+}, Hg^{2+}, Se^{4+}, Sn^{2+}, Te^{4+}
Demethylation	$R\ Hg^{2+}$

methylated forms of many metals. The microbial methylation of mercury, which caused the Minamata disease, is the most well-known methylation phenomenon. The conversion of Hg within an ecosystem is given in Figure 6. The oxidation or reduction can also strongly influence the toxicity of a metal as significantly shown for chromium. In its reduced state (Cr^{6+}) toxicity is the highest.

After the discovery of the methylation of mercury, interest arose over possible methylation of cadmium, lead, selenium, arsenium, and nickel, but not more than indications for these methylations exist.

B. Chlorinated Organic Compounds

Due to the strong accumulation of chlorinated organic carbon compounds in fat tissue and in food chains, the microbial degradation of chlorinated compounds has received serious attention. Pathways of DDT mineralization have

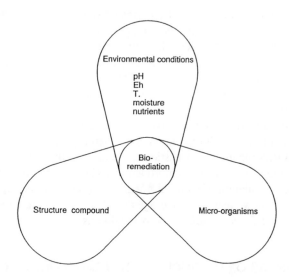

FIGURE 6. Factors affecting mineralization of organic compounds are the structure of the chemical compounds, the presence and function of suitable microorganisms, and the presence of the right environmental conditions.

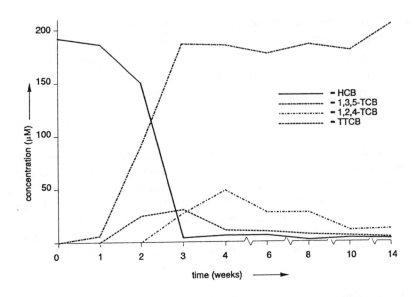

FIGURE 7. The disappearance of hexachlorobenzene and the appearance of intermediates.

been elucidated by Focht in 1972. Nevertheless, DDT can still be detected in many soils.

Many organic compounds and chlorinated organic compounds are highly persistent ("recalcitrant") in soil, notwithstanding the structure of the chemical reveals its potential microbial degradability. Therefore, either the right microorganism and/or the right environmental conditions do not prevail. Generally, anaerobic (reduced) dehalogenation occurs when an organic compound contains four chloride atoms. For bioremediation it is necessary that the triangle is present as shown in Figure 7.

Since mineralization often involves a sequential number of transformation steps, various process conditions and various microorganisms are necessary. For example, hexachlorobenzene, a pesticide, needs initially anaerobic conditions for dechlorination to tri- and di-chlorobenzene (Figure 8). This principle also counts for the degradation of tetra-chloroethylene (Fathepure et al., 1988). This example clearly shows that ecotoxicological and degradation researches demand attention for the effects of various redox conditions.

VII. AWARENESS OF INCOMPLETE BIOTRANSFORMATION

Tetrachloro-ethylene (PER) and trichloro-ethylene (TCE) are well-known organic degreasing solvents and often occurring soil water and groundwater pollutants. TCE and PER can be completely microbially degraded. Figure 9

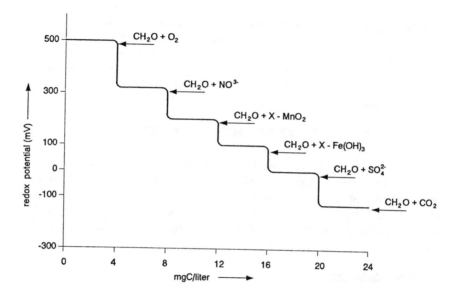

FIGURE 8. The various redox conditions for different electron acceptors.

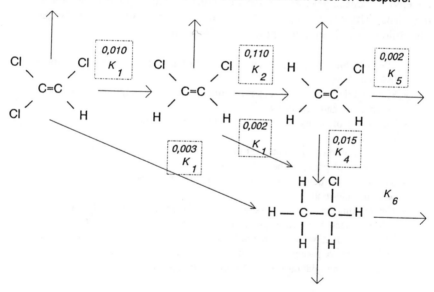

FIGURE 9. The microbial degradation of PER.

gives the biotransformation steps and their kinetics (Jaffe et al., 1988). Since K_4 and K_5 are the smallest rates, accumulation of vinyl chloride may occur. As a consequence an ecotoxicological problem has been created by degradation, since vinyl chloride is far more toxic and carcinogenic than PER and TCE.

Aerobic degradation, primarily resulting in hydroxylating, may increase the solubility and toxicity of the organic compound as shown for the pesticide 2,4D.

As mentioned earlier, heavy metals may inhibit microbial transformations. A combination of heavy metals may result in a synergistic inhibitory interaction as shown in a simple food chain where nematodes were feeding on bacteria, previously adsorbing Pb and Cd (Doelman et al., 1984).

VIII. RECOMMENDATIONS FOR MICROBIAL TESTING

Environmental pollution as well as the application of pesticides causes world-wide conern. Ecotoxicological testing should be satisfying to scientists as well as to governmental decision makers and regulatory bodies. Due to their role in element recycling and in foodwebs, microorganisms in soil could meet this demand. Distinction should be made between prognostic and diagnostic evaluations.

A. Prognostic Tests

Prognostic tests will be used before a compound is introduced into soil and the environment in general in order to predict its potential hazard to soil microorganisms. These tests will be very critical and a variety of tests may be needed. The following tests may be appropriate (Domsch, 1991):

- Microbial biomass measurement: C or N flush after fumigation- or substrate-induced respiration
- Microscopic counting of bacteria, actinomycetes, and fungi, including check on diversity and versatility
- Measuring of nitrogen fixation by *Rhizobia* and blue-green algae (*Cyanobacteria*)
- Nitrification measurements, autotrophic as well as heterotrophic
- Measuring of photosynthesis by algae
- Measuring the development and functioning of mycorrhizas, ectotrophic as well as endotrophic
- Determining the ratio of catabolically/anabolically used carbon when a defined substrate such as glucose, glutamate, or acetate has been added to soil
- The determination of the degradation of defined organic substrates under different redox potentials with their specific electron acceptors
- Measuring the mineralization of nitrogen compounds

As pointed out in section II, some of these tests may be time consuming and therefore not suitable for routine purposes.

B. Diagnostic Tests

Diagnostic research is extremely complex, as often displayed in plant pathological soil research. When unexpectedly an economical crop such as cotton, tomatoes, or wheat declines, viruses, bacteria, fungi, and nematodes, besides nutrient deficiency, may be responsible for poor growth. Searching for the real cause is difficult and time consuming while a single-factor explanation hardly ever occurs.

A diagnostic test should consist of general assessment characteristics (health thermometers) such as the adenylate energy charge (AEC) test and the ratio sensitive/resistant bacteria towards heavy metals or other persistent pollutants. The AEC provides a measure of the metabolic energy stored in cells, which relates to the physiological status of the microbial consortia in soil.

$$AEC = \frac{ATP + 0.5\ ADP}{ATP + ADP + AMP}$$

AEC varies between 1 and 0. For heavy metal-polluted soils the AEC value will be 0.10 to 0.15 lower than for the unpolluted soil (Brooks et al., 1987). For heavy metal-contaminated soil also the sensitivity/resistance index towards heavy metals may be applied. This test may be ecologically very appropriate since heavy metal-sensitive bacteria are more versatile in decomposition of aromatics than resistant bacteria. So loss of sensitive organisms means lowering of metabolic diversity (Doelman et al., 1993).

Another similar approach in ecotoxicological diagnostic research may be found in measuring the activity of enzymes involved in the electron transport system (ETS). Measurement of the ETS system has been recommended to measure activity of sediments (Trevors, 1987) and may also be applied to terrestrial systems.

Preferably, the most elegant way to measure the ecotoxicological effects within a microbial consortium, but which is hardly used, should be a combination of the functional and structural aspect expressed as activity, efficiency, and diversity measurements. This can be achieved by combining (a) activity measurements such as the most sensitive one of the prognostic tests, (b) the adenylate energy charge, (c) the respiration/growth index of a soil as elegantly reported by Killham (1985), and (d) a sensitivity/resistant index. These four diagnostic tests can also be used as prognostic tests.

It is clear that we emphasize the microbial approach, but no full consensus of approaches in soil microbial ecotoxicology has been reached. Therefore, integrated and combined research as recommended is strongly needed worldwide.

REFERENCES

Brooks, P.C., Newcombe, A.D., and Jenkinson, D.S., Adenylate energy charge measurements in soil, *Soil Biol. Biochem.*, 19, 211, 1987.

Brooks, P.C. and McGrath, S.P., Adenylate energy charge in metal contaminated soil, *Soil Biol. Biochem.*, 19, 219, 1987.

Doelman, P. and Haanstra, L., Effect of lead on soil respiration and dehydrogenase activity, *Soil Biol. Biochem.*, 11, 475, 1979.

Doelman, P., Nieboer, G., Schrooten, J., and Visser, M., Antagonistic and synergistic toxic effects of Pb and Cd in a simple foodchain: nematodes feeding on bacteria and fungi, *Bull. Environ. Contam. Toxicol.*, 32, 717, 1984.

Doelman, P., Jansen, E., Michels, M., and van Til, M., Effects of heavy metals in soil on microbial diversity and activity, *Biol. Fertil. Soils*, in press.

Domsch, K.H., Jagnow, G., and Anderson, T.H., An ecological concept for assessment of side-effects of agrochemicals on soil microorganisms, *Res. Rev.*, 86, 65, 1983.

Domsch, K.H., Status and perspectives of side-effect testing, *Toxicol. Environ. Chem.*, 30, 147, 1991.

Eijsackers, H. and Doelman, P., Het biologisch bodemsysteem, als natuurlijke basis van de Wet op de Bodembescherming, *Milieu Recht*, 7, 133, 1981.

Fathepure, B.Z., Tiedje, J.M., and Boyd, S.A., Reductive dechlorination of hexachlorobenzene to tri- di-chlorobenzene in anaerobic sewage sludge, *Appl. Environ. Microbiol.*, 54, 327, 1988.

Focht, D.D., Microbial degradation of DDT metabolites to carbon dioxide, water and chloride, *Bull. Environ. Contam. Toxicol.*, 7, 52, 1972.

Genber, H.R., Anderson, J.P.E., Castle, D., Domsch, K.H., Somerville, L., Arnold, D.J., Werff, H. van der, Verbeken, R., and Vonk, J.W., *Toxicol. Environ. Chem.*, 30, 249, 1991.

Haanstra, L. and Doelman, P., An ecological dose-response model approach to short- and long-term effects of heavy metals on arylsulphatase in soil, *Biol. Fertil. Soils*, 11, 18, 1991.

Jaffe, P.R., Taylor, S.W., Baek, N.H., Milly, P.C.D.M., and Marinucci, A.C., Biodegradation of trichloroethylene and biomanipulation of aquifers, Department of Civil Engineering and Operations Research, Princeton University, Princeton, NJ, 1988.

Kilham, K., A physiological determination of the impact of environmental stress on the activity of microbial biomass, *Environ. Pollut.*, 38, 283, 1985.

Loonen, H., Doelman, P., and Vogels, G.D., Zelfreinigend vermogen van hexachloorcyclohexaan verontreinigde grond voor alpha-HCH under landfarming conditions, RIN-report, 1987.

Ruhling, R., Baath, E., Nordgren, A., and Soderstrom, B., Fungi in metal-contaminated soil near the Gusum brass mill, Sweden, *Ambio*, 13, 34, 1984.

Sheenan, P.J., Miller, D.R., Butler, G.C., and Bourdeau, Ph., Eds., Effects of pollutants at the ecosystem level, Scope 22, John Wiley & Sons, Chichester, England, 1984, 443 pp.

Summers, A.O. and Silver, S., Microbial transformation of metals, *Annu. Rev. Microbiol.*, 32, 637, 1978.

Stigliani, W.M., Changes in valued capacities of soils and sediments as indicators of non linear and time delayed environmental effects, *Environ. Monit. Assess*, 10, 245, 1988.

Trevors, J.T., Electron transport system activity in soil, sediment and pure culture, *CRC Crit. Rev. Microbiol.*, 11, 83, 1988.

Vonk, J.W., Testing of pesticides for site-effects on nitrogen conversions in soil, *Toxicol. Environ. Chem.*, 30, 241, 1991.

Walker, J.D., Effects of chemicals on micro-organisms, *Water Poll. Control Fed.*, 60, 1106, 1988.

CHAPTER 6

A Method for the Ecotoxicological Risk Analysis of Polluted Sediments by the Measurement of Microbial Activities

P. van Beelen and P.L.A. van Vlaardingen

TABLE OF CONTENTS

0-87371-530-6/94/$0.00 + $.50

ABSTRACT

The toxic effects of five toxicants on five anaerobic microbial processes in freshwater sediments were measured. Four ^{14}C-labeled substrates were mineralized to $^{14}CO_2$ at concentrations from 1 to 5 µg/l in fresh sediment microcosms. The half-lives of the substrates acetate, benzoate, and 4-monochlorophenol (MCP) were 12 to 30, 24, and 84 min, respectively. The chloroform mineralization was much slower with a half-life of 4.5 to 12 d. The natural methane production in the sediment was also used to measure the toxic effects.

A dangerous concentration 5% (DC_5) can be calculated from the toxicity data using methods previously developed for single animal species toxicity tests. Above this toxicant concentration more than 5% of the possible mineralization reactions can be partly inhibited. The DC_5 concentrations for benzene, pentachlorophenol, 1,2-dichloroethane, chloroform, and zinc are 1, 0.4, 0.004, 0.0001, and 300 mg/kg sediment dry weight (d.w.), respectively. The chlorinated alkanes are many orders of magnitude more toxic for anaerobic microorganisms than for aerobic fish. This indicates that a sediment quality standard which is derived solely from animal toxicity tests can be erroneous. Only a combination between animal and microbial tests can lead to realistic sediment quality standards.

I. INTRODUCTION

A. How to Derive "Safe" Concentrations From Toxicity Data

Different species have different sensitivities for different toxicants. There is not a single species which is most sensitive for all pollutants.[1] Principally, this means that a number of taxonomically dissimilar species must be tested to assess the effect of a pollutant. A statistical method was described by Kooijman[2] to derive a concentration without effect on any species, when the no observed effect concentration (NOEC) of a toxicant on a number of species can be described by a normal distribution. This method was adapted by Van Straalen and Denneman[3] to derive a concentration which would give no effect on 95% of the species. This method also includes a correction factor for the bioavailability of pollutants in soil. Recently the method was refined to include the statistical uncertainty of the limited number of input data.[4] When more data become available there is a fair chance that the lowest reported NOEC will be lower, since the probability of picking a very sensitive test organism increases when more species are investigated. This will lead to very low allowable concentrations for ecotoxicologically well-studied pollutants and high allowable concentrations for less-studied pollutants. In the latter case a more prudent approach might be desirable.

B. Limitations of the Methods Used For Deriving "Safe" Concentrations

The statistical methods are based on restricted assumptions which can have a large effect on the outcome of the method. The sensitivity of all species is assumed to be distributed symmetrically. Pesticides, for example, are selected to have a stronger effect on target organisms than on nontarget organisms. This can cause an asymmetrical distribution, and therefore the statistical methods might give misleading results. The compensation for the bioavailability in soils and sediments is also a factor which is not easily predicted for all possible soils and organisms. Natural compounds can also give toxic effects, but the species are adapted to normal concentrations. This will also give an asymmetrical distribution of sensitivities, since the most sensitive species will not be present. The estimation whether 95% or more of the species should be fully protected to exclude major effects on a given ecosystem is not answered yet. Investigation of microbial species and activities contributes to the answer how to protect a whole ecosystem with limited experimental tools.

C. The Sensitivity of Microorganisms to Toxicants

Animals, plants, and microorganissm have very different sensitivities for toxicants. Cholinesterase inhibitors affect animals, herbicides intoxicate plants, and antibiotics inhibit certain groups of microorganisms. The physiological differences between microorganisms can be much larger than the differences between plants and animals. In river sediments a large and diverse community of microorganisms is present[5] which plays an important role in the mineralization of natural and xenobiotic substances. In contrast with toxicity tests using animals, tests with microorganisms rarely use a single species. A notable exception is the microtox test with *Photobacterium phosphoreum*. Microbial processes are often performed by a community of microorganisms, each having a different sensitivity for a pollutant. A large range of concentrations with partial effects can often be observed when microbial processes are monitored. This can lead to erroneous conclusions when the mineralization of a high concentration of a substrate is studied. The resistant and fast-growing species in the community will be able to grow and compensate for the effect on the major part of the community.[6] At low substrate concentrations there can be no growth of the resistant species and therefore the toxic effect will not be compensated.[6] A toxicant will decrease the mineralization rate when the degradation is monitored at environmentally relevant low concentrations. The toxicant dose which leads to a 50% decreased mineralization rate is called inhibition concentration 50% (IC_{50}). The IC_{10} inhibits 10% of the mineralization rate and is similar to the NOEC when experiments are performed in duplicate. (The NOEC will be lower when experiments are performed with more replicates.) In practice it is laborious to determine the mineralization rate at each toxicant concentration, since the influence of a toxicant on the percentage of substrate mineralized at one incubation time is more easily determined. From a graph of the percentage of substrate mineralized at a certain incubation time vs. the toxicant concentration the EC_{50} and EC_{10} can be determined. The EC_{50} is the toxicant concentration which decreases the percentage of substrate mineralized at the incubation time by 50%. The corresponding EC_{10} decreases the percentage by 10%. The IC_{50} and IC_{10} can be calculated when the half-life of the substrate and the EC_{50} and EC_{10} at a fixed incubation time are known.[6]

D. How to Derive "Safe" Concentrations From the Results of Microbial Toxicity Tests

This chapter summarizes the results of the effect of five toxicants on the mineralization of four substrates and on the formation of methane. Subsequently, an extrapolation method is used to derive "safe" concentrations of the five toxicants in anaerobic sediments. These concentrations are defined as the concentrations where no effect occurs on 95% of all possible mineralization reactions which might occur in the anaerobic sediment. This concentration is called DC_5 and is analogous to the HC_5 which was derived from single species animal toxicity tests.[3]

II. MATERIALS AND METHODS

Fresh anaerobic (methanogenic) sediment samples were homogenized with distilled water (1 ml/g sediment). The bottles were filled with 20 ml sediment suspension plus toxicant and radiocarbon-labeled substrate in an anaerobic glove box. The bottles were closed with a rubber stopper and the $^{14}CO_2$ formed was trapped in Carbosorb® (Packard®) and counted with a scintillation detector. The remaining labeled substrate was also counted in the scintillation detector.[7] Four ^{14}C uniform labeled substrates were used in a concentration range of 1 to 5 µg/l; acetate, benzoate, MCP, and chloroform. The methods were described in detail elsewhere.[8,9] The methane production was measured by gas chromatography without addition of a substrate.[10]

III. RESULTS AND DISCUSSION

A. The Mineralization of Organic Compounds in Sediment

The anaerobic mineralization of the substrates followed first-order kinetics with half-lives of 12, 24, and 84 min for acetate, benzoate, and MCP. Chloroform was degraded much slower with a half-life of 4.5 to 12 d. Methane was formed with an initial rate of 24 ml methane/d/kg sediment d.w.

B. The Effect of Toxicants on the Mineralization of Substrates

The effects of five toxicants (benzene, pentachlorophenol [PCP], 1,2-dichloroethane [DCE], chloroform, and zinc) were measured on the mineralization of the four substrates and also on the formation of methane. Table 1 shows the IC_{10} concentrations of each toxicant for each process. The effect of chloroform on the mineralization of chloroform could not be monitored, of course. In analogy with the results of animal toxicity tests there are large differences in sensitivity and there is no single test which is most sensitive for all toxicants. This analogy can be elaborated further when a statistical extrapolation method is applied to the data.[3] The method described by van Straalen and Denneman[3] uses the NOEC of single species animal toxicity tests to derive a hazardous concentration 5% (HC_5) which can be expected to protect 95% of all species in the environment. We had to use a slightly different approach for the microbial tests. The IC_{10} values were used instead of NOEC values and no soil correction factor was

Table 1. The IC_{10} Concentrations of Five Toxicants on the Mineralization For Four Substrates and Also on the Formation of Methane

Toxicants	Substrates					DC_5
	Acetate	Benzoate	MCP	Chloroform	Methanogenesis	
Benzene	480	150	150	140	>10000	1
PCP	19	6	3.1	15	140	0.4
DCE	0.7	71	23	0.07	860	0.0004
Chloroform	0.04	0.04	0.05		5.5	0.0001
Zinc	>3500	42+800	7.5+800	11+800	1780+800	300

All concentrations in mg toxicant/kg sediment (d.w.). The background concentration of zinc is 800 mg Zn/kg. The IC_{10} is that concentration of toxicant which gives a 10% decrease in the mineralization rate. The DC_5 is that concentration of toxicant which gives more than 10% inhibition of the 5% most sensible mineralization processes.

applied. A DC_5 is derived from the different IC_{10} values by a similar method as the HC_5 was derived from NOEC values. At the DC_5 concentration of a toxicant, 5% of all possible mineralization processes are expected to be inhibited.

C. Comparison Between the "Safe" Values For Microorganisms and Animals

Table 1 shows the DC_5 which can be calculated from the IC_{10} values. Benzene does not sorb strongly to the organic material in sediments. Fish are sensitive for benzene showing LC_{50} of 5 to 50 mg/l.[11] The anaerobic mineralization reactions shown in Table 1 are relatively insensitive for benzene.

A HC_5 of PCP in water was recently derived from aquatic toxicity data. Using the sorption coefficient, a HC_5 of 0.18 mg PCP/kg sediment was derived.[12] This HC_5 is slightly lower than the DC_5 of 0.39 mg PCP/kg derived from our experiments.

The 96 h LC_{50} of DCE is 480 mg/l for the fish *Lepomis macrochirus* and 550 mg/l for another fish named *L. beryllina*.[13] The mineralization of chloroform is much more sensitive for DCE than the methane production (see Table 1). This might be attributed to competitive inhibition of the chloroform mineralization by the structurally similar DCE. The LC_{50} of DCE for fish is four orders of magnitude higher than the IC_{10} of DCE on the chloroform mineralization. It is evident that the chloroform mineralization test is much more sensitive for DCE than the fish test.

Chloroform is not very poisonous for fish ($LC_{50} = 18$ to 191 mg/l), but more poisonous for the larvae of amphibia.[14] The anaerobic mineralization reactions shown in Table 1 are much more sensitive for chloroform compared to adult fish. This might be attributed to the formation of unstable toxic intermediates which are formed during the anaerobic mineralization of chloroform. Both DCE and chloroform are unreactive narcotic toxicants when applied to aerobic fish toxicity tests, but under anaerobic conditions a different mode of action is employed.

The background concentration of zinc was 800 mg/kg in the sediments used. A part of this zinc might be less bioavailable because it is sorbed to clay or precipitated as ZnS.[15] Table 1 shows that a small enlargement of this background concentration can give toxic effects. The microbial communities in a number of surface soils are also relatively sensitive for additions of a 10 to 50 mg Zn/kg soil.[16] The DC_5 of 250 mg Zn/kg is higher than the background concentration. The extrapolation method used to derive this DC_5 assumes a normal distribution. This is not true, since mineralization reactions which are inhibited by 800 mg Zn/kg cannot be measured in this sediment. Based on the small amount of extra zinc needed to produce an effect, one can assume that the present zinc concentrations in Dutch river sediments cause effects on microorganisms.

IV. CONCLUSIONS

Monitoring the toxic effects of pollutants on microbial mineralization processes in anaerobic sediments can give valuable information about the sensitivity of the local microflora. The difference between the sensitivities of microbial reactions and animal toxicity tests can be many orders of magnitude. Therefore, any attempt to derive sediment quality standards solely from animal toxicity tests is futile. Only a combination between animal and microbial tests can lead to realistic sediment quality standards.

REFERENCES

1. Slooff, W., van Oers, J.A.M., and de Zwart, D., Margins of uncertainty in ecotoxicological hazard assessment, *Environ. Toxicol. Chem.*, 5, 841, 1986.
2. Kooijman, S.A.L.M., A safety factor for LC50 values allowing for differences in sensitivity among species, *Water Res.*, 21, 269, 1987.
3. Van Straalen, N.M. and Denneman, C.A.J., Ecotoxicological evaluation of soil quality criteria, *Ecotoxicol. Environ. Saf.*, 18, 241, 1989.
4. Aldenberg, T. and Slob, W., Confidence Limits for Hazardous Concentrations Based on Logistically Distributed NOEC Toxicity Data, *Ecotoxicol. Environ. Saf.*, 25, 48, 1993.
5. Van Beelen, P. and Fleuren-Kemilä, A.K., Enumeration of anaerobic and oligotrophic bacteria in subsoils and sediments, *J. Contamin. Hydrol.*, 4, 275, 1989.
6. Van Beelen, P., Fleuren-Kemilä, A.K., Huys, M.P.A., van Montfort, A.C.P., and van Vlaardingen, P.L.A., The toxic effects of pollutants on the mineralization of acetate in subsoil microcosms, *Environ. Toxicol. Chem.*, 10, 775, 1991.
7. Van Beelen, P. and van Keulen, F., The kinetics of the degradation of chloroform and benzene in anaerobic sediment from the river Rhine, *Hydrobiol. Bull.*, 24, 13, 1990.
8. Van Beelen, P. and van Vlaardingen, P.L.A., The mineralization of chloroform in river sediments, *Netherlands J. Aquatic Ecol.*, in press.
9. Van Beelen, P. and van Vlaardingen, P.L.A., Toxic effects of pollutants on the mineralization of chloroform in river sediments, *Ecotoxicol. Environ. Saf.*, in press.
10. Van Vlaardingen, P.L.A. and van Beelen, P., Toxic effects of pollutants on methane production of river sediment, *Bull. Environ. Contamin. Toxicol.*, 49, 780, 1992.
11. Slooff, W., Ed., Benzeen Basisdocument. RIVM Report 758476001, 1987.
12. Van de Meent, D., Aldenberg, T., Canton, J.H., van Gestel, C.A.M., and Slooff, W., Desire for levels. Background study for the policy document "Setting Environmental Quality Standards for Water and Soil", RIVM Report 670101002, 1990.
13. Dawson, G.W., Jennings, A.L., Drozodowski, D., and Rider, E., The acute toxicities of 47 industrial chemicals to fresh and saltwater fishes, *J. Hazardous Mater.*, 1, 303, 1975.

14. Van der Heijden, C.A., Speijers, G.J.A., Ros, J.P.M., Huldy, H.J., Besemer, A.C., Lanting, R.W., Maas, R.J.M., Heijna-Markus, E., Bergshoeff, G., Gerlofsma, A., Mennes, W.C., van der Most, P.F.J., de Vrijer, F., Janssen, P.C.J.M., Knaap, A.G.A.C., Huigen, C., Duiser, J.A., and de Jong, P., Criteria document Chloroform. RIVM Report 738513004, 1986.
15. Di Toro, D.M., Mahony, J.D., Hansen, D.J., Scott, K.J., Hicks, M.B., Mayr, S.M., and Redmond, M.S., Toxicity of cadmium in sediments: the role of acid volatile sulfide, *Environ. Toxicol. Chem.,* 9, 1487, 1990.
16. Doelman, P. and Haanstra, L., Short- and long-term effects of heavy metals on phosphatase activity in soils: an ecological dose-response model approach, *Biol. Fertil. Soils,* 8, 235, 1989.

CHAPTER 7

Biosensors for Pollution Monitoring and Toxicity Assessment

A.L. Atkinson and D.M. Rawson

TABLE OF CONTENTS

0-87371-530-6/94/$0.00 + $.50

ABSTRACT

The development of bioassays during the last decade has moved away from the use of higher organisms for toxicity assessment in favor of microorganisms offering the potential for rapid, continuous monitoring. The increased demand for toxicity assessment and pollution monitoring of surface waters and water supplies has led to the development of mediated, whole cell biosensors. The success of biosensors incorporating both photoautotrophic (*Synechococcus*) and heterotrophic (*Escherichia coli, Paracoccus denitrificans*) microorganisms has enabled broad-spectrum, continuous monitoring systems to be established. The biosensors are sensitive to a wide range of potential pollutants, including heavy metals, organometallics and phenolics, and the Synechococcal biosensor exhibits sensitivity to a variety of herbicides. The spectrum of sensitivity may be extended by using an array of biosensors in a single system. The technology may have a role to play in other areas of environmental monitoring, and whole cell biosensors are already planned for monitoring aqueous extracts from air and soil samples.

I. INTRODUCTION

The range of possible pollutants in the environment grows continually as new chemicals are developed. The priority of those involved in the production or disposal of chemicals or in environmental monitoring must be to protect all living systems from chemicals which are potentially harmful. Therefore, methods are required to determine the toxicity of these compounds and to monitor their levels in the environment, and procedures must be employed to ensure that toxic levels are avoided or that corrective measures are implemented as soon as possible following their detection.

Toxicity studies using living systems are used to assess the risk of chemicals to organisms and ecosystems and provide the data for establishing safety thresholds. Analytical instrumentation is generally used for pollution monitoring and involves the qualitative and quantitative determination of those chemicals identified as a potential threat from the bioassays. Previously, pollution monitoring and toxicity assessment have been regarded, and dealt with, as separate issues, but the potential for exploiting bioassays in both toxicity assessment and detection/monitoring systems has gained interest over the past decade.

The ability to monitor bacterial metabolic activity using transducer technology has enabled the development of bacterial biosensors capable of toxicity assessment and pollution monitoring. To date, detection of freshwater pollution has been the most commonly addressed application,[1-3] with systems being developed as a direct result of the increased demand for information on the threat of contamination to waterways and water supplies. In addition, there is a need for systems for prescreening and assessing the levels of toxicants in water samples and effluent.

Both molecular- and cellular-based biosensors can, with little or no sample preparation, be made to operate in the aqueous phase. The analysis of sediment or soil samples would require a phase change prior to presentation to the biosensor, and therefore poses the additional problem of sample pretreatment, but there is no reason why systems could not be developed for the detection and monitoring of pollutants in sediments and soils.

II. DEVELOPMENT OF A SYSTEM FOR POLLUTION MONITORING AND TOXICITY ASSESSMENT

A. Rationale For the Use of Microorganisms

There is a need in environmental protection for rapid and potentially continuous monitoring systems which combine toxicity assessment and pollution monitoring for applications such as the protection of surface water intake sites. Toxicity assessment is a marriage of interpretation and prediction, using bioassays to evaluate the risk of exposure of particular sections of the biosphere to a chemical and to predict the likely adverse effects if and when critical concentrations of chemicals are released into the environment. The toxicity data obtained for a particular chemical from bioassays performed on a limited number of living systems are often extrapolated to predict the effect of the chemical on individual species, e.g., humans, and entire ecosystems.

Therefore, if a system can be devised which can incorporate organisms which mimic the biological system under threat, then both risk assessment and prediction can be achieved with a single test. If the system can also be calibrated for a range of concentrations of a chemical, then the criteria for pollution monitoring may also be met.

The use of microorganisms allows the exposure of relatively large numbers of organisms to a toxicant simultaneously, compared to the inherently small populations which may be used in bioassays using higher life forms, giving good resolution. The enormous wealth and diversity of microbial species makes them a potentially valuable source of bioassay materials encompassing a range of metabolic types which could be exploited in monitoring systems. They have many of the biochemical pathways present in higher organisms, allowing some extrapolation of results to higher organisms. In addition, microorganisms are generally easier and cheaper to maintain than higher organisms.

B. Microbial Assays For Toxicity Assessment

The toxic effect of a pollutant on a living system may be cytotoxic or genotoxic, and lead either to lethal or sublethal damage. The development of the Ames test[4] for the specific detection of mutagens and carcinogens stimulated research into the use of bacteria for rapid and inexpensive screening of chemical toxicity. The success of *Salmonella typhimurium* in the Ames test highlighted the fact that bacteria offer the potential for exploitation in the monitoring of chemicals with different modes of action as well as allowing a comparison to be made of the effects on general cellular activity or growth between species.

Toxicity screening tests that use bacteria[5,6] fall into two categories:

1. Assays based on measurements of viability or growth of a selected bacterial species: growth is a summation of cellular processes and will have an integrating effect on any challenge to the numerous biological functions within the cell. The major drawback of determining growth inhibition which prevents its potential use as a monitoring system is time, as the assays require the time taken for several cell divisions and usually for cells to reach logarithmic growth before reliable results can be obtained. Detailed analysis of the data can also be time consuming and a potentially expensive stage.

2. Assays based on measurements of the effect of toxic chemicals upon general or specific biochemical/physiological activity of selected bacterial species or mixed populations: a variety of substrates and products have been used as indicators of specific physiological function, including oxygen consumption or production;[7,8] the production of metabolites such as carbon dioxide[8] and hydrogen peroxide;[9] and luminescence.[10]

The success of the Microtox® bioassay[10] highlighted the potential for exploiting unusual metabolic activity for assessing acute toxicity in aqueous samples. The luminescent marine bacterium *Photobacterium phosphoreum* emits light under normal metabolic activity, and stimulation or inhibition of its metabolism affects the intensity of light output due to changes in the metabolic pathways fueling this process. A comparison of light emission from control and toxicant-treated samples after 15 min exposure enables toxicity assessment of the aqueous samples. The system is inherently sensitive to a wide range of toxicants by virtue of the complex nature of the process involved in light production. However, the relevance of the system for environmental monitoring has been questioned, since it uses a marine bacterium, but is employed for assessing the quality of fresh waters. Differences in the toxicity of sediment samples eluted with deionized water or seawater have been observed,[11] and it has been suggested that these differences could be due not only to changes in bioavailability, but also to the physiology of the bacterium in the different waters.

The Microtox® instrument was developed for centralized, routine laboratory use for toxicity assessment rather than on-line monitoring. Although the system is fairly compact, allowing on-site monitoring, it is not portable and the short monitoring period of the microbial component makes on-line, continuous monitoring difficult. The use of bacterial assays for toxicity assessment of aqueous pollutants has become increasingly popular, but the assays still have limitations which prevent their use as aqueous pollution monitoring systems.

C. Bioassays to Pollution Monitoring

The main challenge in the development of a combined toxicity assessment and pollution monitoring system has been to achieve real-time, continuous monitoring. One successful approach in developing a real-time monitoring system has been the use of biosensor transducer technology for direct monitoring of cellular activity. Biosensors exploit the fact that selected biochemically generated signals can be converted into quantifiable electrical signals using a transducer and, in general terms, consist of a biological component such as an enzyme, tissue slice, or population of whole cells held in close proximity to the surface of a transducing element.[12] The optimal choice of transducer will depend on the products and substrates involved in the biocatalytic process, with the main requirements being high specificity and rapid response within a designated concentration range. The majority of biosensors use electrochemical devices and, of these, potentiometric (where steady current is maintained and changes in potential followed) and amperometric (where current change at a constant potential is measured) are most commonly used.[13]

Biosensors were initially developed as analyte-specific detectors[14] and, in particular, the development of analyte-specific enzyme sensors has been the largest area of biosensor research and development in response to the demands of the health care industry.[15] However, environmental monitoring has requirements for both analyte-specific biosensors and broad-band detectors. Sensors for pollution monitoring must be able to respond to a wide range of compounds both individually and in complex mixtures, and, as a result, the emphasis has changed from the use of isolated biomolecular to multireceptor cellular biocatalysts.[16] One class of multireceptor biosensors has emerged as a potential dual-purpose system, namely, whole cell biosensors, which can incorporate both respiratory and photosynthetic bacteria as the biocatalyst.[2,3,19-21]

The majority of applications targeted for both biomolecular and cellular biosensors have involved analysis of aqueous samples, since water is the natural medium of the biocatalyst and often little or no sample pretreatment is required.

III. BACTERIAL AND CYANOBACTERIAL BIOSENSORS FOR POLLUTION MONITORING

A. Introduction

Biosensors incorporating enzymes as the biocatalyst fulfill the criterion of rapid monitoring, but their inherent specificity rules out their use as broad-band detectors. Ideally, a multienzyme system *in vitro* could offer the required sensitivity and speed of response, but would be practically impossible.

An alternative approach to the measurement of substrate consumption or product formation is to monitor the inhibition of enzymes in the electron transport chain (ETS) of microorganisms. The metabolic activity of both respiratory and photosynthetic organisms can be monitored indirectly by diverting electrons from the ETS to exogenous redox couples, termed mediators, and can be applied to both amperometric and potentiometric systems.[20]

Biosensors require intimate contact between the biocatalyst and the transducer element, and construction is generally based on immobilization of the biocatalyst. Several methods of immobilizing whole cells have been investigated to obtain a method that optimizes sensitivity, stability, and operational life of the biocatalyst. Successful methods used for immobilizing microbial cells have been limited to physical methods (adsorption, entrapment in supports), since chemical methods (e.g., covalent binding) generally cause a loss of biological activity.[22] Gel membranes (biological and chemical) and filters have been used successfully for immobilizing bacteria by providing mechanical support for the cells with virtually no chemical interaction between the cells and the support, thereby reducing the risk of damage to the cells.

A key problem of bioassay development is the choice of the biological component, and biosensors are no exception. In addition to being an appropriate organism for the purpose intended, the biocatalyst should be low cost; easy to grow, harvest, and immobilize; generate a good bioelectrochemical signal as a result of a high metabolic rate; and be stable in the presence of a mediator. The initial choice of both respiratory and photosynthetic microorganisms for biosensor development should be seen as leading to a model system rather than being of particular ecotoxicological relevance. However, a wide range of bacterial species are now under investigation for incorporation into biosensors, and here, selection is based on data from toxicity tests to obtain ecologically relevant species which also exhibit sensitivity to selected pollutants.

B. Types of Whole Cell Biosensor

The basic design of the biosensor unit developed for the continuous monitoring of respiratory and photosynthetic bacteria is shown in Figure 1. The immobilized cells are held against a working carbon electrode polarized against a silver/silver chloride reference electrode and irrigated with the sample solution.

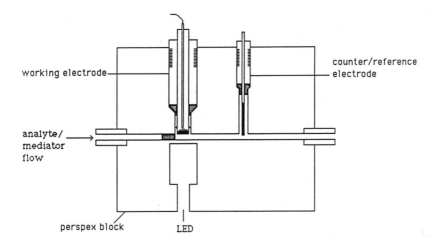

FIGURE 1. The electrochemical flow cell.

Stimulation of the cells can be achieved in different ways, depending on the metabolic activity to be monitored, but a pulsed regime is generally employed so that the activity of the cells can be determined from the difference in current between active (stimulation and/or addition of mediator) and resting (removal of stimulation and/or mediator) states (Figures 2 and 3).

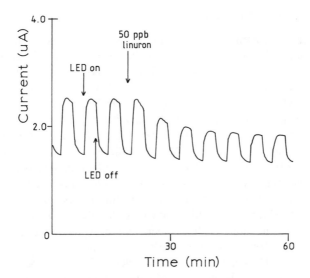

FIGURE 2. The mediated response of the synechococcal biosensor before and after exposure to 50 ppb linuron.

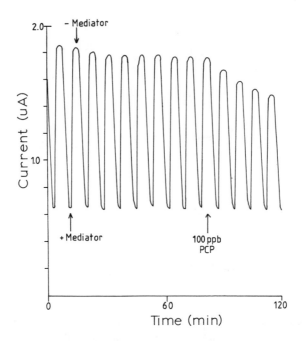

FIGURE 3. The mediated response of the *E. coli* biosensor before and after exposure to 100 ppb pentachlorophenol (PCP).

1. Photosynthetic Organisms

The potentially harmful presence in natural waters of herbicides which are known to inhibit photosynthesis prompted the development of biosensors incorporating photosynthetic organisms. Stimulation of photosynthesis can be achieved by pulsing light directed at the immobilized cells from an on-board light-emitting diode (LED) while irrigating the electrodes in the dark with the sample medium containing the mediator. A range of photosynthetic organisms, both cyanobacteria and algae, was tested to compare the magnitude and stability of response, sensitivity, and operational life of the immobilized cells using a variety of mediators. Sensor stability in the absence of pollutant poisoning has been found to be greatly influenced by the choice of mediator. Membrane-penetrating mediators such as *p*-benzoquinone caused a steady decline in response with continued exposure, and reduced sensor life, indicating that the mediator is toxic to the cells.[2,3] The nonpenetrating mediator potassium ferricyanide was generally used, since this proved less toxic to the cells. The cyanobacterium *Synechococcus* proved to be a good biocatalyst as further problems of loading, speed of response, and lack of response in the presence of nonpenetrating mediators ruled out the use of algal cells.

The synechococcal biosensor was used for toxicant studies to determine its range and magnitude of sensitivity. For laboratory purposes, toxicant additions were made directly into the sample reservoir located upstream of the flow cell and any perturbations in the current/time curve following the additions (Figure 2) were used to indicate sensitivity of the biosensor to the chemical. The biosensor exhibited the expected sensitivity to herbicides which interfere with photosynthesis, e.g., linuron and atrazine (Table 1).

2. Respiratory Organisms

Respiratory activity was monitored using the apparatus shown in Figure 4 in which respiratory substrates replaced the light stimulation used in the synechococcal biosensor. The biosensors were continuously irrigated with respiratory substrates and electron flow was only monitored periodically by repeated introduction and removal, typically every 5 min, of the mediator. This pulsed regime produced a rapid cessation in current when the mediator was removed and a series of sharp peaks, similar to those obtained for the synechococcal biosensor, when the mediator was added (Figure 3).

A wide variety of substrates can be used to stimulate respiration, but a simple cocktail of glucose, succinate, and lactate was found to produce good, reproducible stimulation and is generally used. Investigations into the stimulation achieved by the individual sugars revealed that it is possible to saturate the biocatalyst with one substrate, but increase current output upon addition of a second and third substrate, confirming that the biocatalyst is performing a multi-enzymic detection.[17]

The *Escherichia coli* and *Paracoccus denitrificans* biosensors produced the most stable responses, and an operational life of up to 5 d was achieved. The sensitivity of the biosensors to a range of potential pollutants, including heavy metals, organometallics, and phenolics (Table 1) was determined, and revealed differences not only between species, but also between strains.[23] The levels of inhibition (EC_{50}) and limits of detection obtained with the biosensors compared favorably with those reported for established laboratory-based toxicity tests and bioassays (Table 2), demonstrating the potential of the respiratory biosensor as a rapid toxicity assessment and pollution monitoring system.

C. Potential Uses

The development of bacterial biosensors offers a potential solution to the problem of real-time, continuous monitoring of aqueous samples. The screening of potential biocatalysts, both bacterial and cyanobacterial, produced several biosensors with different activities and sensitivities, and highlighted the potential for producing a field-based system incorporating a battery of biosensors to permit broad-spectrum monitoring. The need for such a system to protect surface water intake sites was the driving force in the biosensor development and is still the

Table 1. Detection Limits For Whole Cell Bacterial Biosensors Obtained at Luton College

	Organism	
Pollutant	*Synechococcus* 6301 (ppb)	*Escherichia coli* (ppb)
Adicarb	n/t[a]	>1000[b]
Ammonia N (total)	n/t	575
Ammonia N (unionized)	n/t	163
Aniline	>1000	>1000
Atrazine	25	>1000
Bromoxynil	12	n/t
Cadmium	>1000	>1000
Carbaryl	>1000	n/t
2-Chlorophenol	n/t	50
Copper	>1000	500
Dibutyltin dichloride	n/t	72
2,5-Dichlorophenol	n/t	1200
2,4-Dichlorophenoxyacetic acid	>1000	1000
3,5-Dichlorophenol	n/t	1000
Dichromate	>1000	n/t
2,6-Dinitro-*o*-cresol	n/t	500
Ferbam	n/t	550
Formaldehyde	n/t	4600
Ioxynil	10	n/t
Isoproturon	25	n/t
Lindane	>1000	n/t
Linuron	50	n/t
MCPA	1000	150
Mecoprop (MCPP)	1000	n/t
Mercury chloride	n/t	27
Methyl parathion	>1000	n/t
Metoxuron	50	n/t
Paraquat	1000	n/t
Pentachlorophenol	100	9
Phenol	>1000	>1000
Propanil	10	n/t
2,3,5,6-Tetrachlorophenol	n/t	4
Thiram	n/t	480
Toluene	>1000	n/t
Tributyltin	n/t	11
Tributyltin chloride	n/t	11
2,4,5-Trichlorophenol	n/t	160
3,4,5-Trichlorophenol	n/t	250
Ziram	n/t	260

[a] Not tested.
[b] No detection at this level.

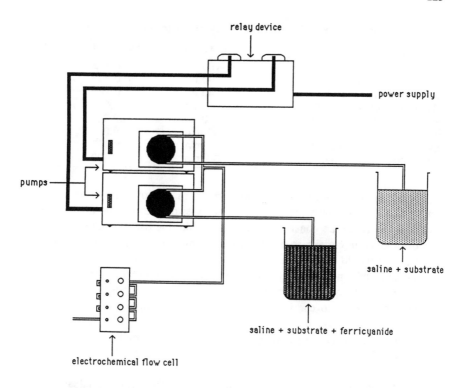

FIGURE 4. Continuous monitoring system for bacterial biosensors.

major potential site of application. However, the technology has raised interest from other areas of environmental monitoring, including soil and sediment analysis, and bacterial biosensors may also have a role to play here.

With the exception of enzyme biosensors that are capable of operating in organic solvents,[24] all biosensors require the analyte to be presented to the biological component in aqueous solution. Analysis of soil or sediment samples, therefore, would require a phase change prior to presentation to the biosensor. Two possible approaches could be used.

1. Analyte elution from soil/sediment samples into aqueous solution for presentation to the biosensor
2. *In situ* exposure of the biosensor in soil/sediment samples, either irrigated with sufficient water to allow analyte solution formation at the biosensor surface, or following modification of the sample to a slurry prior to biosensor insertion

Table 2. Sensitivity of a Range of Toxicity Tests to Pentachlorophenol

Toxicity test	Test time	EC$_{50}$ (ppm)	Ref.
Rainbow trout lethality	96 h	0.115	25
Fathead minnow lethality	96 h	0.266	25
Daphnia magna motility	48 h	0.145	25
Microtox® bioluminescence	5 min	1.5	26
Activated sludge respiration	3 h	22.0	27
Resazurin reduction	30 min	8.0	28
Escherichia coli biosensor	30 min	0.400	
Paracoccus denitrificans biosensor	30 min	0.600	

While both elution, irrigation and slurry formation of samples prior to biosensor exposure will only make available analytes that readily dissolve in water, and could therefore be considered as bioavailable, the use of nontoxic solvents and detergents may make available material which would not normally be bioavailable. It would be very difficult to reproduce in a rapid biosensor toxicity test the range of conditions that might govern bioavailability of chemicals. However, the use of aqueous extraction, irrigation, and slurry formation may be sensible approaches to use in screening tests for contaminated soils and sediments.

Bacterial biosensor assessment of toxicity could involve two different approaches.

1. Monitoring of the biosensor during its exposure to the environmental sample
2. Monitoring of the biosensor in a controlled medium following its exposure to an environmental sample

The former approach would be possible in most cases where the environmental sample was an eluent. The latter approach may be more appropriate where the biosensor was to be exposed to soil/slurry preparations that may not allow reliable monitoring, for example, where modification of the monitoring medium with chemical mediators was required.

The vast array of soil and sediment bacteria could offer a new pool of ecologically relevant biocatalysts. One limitation of existing bacterial biosensor systems is the inability to use anaerobic organisms as biocatalysts, and this will reduce the number of sediment bacteria which could be employed. On a more positive note, the isolation of organisms from different soils may provide biocatalysts with a range of sensitivities or resistances to selected pollutants, thereby aiding the development of battery arrays of bacterial biosensors for monitoring soils and sediments.

REFERENCES

1. Evans, G.P., Briers, M.G., and Rawson, D.M., Can biosensors help protect drinking water?, *Biosensors*, 2, 287, 1986.
2. Rawson, D.M., Willmer, A.J., and Cardosi, M.F., The development of whole cell biosensors for on-line screening of herbicide pollution of surface waters, *Toxicity Assessment*, 2, 325, 1987.
3. Rawson, D.M., Willmer, A.J., and Turner, A.P.F., Whole-cell biosensors for environmental monitoring, *Biosensors*, 4, 299, 1989.
4. McCann, J. and Ames, B.N., The Salmonella/microsome mutagenicity test: predictive value for animal carcinogenicity, in *Origins of Human Cancer*, Hiatt, H.H., Watson, J.D., and Winsten, S.A., Eds., Cold Spring Harbor Laboratory, Cold Spring Harbor, NY, 1977, 1431.
5. Bitton, G., Bacterial and biochemical tests for assessing chemical toxicity in the aquatic environment: a review, *CRC Crit. Rev. Environ. Control*, 13, 51, 1983.
6. Berkowitz, D., Potential uses of bacteria in toxicology, *Vet. Human Toxicol.*, 21, 422, 1979.
7. International Standards Organization, Activated Sludge Inhibition Test, ISO/TC 147/SC 5/WG1, 1981.
8. Tam, T.-Y. and Trevors, J.T., Toxicity of pentachlorophenol to *Azotobacter vinelandii*, *Bull. Environ. Contam. Toxicol.*, 27, 230, 1981.
9. Kulys, J.J., Amperometric enzyme electrodes in analytical chemistry, *Fresenius' Z. Anal. Chem.*, 335, 86, 1989.
10. Beckman, Inc., *Microtox System Operating Manual*, Beckman Instruments Inc., Microbiotics operations, Carlsbad, CA, 1982.
11. Ankley, G.T., Hoke, R.A., Giesy, J.P., and Wi..ger, P.V., Evaluation of the toxicity of marine sediments and dredge spoils with the Microtox bioassay, Microbics Corp., Reference #163.
12. Lowe, C.R., An introduction to the concepts and technology of biosensors, *Biosensors*, 1, 3, 1985.
13. Schmid, R.D. and Karube, I., Biosensors and "bioelectronics", *Biotechnology*, 6b, 317, 1988.
14. Tran-Minh, C., Immobilized enzyme probes for determining inhibitors, *Ion-Selective Electrode Rev.*, 7, 41, 1985.
15. Schramm, W., Yang, T., and Midgley, A.R., The commercialization of biosensors, *Medical Device and Diagnostics Industry*, Nov., 52, 1987.
16. Riedel, K., Biochemical fundamentals and improvement of the selectivity of microbial sensors — a minireview, *Bioelectro. Bioenerg.*, 25, 19, 1991.
17. Richardson, N.J., Gardner, S., and Rawson, D.M., A chemically mediated amperometric biosensor for monitoring eubacterial respiration, *J. Appl. Bacteriol.*, 70, 422, 1991.
18. Karube, I. and Tamiya, E., Biosensors for environmental control, *Pure Appl. Chem.*, 59, 545, 1987.

19. Hansen, P.-D., Pluta, H.-J., and Beeken, J.A., Biosensors for on-line monitoring of the waterways and for sample taking aimed at relieving disturbances, in *Biosensors: Applications in Medicine, Environmental Protection and Process Control,* Schmid, R.D. and Scheller, F., Eds., 113, 1989.
20. Rawson, D.M., Whole cell biosensors, *Intern. Indust. Biotechnol.,* 8, 18, 1988.
21. Gaisford, W.C., Richardson, N.J., Haggett, B.G.D., and Rawson, D.M., Microbial biosensors for environmental monitoring, *Biosensors,* 19, 15, 1991.
22. Reidel, K., Renneberg, R., Wollenberger, U., Kaiser, G., and Scheller, F.W., Microbial sensors: fundamentals and application for process control, *J. Chem. Tech. Biotechnol.,* 44, 85, 1989.
23. Richardson, N.J., Eubacterial Biosensors for Pollution Monitoring and Surface Water Intake Protection, Ph.D. thesis, Cranfield Institute of Technology, 1991.
24. Saini, S. and Turner, A.P.F., Biosensors in organic phases, *Biochem. Soc. Trans.,* 19, 28, 1991.
25. Thurston, R.V., Gilfoil, T.A., Meyn, E.L., Zajdel, R.K., Aoki, T.I., and Veith, G.D., Comparative toxicity of ten organic chemicals to ten common aquatic species, *Water Res.,* 19, 1145, 1985.
26. Vasseur, P., Bois, F., Ferrard, J.F., and Rast, C., Influence of physico-chemical parameters on the Microtox test response, *Toxicity Assessment,* 1, 283, 1986.
27. King, E.F. and Painter, H.A., Inhibition of respiration of activated sludge: variability and reproducibility of results, *Toxicity Assessment,* 1, 27, 1985.
28. Thompson, K., Lui, D., and Kaiser, K.L.E., A direct resazurin test for measuring chemical toxicity, *Toxicity Assessment,* 1, 407, 1986.

CHAPTER 8

The Effects of Heavy Metal-Contaminated Sewage Sludge on the Rhizobial Soil Population of an Agricultural Field Trial

J.P. Obbard, D.R. Sauerbeck, and K.C. Jones

TABLE OF CONTENTS

ABSTRACT

Soils from a well-controlled field experiment were screened for the presence and number of effective cells of *Rhizobium leguminosarum* biovar *trifolii* capable of nodulating the host plant, white clover (*Trifolium repens*). Soils had been amended with anaerobically digested and undigested sewage sludge at rates of 0, 100, and 300 m^3/ha/yr on soils of differing pH since 1980 and up to the present. Applications of anaerobically digested sludge included additions with and without heavy metal salts. *Rhizobium* were found to be present in all of the treatments, apart from the most contaminated treatment in the soil of lower pH, despite the absence of the host plant from the field sward. Lack of nodulation and nitrogen fixation (acetylene reduction activity) for *T. repens* growing in soils, in some cases, was probably caused by the high levels of extractable nitrate present as plants subsequently grown in N-free media were effectively nodulated. Important effects on the size of the effective rhizobial population were apparent in relation to the soil pH, sludge type and addition rates and the concentration of heavy metals present.

I. INTRODUCTION

The addition of sewage sludge to agricultural soil improves physical characteristics and acts as a valuable source of nutrients to growing crops. However, there is increasing concern over the use of sewage sludge which has been contaminated by heavy metals. Heavy metals are known to be persistent in the soil over long time periods, and are known to have ecotoxicological effects on plants and soil microorganisms.[1-6] Investigations up until recently have concentrated on the uptake and transfer of heavy metals into the food chain via crops.[7-9] Now, however, there is increasing evidence of an adverse effect on microbial processes related to nutrient cycling in these types of soils.[10]

The soil microbial ecosystem is functionally complex. It contains key groups of microorganisms which have an integral role in maintaining soil fertility in relation to plant nutrition. This can either take place via general decomposition activity, whereby organic substrates are broken down into their inorganic constituents, or more specifically, via symbiotic relationships involving the direct exchange of inorganic nutrients for photosynthetic carbohydrates between the microorganism and its host plant, respectively. Arguably, the most important symbiotic relationship occurring in the agriculture is that between the genus *Rhizobium* and their leguminous host plants. In this symbiosis atmospheric nitrogen is fixed by the *Rhizobium* and supplied to the host plant. The process can assume great significance in marginal areas where artificial N inputs are uneconomical. For example, the symbiosis between *R. leguminosarum* biovar *trifolii* and white clover (*Trifolium repens*) can result in fixation rates of up to 200 kg N/ha/year under favorable conditions.[11]

Table 1. Maximum Permissible Concentrations of Elements in U.K. Arable Soil After Application of Sewage Sludge

Element	Maximum permissible concentration of element in soil (µg/g dry soil)			
	pH 5.0 < 5.5	**pH 5.5 < 6.0**	**pH 6.0 < 7.0**	**pH > 7.0**
Zinc	200	250	300	450
Copper	80	100	135	200
Nickel	50	60	75	110
	For pH 5.0 and above			
Cadmium	3			
Chromium	400 (provisional)			
Lead	300			

Nitrogen fixation by *R. leguminosarum* bv. *trifolii,* in symbiosis with white clover (*T. repens*), has been found to be affected by the presence of heavy metals in sludged soils from the Woburn Experiment at the Rothamsted Experimental Station, U.K.[2-4] Field trials resulted in yield reductions of up to 40% of *T. repens* in sewage sludge-amended soils compared to those amended with farmyard manure (FYM). This was despite organic applications having ceased over 20 years prior to the study. Host plant nodules from the sludge-grown plants were small and white (indicating a lack of leghemoglobin) and were ineffective at fixing nitrogen, despite concentrations of heavy metals in the soils approximating to maximum permissible limits, as specified in the current European Commission (EC) Directive[12-13] (see Table 1). Isolation of the ineffective *Rhizobium* strain indicated that the inability of the cells to fix nitrogen was a genetic trait and the plasmid conformity of these cells contrasted with the normal diversity found in FYM isolates.[2,4] Inoculation of the contaminated soil with an effective strain of *R. leguminosarum* bv. *trifolii* resulted in nitrogen fixation, but this ability was lost over a 2-month period in the absence of the host plant, unless very large inoculum densities (10^{10} cells/270 g dry soil) were used. This indicated that effective *Rhizobium* strains were unable to survive, or at least remain effective, in the presence of sludge-derived heavy metals.

The purpose of the further research which has been conducted at Lancaster University has been to extend microbial investigations to other agricultural sites which have also received sewage sludge. This has been in order to determine the prevalence of the above described phenomenon and to understand more of the behavior of *R. leguminosarum* bv. *trifolii* populations in the presence of heavy metals. Initial work involved the screening of a large number of soil samples, from a range of sites, to determine the presence of rhizobial cells capable of forming a symbiosis with *T. repens* and fixing atmospheric nitrogen.[14] It was shown that the presence of the host plant in contaminated soils, irrespective of metal concentration, resulted in the presence of effective cells, both within the host plant nodule and the rhizosphere. Other sites, where the host plant was absent, demonstrated that the introduction of the host to the soil did not result in the establishment of a nitrogen fixing symbiosis. This indicated that the microbial symbiont was absent or ineffective, thereby supporting earlier work.[2-4] The work at Lancaster has also highlighted the role of the host plant in maintaining an effective rhizobial population under contaminated conditions. In certain cases where rhizobial cells were able to establish an effective symbiosis upon the introduction of the host plant, this ability was invariably lost when cells were diluted and inoculated onto plants growing in artificial growth media. This suggested that heavy metals were having a numeric and not a genetic effect on cells, which is in contrast to work conducted at the Rothamsted Experimental Station.[2-4]

This quantitative ecotoxicological effect has been further elucidated by the enumeration of effective *R. leguminosarum* bv. *trifolii* cells found in an experimental agricultural site. The work has revealed apparent effects on the size of the effective rhizobial population in relation to soil pH, the concentration of heavy metals present, and sludge type and addition rates.

II. MATERIALS AND METHODS

A. Experimental Site

Soils from the Braunschweig Experimental site in Germany were used for experimental purposes. Soils have been amended from 1980 up to the present with anaerobically digested sludge at rates of 0, 100, and 300 m³/ha/year on two soil trials of the same texture (loam), but of differing pH ("low" and "high" pH). Sludge was digested with or without heavy metal salts to give two treatments for each addition rate, i.e., a "metal-spiked" treatment and a "nonspiked" treatment. Each trial also included a plot which was untreated with sludge. The "high" pH soil trial also received additions of "nonspiked" undigested sewage sludge at the two application rates over the same period. Soil was cultivated to a depth of 23 cm following sludge application. *T. repens* was not indigenous to the site at the time of sampling and there is no history of the plant growing on the site.

The site has therefore provided the opportunity to study populations of *R. leguminosarum* bv. *trifolii* in sludged soils of differing pH with and without the addition of heavy metals added at specific application rates, as well as allowing the comparison between different sludge types with untreated soils.

B. Sampling and Analysis

Four replicate samples were removed to a depth of 12 cm from each of the representative treatments and were then bulked and homogenized before removing separate aliquots for experimental purposes. Soil was kept moist throughout this process and for cell enumeration tests. Soil samples for chemical analysis were air dried in the laboratory and sieved (<2 mm). Analytical methods for characterizing soil chemistry were based on standard procedures.[14,15]

The presence of effective rhizobial cells in the soils was tested by growing the host plant (using sterilized seed) on moist soil in 150 × 25 mm sterilized boiling tubes in a Fisons growth cabinet (20°C, 16/14-h light/dark cycle). After an appropriate growth period (4 weeks) the acetylene reduction assay was used to test for the symbiotic fixation of atmospheric nitrogen on triplicate samples in conjunction with unseeded control soils.[14,16] Where acetylene reduction activity (ARA) was not detected, soil suspensions were inoculated onto plants growing in N-free sterile agar, under the same conditions, to test for nodulation.[14,17] This allowed for the detection of symbiotic nitrogen fixation in cases where activity may have been suppressed due to soil toxicity or high levels of plant-available nitrogen.

In most cases, effective rhizobial cells were found despite the absence of the host plant. This provided the basis for determining quantitative effects on the effective rhizobial population in relation to ecotoxicological effects of heavy metals and other soil properties. Enumeration of effective cells was carried out using a most probable number (MPN) technique.[17] Pregerminated *T. repens* seeds (two per agar growth tube) were inoculated using a dilution series from an initial soil suspension prepared at a given soil ratio (e.g., 1 part soil dry weight equivalent: 25 parts sterilized deionized distilled water). The initial dilution was decreased where rhizobial populations were suspected to be low in order to increase assay sensitivity. Four replicate tubes for each dilution were used after homogenization of the suspension. Dilutions of suspension were prepared up to 3125 times that of the initial suspension, representing 24 inoculation tests for each of the 12 soil treatments. Plants were grown for 4 weeks (as above) before each replicate tube, for each dilution, was inspected for the presence of red nodules on the host plant root system. This procedure was used to indicate the presence of an effective N_2 fixing symbiosis between the host plant and microsymbiont. The presence of nodules counted as a positive result, where the replicate results were used to construct an infection profile for the dilution series. This profile was used to determine the number of effective cells present in the initial soil sample using standard statistical tables.[17]

III. RESULTS

Total concentrations of heavy metals for the Braunschweig soil treatments are shown in Table 2, together with data for other characteristic soil properties. Organic matter and the concentrations of extractable nitrate, were, as may be expected, dependent on the amount and type of sludge added to the soils. Soil pH was found to decline in both soil trials with increasing sludge additions, probably due to the nitrification of organic residues which causes a concomitant

Table 2. Physicochemical and Biological Data For the Braunschweig Experimental Site

| Sample | Total metals (µg/g dry soil) | | | | | | pH | Organic matter (%) | Extractable nitrate (µg/g dry soil) | Extractable nitrate Sd. |
	Zn	Cu	Ni	Cr	Pb	Cd				
					"High" pH trial					
1	48	12	6	13	42	0.18	6.8	2.97	30	3
2	90	21	8	15	43	0.36	6.7	3.43	105	14
3	138	38	12	37	64	1.0	6.8	3.17	148	7
4	168	33	11	22	44	0.8	6.3	4.94	1013	97
5	341[a]	**99**	25	98	107	2.7	5.4	4.56	1398	180
6	84	20	6	13	40	0.3	7.0	3.71	83	5
7	198	43	12	24	55	0.6	6.4	4.27	248	62
					"Low" pH trial					
1	36	7	5	11	33	0.2	5.8	3.79	68	2
2	77	14	7	13	34	0.4	5.5	4.47	155	17
3	144	36	12	42	55	1.2	5.7	4.08	187	12
4	175	30	11	21	40	0.9	4.9	5.93	1241	163
5	**335**	**93**	24	96	107	2.9	4.8	7.20	1337	180

Key: *Anaerobically digested sludge* 1. untreated 2. "nonspiked" 100 m³/ha/year. 3. "metal spiked" 100 m³/ha/year. 4. "nonspiked" 300 m³/ha/year, 5. "metal spiked" 300 m³/ha/year. *Undigested sludge* 6. "nonspiked" 100 m³/ha/year. 7. "nonspiked" 300 m³/ ha/year.

[a] Total metal figures underlined indicate U.K. limit is exceeded.

reduction in pH.[18] Total metal concentrations only exceeded statutory limits in the soils which had received "metal-spiked" sludge at the highest addition rate. For these treatments, concentration exceeded EC maximum permissible concentrations by 141 and 135 μg/g for zinc and by 19 and 13 μg/g for copper in the "high" and "low" pH trials, respectively. It can also be noted that the addition of "nonspiked" sludge resulted in an elevated concentration of metals compared to the untreated soils in each pH trial, and that the addition of undigested sludge resulted in lower soil contamination than did the anaerobically digested form.

Plant characteristics and ARA results for the *T. repens* growth trial, together with the MPN results for soil samples, are given in Table 3. Plant yields were variable, as were nodule numbers, but it is important to note that nodulation and ARA were detected in all soil treatments except for those amended with highest application rates of "nonspiked" and "metal-spiked" sludges on the "high" pH trial and the "metal-spiked" of the "low" pH trial. Growth of host plants inoculated with soil suspensions and grown on N-free media from these treatments resulted in nodulation for all treatments except that from the "low" pH trial. This indicated that the expression of nodulation was being suppressed at the higher application rate in the "high" pH trial and that rhizobial cells capable of forming an effective symbiosis were indeed present.

Intra- and intertreatment differences in rhizobial population MPN estimates occurred between the different soil treatments. While MPN data represent only a single infection profile for each treatment, this was composed of four replicates for each of six dilution series. It is difficult to attribute MPN differences to specific soil properties, due to the inherent variability in soil properties generally between treatments, but variations in effective rhizobial population size were apparent in relation to sludge application rate and type, heavy metal concentrations, and soil pH. These differences can be described as follows:

1. In both the "low" and the "high" pH trials the addition of "nonspiked" sludge at the lowest rate (100 m³/ha/year) enhanced the size of the effective population relative to the untreated soils. The MPN of cells in the untreated soil of the "low pH" trial was greater than that of the "high" pH trial (1.366 × 10⁴ cells/g dry soil vs. 9.200 × 10² cells/g) and was also enhanced to a greater degree with the addition of the "nonspiked" sludge at the 100 m³/ha/year application rate (6.437 × 10⁴ cells/g vs. 1.586 × 10⁴ cells/g).

2. The addition of greater amounts of "nonspiked" sludge (300 m³/ha/year) on both trials, irrespective of soil pH, resulted in a reduction in rhizobial population size by an order of magnitude to 3.70 × 10² cells/g in the "high" pH soil and 2.60 × 10² cells/g in the "low" pH soil. This effect was also noted in samples from the "high" pH soil which had received applications of undigested sludge, where MPN estimates decreased from 4.567 × 10⁴ cells/g to 3.50 × 10² cells/g at the higher rate of application. The difference in MPN estimates for soil which had received the undigested sludge, compared to anaerobically digested sludge, indicates that the effect of sludge applications on effective cell numbers is dependent on sludge

Table 3. Plant and Nodulation Characteristics of *T. repens* Grown in Braunschweig Experimental Soils

Sample	Mean yield (mg)	Yield Sd	Mean nodule number[a]	Nodule number Sd	ARA (+/−) Y/N?[b]	Nodulation in agar	MPN of effective cells (cells/g dry soil)
			"High" pH trial				
1	8.6	0.2	3.0	0.6	+(3)	Y	920
2	5.1	0	6.2	4.3	+(3)	Y	15860
3	4.9	0.3	3.8	0.7	+(3)	Y	9160
4	9.6	0.5	<1	0.1	+(3)	Y	370
5	5.9	1.0	0	—	—	Y	130
6	12.9	0.5	4.8	1.3	+(3)	Y	45670
7	7.9	1.7	1.8	0.7	+(3)	Y	350
			"Low" pH trial				
1	8.2	1.8	4.8	0.3	+(3)	Y	13660
2	3.9	1.2	3.2	1.5	+(3)	Y	64370
3	6.0	0.7	2.3	0.5	+(3)	Y	1000
4	4.9	0.9	0	—	—	Y	260
5	1.7	0.4	0	—	—	N	0

Key: Anaerobically digested sludge: 1. untreated 2. "nonspiked" 100 m³/ha/year, 3. "metal spiked" 100 m³/ha/year, 4. "nonspiked" 300 m³/ha/year, 5. "metal spiked" 300 m³/ha/year. *Undigested sludge:* 6. "nonspiked" 100 m³/ha/year, 7. "nonspiked" 300 m³/ha/year.

[a] Maximum values in set of four replicates.
[b] Figures in brackets refer to the number of replicates (out of three) where positive acetylene reduction activity (ARA) was detected.

type. At the higher rates of addition, nitrate levels were an order of magnitude greater than at the lower rates. This and other soil properties were probably inhibitory to cellular activity and hence population size.[19,20]

3. Comparing treatments from each trial which have received equal amounts of either "metal-spiked" or "nonmetal-spiked" sludges, there is evidence for a metal-related effect on rhizobial population size. It is important to note, however, that other differences in soil properties between treatments may also partially account for the differences observed. In soils which had received "metal-spiked" sludge at the lower application rate in the "high" pH soil the population was reduced by 42% (1.586×10^4 cells/g to 9.160×10^3 cells/g) compared to its equivalent "nonmetal-spiked" sludge treatment. In the "low" pH soil this effect was even more severe. Effective population size was reduced by 98% compared to the equivalent treatment with "nonmetal-spiked" sludge (i.e., 6.437×10^4 cells/g vs. 1.00×10^3 cells/g) at the lower rate of application. Soil samples which had the highest concentrations of heavy metals in the most contaminated soil of the "low" pH trial were completely devoid of effective rhizobial cells.

IV. DISCUSSION

Effective symbiosis was determined for host plants growing in the majority of the soil samples collected from the Braunschweig experimental site. This was despite the host plant being absent from the field sward. This contrasts with the majority of field sites discussed in a previous paper,[14] where the presence of effective symbiosis was dependent on the presence of the host plant in the indigenous field sward. In these soils, sludge applications had ceased several years prior to study. As the Braunschweig experimental site is still freshly amended with sewage sludge on an annual basis it is possible that other factors, such as the enhanced organic and nutritional status of the soil, are maintaining the rhizobial population in this instance. The reduction in organic matter and nutrient status of soils following the termination of sludge application, in contrast to the long-term persistence of heavy metal toxicity,[1] is likely to be the reason for this apparent contradiction in results. The contrast highlights the importance of considering ecotoxicological effects in sewage sludge-amended soils in the long-, as well as the short-term time period following application.

The enumeration of rhizobial populations in soils where *R. leguminosarum* bv. *trifolii* cells have been found to be effective in forming an N_2 fixing symbiosis with *T. repens*, as at Braunschweig, has proved useful in understanding the ecotoxicological effects of heavy metals in the context of other soil properties altered by sludge addition. An adverse effect on the size of the rhizobial population by heavy metals in isolation from other soil properties was apparent, although the confounding effect of other variable soil properties, despite the design of the field experiment, precluded precise conclusions. The fact that large effects were apparent in some soils which had heavy metal concentrations below their statutory limits warrants concern for agricultural sludge utilization practices. In the extreme case in the "low" pH soil, where concentrations of zinc and copper exceeded their limits by 135 and 13 μg/g, respectively, effective *R. leguminosarum* bv. *trifolii* cells were eliminated completely.

Despite the potential of sewage sludge to have harmful effects on the microbial community, it is important to note that the beneficial effects of sludge on rhizobial population size are also apparent from this study. The application of "nonmetal-spiked" sewage sludge at 100 m³/ha/year enhanced population size considerably compared to the untreated soil, the degree of which was dependent on soil pH and sludge type. The conclusion of the research is, therefore, that careful consideration is required to promote the sustainable use of sewage sludge in agriculture in order to promote soil fertility and to avoid potentially adverse ecotoxicological effects on the soil microflora.

REFERENCES

1. McGrath, S.P., Long-term studies of metal transfers following applications of sewage sludge, in *Pollutant Transport and Fate in Ecosystems,* Coughtrey, P., Martin, M.H., and Unsworth, M.H., Eds., Special Publication No. 6, The British Ecological Society, London, 1987, 301.
2. Giller, K.E., McGrath, S.P., and Hirsch, D.R., Absence of nitrogen fixation in clover grown on soil subject to long term contamination with heavy metals is due to survival of only ineffective *Rhizobium, Soil Biol. Biochem.,* 21, 841, 1989.
3. McGrath, S.P., Hirsch, P.R., and Giller, K.E., Effect of potentially toxic metals in soil derived from past applications of sewage sludge on nitrogen fixation by *Trifolium repens* L., *Soil Biol. Biochem.,* 20, 415, 1988.
4. McGrath, S.P., Hirsch, P.R., and Giller, K.E., Effect of heavy metal contamination on the genetics of nitrogen fixing populations of *Rhizobium Leguminosarum* bv. *trifolii,* in *Environmental Contamination,* Oriio, A.A., Ed., CEP, Venice, 1988, 164.
5. Witter, E., Agricultural Use of Sewage Sludge: Controlling Heavy Metal Contamination, National Environmental Protection Board, Sweden, 1989.
6. Smith, S., Effects of Sewage Sludge Application on Soil Microbial Processes and Soil Fertility, Water Research Centre Report No. FR0034, Foundation of Water Research, 1990.
7. Logan, T.J. and Chaney, R.L., Metals, in *Utilization of Municipal Wastewater and Sludge on Land,* Page, A.L., Gleason, T.L., Smith, J.E., Iskander, I.K., and Sommers, L.E., Eds., University of California, Riverside, 1983, 235.
8. Davis, R.D., *Sludge Utilisation to Farmland, Part 2, Operational Aspects,* Water Research Centre, Medmenham, England, 1983.
9. Chang, A.C., Page, A.L., and Bingham, F.T., Heavy metal absorption by winter wheat following termination of cropland usage of sludge applications, *J. Environ. Qual.,* 11, 705, 1982.

10. Obbard, J.P., The Effect of Heavy Metals on Microbial Processes Related to Nutrient Cycling in Sewage Sludge Amended Soils, Ph.D. thesis, University of Lancaster, England, 1991.

11. Robson, M.J., Parsons, A.J., and Williams, T.E., Herbage production: grasses and legumes, in *Grass: Its Production and Utilisation,* Holmes, W., Ed., Blackwell Scientific, Oxford, 1989, 7.

12. C.E.C. (Commission of the European Communities), Council Directive of 12 June 1986 on the protection of the environment, and in particular of the soil, when sewage sludge is used in agriculture, Official Journal of the European Communities. No. L 181/6 (86/278/EEC), 1986.

13. Department of the Environment, The Use of Sewage Sludge in Agriculture. A National Code of Practice, Her Majesty's Stationery Office, London, 1989.

14. Obbard, J.P. and Jones, K.C., The effect of heavy metals on Di-Nitrogen fixation by Rhizobium-white clover in a range of long-term sewage sludge amended and metal contaminated soils, *Environ. Pollut.,* 79, 105, 1993.

15. Ministry of Agriculture Fisheries and Food, The Analysis of Agricultural Materials, Reference Book 427, 3rd ed., Her Majesty's Stationery Office, London, 1986.

16. Stewart, W.D.P., Fitzgerald, G.P., and Burris, R.H., In-situ studies on N_2 fixation using the acetylene reduction technique, *Biochemistry,* 58, 2071, 1967.

17. Vincent, J.M., *A Manual for the Practical Study of Root-Nodule Bacteria,* IBP Handbook No. 15, Blackwell Scientific, Oxford, 1970.

18. King, L.D. and Morris, H.D., Land disposal of liquid sewage sludge. 2. The effect on soil pH, Mn, Zn, and growth and chemical composition of Ryegrass, *J. Environ. Qual.,* 1, 425, 1972.

19. Vincent, J.M., The genus Rhizobium, in *The Prokaryotes. A Handbook on Habitats, Isolation, and Identification of Bacteria,* Syars, M.P., Stolp, H., Truper, H.G., Balows, A., and Schlegel, H.C., Eds., Springer-Verlag, Berlin, 1980, 818.

20. Postgate, J.R., *The Fundamentals of Nitrogen Fixation,* Cambridge University Press, Cambridge, 1986.

CHAPTER 9

Effects of Heavy Metals, Organic Substances, and Pesticides on Higher Plants

J.A.C. Verkleij

TABLE OF CONTENTS

0-87371-530-6/94/$0.00 + $.50

I. INTRODUCTION

A. Pollutants

An increasing number of substances are produced and introduced into the environment (environmental chemicals). These chemicals are potentially hazardous to biota (animals, plants, and microorganisms) and humans, and, in addition to acidification, eutrophication and drought may have contributed, directly or indirectly, to a strong impoverishment of the flora and fauna. Although every year new chemicals are developed, the acute toxicity and long-term toxicological effects are only known for a fraction of this large number (about 60,000); this applies especially for drugs, food additives, and pesticides. Ecotoxicological risk assessment studies have been carried out for a very small part of the total number of chemicals.

The hazardous effects of various environmental pollutants for higher plants are less or hardly known in comparison to other soil organisms. However, during the last 5 to 10 years there has been a growing interest in studying the potential toxic effects of several pollutants on agricultural and horticultural crops and wild species. In this chapter I will discuss the effects of three main groups of pollutants: heavy metals, pesticides (herbicides and nonherbicides), and a third group, comprising polychlorinated biphenyls (PCBs), polycyclic aromatic hydrocarbons (PAHs), and other organic chemicals. I have left out of consideration the effects of gaseous air pollutants (SO_2, NO_x, PAN, O_3) on plants, as this is beyond the scope of the review.

B. Ecotoxicology of Higher Plants

In studies for environmental toxicity testing, regulatory agencies routinely use three main species, the fish (fathead minnow, *Pimephales promelas*), the macroinvertebrate *Daphnia magna*, and the green alga, *Selenastrum capricornutum*. In many reports herbicide toxicity has been assessed merely on the ground of a faunal species test. If, however, a particular herbicide turns out to be practically nontoxic to daphnids, one cannot deduce from this result that the compound is also safe for floral species. More specifically, herbicide toxicity testing must not be assessed only on the basis of toxic effect on faunal species.

Aquatic and terrestrial plants are essential components of the ecosystem, they are primary producers of O_2 and primary food sources for all heterotrophic organisms whether they are animals, fungi, or bacteria. Assessment of the likelihood of risks to humans, livestock, and wildlife from potentially toxic constituents in water and soil requires a knowledge of the toxic effects on plants. Although higher plants and algae both belong to the plant kingdom, criteria developed to assess water quality by means of algae tests may not be sufficient to protect higher plants. It is therefore of importance to include representative members of this group in risk assessment studies.

In contrast to most animals, plants do not migrate and are immobile at a certain site and exposed to potentially toxic compounds at that place throughout the life cycle. Because of this, it is possible and advantageous to perform integrated environmental risk assessment studies.

In studying the potential ecotoxicological effects of pollutants on plants, it appears that plants often modify these pollutants (particularly heavy metals, but also organic substances) by biotransformation. If plants are used as a food source for ecotoxicological studies on animals (by adding toxic compounds, e.g., heavy metals to the food), one must be aware of the fact that, through biotransformation, the speciation of the compound may be changed and as a consequence potential toxicity.

As effect parameters in toxicity tests, the root-elongation test for assessing toxicity of heavy metals, organic compounds, and pesticides is the most widely used method (see Section II). In addition, various life history parameters have been applied, such as germination, seedling growth, relative growth rate (RGR), yield, and reproduction (pollen tube growth, seed production).

In recent years physiological parameters have been developed for a number of pollutants (heavy metals, organic compounds, herbicides) which are mostly aspecific (responding to various kinds of pollutants) and sometimes specific (e.g., phytochelatins as heavy metal-binding peptides; enzymes of which the

Table 1. Sources of Heavy Metal and Metalloid Contamination in the Environment

Source	Important elements
Industry	
Ore outcrops	Zn, Pb, Cu, Cd, Ni
Metal smelters	Zn, Cu, Cd, Pb, Sn, Ni
Blast furnaces	Fe, Zn, Mn
Electrolysis	Hg
Cement industry	Tl
Traffic	
Leaded gasoline	Pb
Catalysts	Al
Metal emission from tires	Ni
Household	
Waste	Cd, Pb, Cu, Zn, Ni
Sewage sludge	Cd, Pb, Cu, Zn, Ni
Energy supply	
Coal-burning power stations	Se, B, Mn, Cd
Petroleum combustion	V, Ni
Nuclear power plants	U, Cs, Pu
High-tension lines	Cu, Zn
Agriculture	
Food additives	Cu, Zn
Phosphate fertilizer	Cd
Pesticides	Cu, Hg, As, Sn

From Ernst, W.H.O. and Joosse, E.N.G., *Umweltbelastung durch Mineralstoffe; Biologische Effecte,* VEB Gustav Fischer Verlag, Jena, 1983. With permission.

activity is specifically inhibited by single herbicides, etc.) for the particular pollutant. A very specific aspect is the phytotoxic effect on plants due to inhibition or inactivation of photosynthesis by heavy metals (Cd) and herbicides. The mode of action of the chemical toxicant and the target molecules are often well documented and contributes to a better physiological understanding.

II. HEAVY METALS AND METALLOIDS

A. Diffuse and Point Sources of Heavy Metal Pollution

Heavy metals and metalloids pollute the environment from various diffuse and point sources[1] (Table 1). The contamination of the terrestrial and aquatic environment by a large number of different heavy metals and metalloids is considerable, and the adverse effects of this type of contamination on vegetation has been observed in several habitats and will probably be observed in others in the near future. The most severe effect on the vegetation can be seen near metal smelters and on mine spoilings due to the extremely high contamination of the soil.[2] The resulting specific vegetation is characterized by endemic species (species confined to a particular area) and metallophytes defined as plant species, which can maintain populations in a metal-enriched environment.[3] Essential heavy metals such as copper, zinc, manganese, and iron, and nonessential ones such as cadmium, lead, aluminum, chromium, and thallium, may occur at toxic levels in metalliferous soils.[4]

The effects on the vegetation and individual plants due to other sources of heavy metal contamination are less pronounced. In these cases, total amounts of heavy metals in the soil and the fraction of heavy metals that is available and toxic to the plant are less compared to mine soils. This holds for the use of metal-containing herbicides such as lead arsenate used in fruit farming, the disposal of sewage sludge and pig manure on arable and grassland, the erosion of zinc and copper from high-tension lines, and the accumulation of heavy metals in marine and fluviatile sediments.

At increasing metal concentrations in the environment (i.e., the metal fraction in the soil which is available and toxic to the plant) heavy metal-sensitive species disappear. A few angiosperms (for instance, in the temperate zone of Europe there are about 40 species known to date) can still be found on soils with toxic heavy metal concentrations (Zn up to 60,000 mg kg^{-1}, Pb 30,000 mg kg^{-1}, Cu 7,000 mg kg^{-1}, and Cd 500 mg kg^{-1}) and an ability to grow on these soils is related to evolution of heavy metal tolerance.[5] In habitats where the metal contamination could be potentially toxic to most plants (Sections II.C and II.D), it is important to know which techniques and parameters can be used in trying to assess metal toxicity and tolerance. The most commonly applied technique is the root-elongation method, but other parameters will be discussed, c.q. RGR, biomass production, other life history characteristics, metal content, and physiological parameters (biomarkers).

B. Plants Occurring on Soils Highly Contaminated by Heavy Metals

On metalliferous soils (ore outcrops, spoil heaps), a specific vegetation is found which consists of some characteristic metal indicator species and metal-resistant populations of a small number of other species.[1] Such indicator species found in Europe are *Viola calaminaria, Thlaspi caerulescens,* and *Armeria maritima.* A few other species, having evolved metal-resistant populations, are also successful, c.q. *Agrostis capillaris, Silene vulgaris, Festuca ovina, F. rubra,* and some other grasses and herbs.[5] This evolutionary process has taken place on undisturbed metalliferous soils since the last glaciation. Metalliferous areas were further enlarged by smelting and washing of ores, where sedimentation of metals took place on river banks accompanied by seed transport. As a result of these processes, metal vegetations are found on the floodplains of rivers in a number of countries in Europe.[6] However, for a number of reasons these heavy metal vegetations are now under pressure. The transport of metals by rivers has been reduced strongly because of mine closures, and the metal content of the upper soil layer has been drastically lowered, allowing species with low metal resistance a better opportunity for establishing and developing. Less competitive species, such as the zinc violet (*V. calaminaria*), are driven out and the metal vegetation will finally disappear due to the increasing use of fertilizers.

The impact of metal emission by metal smelters, refineries, and metal processing industries on the surrounding environment is complex and depends on period and intensity of the emission and the presence of other pollutants (sulfur dioxide, nitrogen oxides). In most cases the effect on the vegetation is drastic and within a few years after the start of the smelting process a species-poor grassland containing metal-tolerant genotypes of a few grasses (*A. capillaris, A. canina*) remains, as trees die back and disappear.[7] Such a typical situation is found near a zinc-cadmium smelter in the Kempens area (Belgium/The Netherlands), where the surrounding vegetation has been reduced to a monotonous stock of *Molinia caerulea* and *A. capillaris*.[8,9] The smelter population of *M. caerulea* showed low metal tolerance and did not differ from the control population, in contrast to the *A. capillaris* population, which shows high tolerance. Due to the fact that heavy metals have accumulated strongly in the topsoil and have not leached to deeper layers of the soil, the deep-rooting grass *M. caerulea* is able to survive without developing metal tolerance.[9,10] As well as the evolution of metal-resistant *A. capillaris*, vesicular arbuscular mycorrhiza (VAM) fungi, living in symbiosis with this grass, have also developed metal resistance.[11]

In the literature on the evolution of metal tolerance in plants there are strong indications that metal tolerance is a constitutive character, i.e., always present. In a study on Cd tolerance in *Holcus lanatus* from an aerially polluted site, however, Baker et al.[12] determined that a part of the original tolerance was lost after transplantation into a noncontaminated soil in the greenhouse. This reduction in tolerance was not observed in populations of the same species from a mine site nor in populations of other metal-tolerant species (*A. capillaris, Deschampsia caespitosa,* and *F. rubra*). It has been suggested that during the early stages of evolution, selection mainly occurs on inducible tolerance while in later stages constitutive tolerance prevails under constant and strong selection pressure. It is as yet unclear what the physiological and genetical basis is of such phenomenon, and convincing evidence for phenotypic plasticity of metal tolerance in plants is still not available.

Evolution of heavy metal tolerance does not necessarily take place over wide geographic ranges. If the metal stress is localized, adaptation to this can be equally localized. It is often stated that this evolution to man-made stress can occur over very short periods of time, but there is a lack of well-documented studies of this phenomenon in contrast to the evolutionary rate of pesticide resistance in insects. A nice example of such a metal stress, which is usually underestimated, are the steel pylons of power lines, which are initially galvanized to prevent corrosion. With time, this galvanized covering is corroded away by rainwater, enriching the soil beneath the pylons with metals, in particular, zinc. *A. capillaris* and, to a lesser degree, other grasses growing beneath these 20- to 30-years old pylons, have evolved zinc tolerance in a very limited area (10 m square).[13,14]

The effect of corrosion of copper and zinc due to acid deposition from high-tension lines, especially on the underlying vegetation, is less clear. Enhanced levels of copper were found in soil and plant species, but this did not result in a change in the genetic make-up of the various plant species and evolution of copper tolerance.[15]

The increasing enrichment of marine and fluvial sediments in salt marshes and floodplains by heavy metals and arsenic is a potential threat to the existing vegetation.[16] Although the contamination in these areas is not as high as in industrial and mining areas, it is considerable compared to background concentrations.[2,16] Total amounts of heavy metals in sediments of a former Rhine estuary in The Netherlands (Biesbosch) are close to the amounts in the floodplains of the Geul and within the range of concentrations below pylons.[17] Genetic changes in heavy metal tolerance have, however, not (yet) been observed in *Urtica dioica* and *A. stolonifera*.[18] The most obvious explanation for the observed differences in effect is the low availability of the contaminants. Other explanations are that this estuarine area is located too far from naturally occurring tolerant plant populations, or that tolerant plants were not able to become established due to the former high flooding frequency in this area (before closing of the estuary).[17]

An elevated concentration of heavy metals and arsenic of the roots in a number of plant species (e.g., *U. dioica*) occurring in this area was observed. Under waterlogged conditions, several plant species are capable of forming an iron plaque on their roots (by oxidizing $Fe[II]$ and $Mn[II]$ to $Fe[III]$ and $Mn[IV]$). Due to the adsorption of appreciable amounts of zinc, copper, and arsenic to this iron plaque, a change in the chemical speciation and mobility occurs, which leads to an increased uptake of these heavy metals and arsenic.[18]

C. Effects of Heavy Metal Pollution on Uptake and Translocation in Plants

When plants take up, translocate, and assimilate toxic amounts of metals, growth becomes inhibited and biomass production decreases.[19] The nature and magnitude of this response will depend on the sensitivity of the individual, the intensity (concentration and duration) of exposure, the specific metal, and the form in which it is present.[20] It is possible to construct yield-dose response curves based on experimental studies, where yield can represent a growth parameter such as biomass production in the long term or inhibition of root elongation in the short term. In order to assess the hazardous effects of heavy metals on plants and crops, biomass production (yield) after long-term field experiments is a commonly used parameter,[21] while for direct toxicity testing methods involving root elongation, germination, or seedling development are more frequently used.[22,23]

The yield of test organisms (crops), or their growth rate give rather ambiguous measures of toxicity; however, since they are affected by many other environmental conditions[24] (light, temperature, soil structure, soil composition, nutrient status, etc.). Also, die off or reductions in yield do not indicate which component is primarily responsible for the toxic effects. Furthermore, soil analyses of the various heavy metals (total, water extractable, or other fractions) are often ambiguous, since the relationship between the availability of an element in soil to plants and its extractable fraction can vary according to soil type and combinations of different metals.

In view of these uncertainties, many authors[24,25] have applied diagnostic tissue analyses to assess toxicity. Their results showed that, although uptake of heavy metals and dry matter yield were affected by environmental conditions and nutrient status, the upper, critical levels of these metals (i.e., the lowest tissue concentration of a metal at which it reduced the production of dry matter) were remarkably independent of these variables[24] and relatively uniform over the range of species studies.[25] Other studies demonstrated that the levels of heavy metals in plants are not only affected by soil properties and climatic conditions, but are also dependent on plant factors (species, cultivar), and marked differences were found with regard to heavy metal accumulation and tolerance by plant species and cultivars.[26]

D. Effect Parameters in Metal Toxicity and Tolerance Tests

1. Root-Elongation Method

Metal toxicity and tolerance tests using the root-elongation method have been widely reported for a number of heavy metals and plant species.[22,23] In (genetic) studies of metal tolerance, particularly, investigators have used the tolerance index (TI) by comparing root growth at a single concentration of metal solution with a control solution during a certain exposure time (days or weeks). The TI can be calculated:

$$TI = \frac{\text{root growth in metal solution}}{\text{root growth in control solution}}$$

This index is only satisfactory when large differences in tolerance between plants are being compared, but is inaccurate because of an inherently high level of statistical noise and the fact that root elongation is determined by many genes unrelated to tolerance. Alternative methods have therefore been developed for exposing individual plants to a multiple concentration test.[27] The sensitivity of the tested population is quantified using the parameters of the concentration effect relationship, i.e., no observed effect concentration (NOEC), EC_{50}, and EC_{100}. (NOEC = no observed effect concentration, i.e., the highest concentration that does not inhibit root growth; EC_{50}, the concentration that inhibits root growth to 50% of the control rate; EC_{100}, the lowest concentration that inhibits root growth completely). In this way Schat and ten Bookum[27] were able to assess

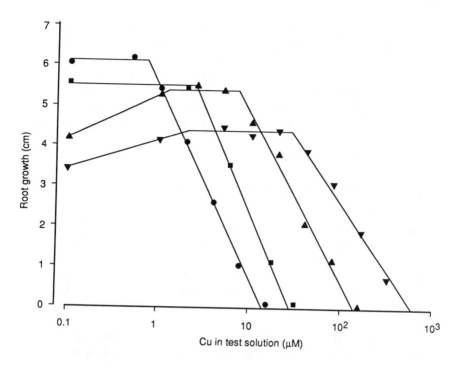

FIGURE 1. Increase in root length during a 72-h exposure to different copper concentrations in a nontolerant population (Amsterdam, ●) and three distinctly copper-tolerant populations (Harlingerode, ■; Marsberg, ▲; Imsbach, ▼). After Schat, H. and ten Bookum, W.M., *Heredity*, 68, 219, 1992. With permission.

large differences in copper toxicity between populations of *S. vulgaris* (Figure 1). Besides the difference in sensitivity to copper, two other conclusions can be drawn:

1. Copper-tolerant populations showed suboptimal growth at the lowest copper concentration, 0.1 μM, whereas the sensitive one did not, implying a copper deficiency for tolerant plants in the control solution.
2. The maximum root elongation of the copper-tolerant populations was lower than that of the sensitive population, and suggests a "cost" of tolerance.[2,27]

In contrast to the essential metals zinc and copper, such an optimum dose-response curve does not apply for non-essential trace elements such as cadmium and lead.[20,28]

Several agencies, such as the U.S. Environmental Protection Agency (EPA), Food and Drug Administration (FDA), and the Organization for Economic Co-operation and Development (OECD), have recommended a number of crop, forage, tree, and other species for ecotoxicity testing using the root-elongation method. This method can be used for testing a number of heavy metals within a whole range of species and seems to be valid as a sensitive indicator of environmental toxicity. It is interesting to notice that in some studies, the root-elongation test detected toxicity in some well water samples from hazardous sites, whereas algal and daphnid tests showed either no effect or stimulation.[29] Data from root-elongation tests by Godbold and Hüttermann[30] suggest that metal concentrations in forest soils are sufficiently high to influence root growth and might be a contributing factor to forest die-back.

2. Other Life History Characteristics

As previously mentioned, the root-elongation method is the most widely applied method for testing metal toxicity or tolerance. The method is very reproducible and sensitive under standard conditions (within 3 d, clear differences in response can be obtained) and has a positive signal to noise ratio. However, root growth responds much faster to copper and other redox-active metals (Hg, Ag) than to other metals[31] (mostly non-redox active ones), and for long-term exposure to low toxic levels of heavy metals, other parameters might be more important.

By using the RGR a clear difference in response was found between cadmium-sensitive and tolerant populations of *S. vulgaris* at a relatively low cadmium concentration[28,32] (1 μM). By contrast, at the early stage of development (germination) increasing concentrations of heavy metals for Cd (up to 10 mM) did not discriminate between the percentages of germination of metal-tolerant and -sensitive populations.[33] It was concluded that the germination phase does not seem to be critical in the process of selection. For the following stage, development of the radicula and the young seedling, significant differences in response have been observed and have also been found for other plant species.[23,33]

Long-term exposure to low doses of heavy metals in combination with gaseous air pollutants (NO_x, SO_2) reduced the number of flowers per plant, which could affect the potential seed biomass.[34] However, the metal load of pollen is low.[35] The effect of low external metal concentrations on pollen tube growth is, however, quite clear and differs between metal-tolerant and non-tolerant populations.[36]

3. Physiological Parameters (Biomarkers)

Upon exposure, higher plants take up heavy metals to a degree that depends on internal and external factors. However, the relationship between metal concentration in the soil (total, plant-available fraction) and metal uptake and tissue concentration (root, shoot) with tolerance and toxicity is not very clear.[20] Therefore, it is very difficult, if not impossible, to assess sensitivity on the basis of heavy metal contents in root and/or shoots of plant species.

A better approach to this problem seems to be to look at physiological parameters which are more closely related to heavy metal stress syndromes.[37]

In plants, both essential and nonessential heavy metals can induce metal binding thiorich peptides (γ-glutamylcysteinyl)$_n$ glycines with n = 2 to 11, also known as phytochelatins.[38,39] Phytochelatins are considered to play a key role in cellular metal homeostasis and metal detoxification. Although it has been argued that these compounds may be involved in heavy metal tolerance, evidence for a significant role is not convincing.[37,40] There are strong indications that production of phytochelatins is merely a measure of the degree of metal toxicity (at least for copper and cadmium) experienced by the plant.[40,41]

Other physiological metal toxicity parameters, which can be considered to be more indirect measures of the toxic heavy metal concentrations, have also been exploited. Van Assche and Clijsters,[42] measuring distinct enzyme activity and metal-specific changes in isoperoxidase patterns, have argued that their parameters can be used as diagnostic criteria to evaluate the phytotoxicity of soils contaminated by several metals. However, other environmental (air pollutants, shortage or excess of water, salt, etc.) and ontogenetic (plant development and differentiation) factors may also bring about a change in peroxidase isoenzyme pattern and/or increased peroxidase activity[43,45] and it will be very difficult, if not impossible, to single out the effects of the individual factors.[45] A similar argument can be raised to other physiological stress components such as proline, the accumulation of which is not only restricted because of metal toxicity experienced by the plant.[46]

E. Conclusions

1. Due to heavy metal pollution in the environment and metal speciation processes in the soil, the availability of metals to the plant will be increased.
2. Heavy metal-sensitive species will disappear from the vegetation, and other species will evolve with large genetic variability in respect to heavy metal tolerance. As long as emissions of metals occur, both processes, species loss and change in genetic structure of surviving species, will take place.
3. The direction and extent of these processes are determined by abiotic conditions (organic matter, pH, redox potential) as well as by the mechanisms of resistance at the level of the cell, organ, and whole plant.[37]
4. It is necessary to develop an ecotoxicological risk assessment program for metal contamination in plants.
5. In addition to commonly applied techniques such as the root-elongation method, other more physiological parameters (enzyme activities, phytochelatins) appear to be appropriate parameters for assessing metal toxicity.

III. ORGANIC SUBSTANCES

A. Polynuclear Aromatic Hydrocarbons (PAHs)

Polycyclic (polynuclear) aromatic hydrocarbons (PAHs) form a large group of organic compounds, characterized by the presence of two or more condensed aromatic rings. Emission of these compounds into the environment is the result of human activity. PAHs are produced by incomplete combustion of organic material such as coal, oil, wood, etc., and are typically present in sewage sludge in the range 1 to 10 mg Σ PAH kg^{-1}, which is significantly higher than found in normal agricultural soils.[47]

A number of PAHs, such as benzo(a)pyrene, are proven or suspected of being carcinogenic and/or mutagenic. Therefore, it is important to investigate whether or not plants are capable of taking up these organic chemicals and, hence, could contaminate human food. According to a few studies it seems clear that PAHs are not readily taken up by plants. These compounds are strongly adsorbed onto soil organic particles and foliar and/or root uptake is very inefficient. Concentrations of PAHs in the range of 0 to 100 μg kg^{-1} have been detected in plant tissues, suggesting that some PAHs (particularly the lower-molecular weight compounds) have been taken up from the soil.[48] However, in a recent study on PAHs in crops from long-term field experiments amended with sewage sludge there was no evidence for this.[49] These authors could not consistently detect increased PAHs concentrations in plant tissues of crops growing on an increased PAHs soil burden relative to the unsludged controls.

There were strong indications that PAHs detected in above-ground plant parts are derived mainly from atmospheric inputs, while PAHs detected in root crops probably arise from adsorption to the root surface. So, atmospheric deposition seems to provide an important input of PAHs onto plant leaves and to dominate over inputs via plant root uptake and translocation.[50] Extremely high concentrations of PAHs (higher than 10,000 μg kg^{-1}) were detected in moss bags (*Sphagnum* spp.) near a plant manufacturing electrodes in a biomonitoring study in The Netherlands, indicating the strong adsorption capacity of mosses.[51]

B. Polychlorinated Biphenyls (PCBs)

Polychlorinated biphenyls (PCBs) are very persistent, nonionic, lipophilic compounds of very low water solubility ranging from 0.95 μg L^{-1} to 5.9 mg L^{-1}.[52] They are or have been used in, among other things, hydraulic systems, as dielectric liquids, softeners, paints, and inks. The emission of PCBs into the environment is due solely to human activity and they are found in widespread areas of the aquatic and terrestrial environment. PCBs appear to be rather toxic (particularly the PCBs in which the meta- and para-positions are substituted, e.q. 3.3', 4.4'-tetrachlorobiphenyl) to animal and man and are probably teratogenic and carcinogenic. The accumulation of these compounds in the fat tissue seems to be dangerous for organisms at the top of the food chain and one of the main causes of the reduction in seal numbers.[53] The concentrations of PCBs in relatively clean agricultural soils are less than 0.5 mg kg^{-1} d.w., but in sewage sludge-amended soils, concentrations up to 25 mg kg^{-1} d.w. have been detected in a depth of 15 cm.[54] However, because of their strong affinity to soil particles, PCBs are very immobile and do not leach to any appreciable degree.

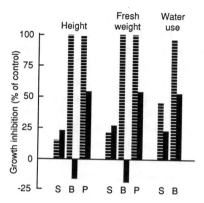

FIGURE 2. Effect of 1000 mg kg^{-1} soil (striped bars) and 100 mg kg^{-1} soil (black bars) of soil-applied Aroclor® 1254 (mixture of PCBs) on soybean (S), beet (B), and pigweeg (P). Growth expressed as % growth inhibition. After Strek, H.J. and Weber, J.B., *Environ. Pollut.*, 28, 291, 1982. With permission.

Although relevant data concerning the effects of PCBs on plants and the vegetation are very scarce (in sharp contrast to the vast amounts of literature on animal and man), these data indicate that plants are capable of taking up these compounds and transferring them into polar metabolites or insoluble macro-molecules.[55] It is not very probable that uptake and transformation of these compounds occur to any great extent, because a very large part (>95%) will adsorb to the root surface.

Some PCBs will reach the plant via atmospheric deposition and will be bound to the wax layer of the cuticle.[54] Some authors claimed that the uptake of PCBs by the leaves from the atmosphere is many times as high as uptake by translocation (less than 10%).[56] However, uptake of PCBs from fallout is unlikely to occur to any great degree because the chemical may merely adsorb to the outer surfaces of the plant and may not be truly present inside the plant.[54] There are no reports on the negative effects of PCBs on plants in field situations due to the relatively low PCBs concentrations. However, in soybean, beet, and pigweed exposed to high doses of soil-applied PCBs, a strong reduction in plant height and water content was observed (Figure 2). The ecological relevance of this study is less clear and limited because of the extremely high concentrations applied (100 and 1000 mg kg^{-1}) to soil. It seems, on the other hand, worthwhile to investigate long-term (several months to years) effects of low concentrations of PCBs (1 to 10 mg kg^{-1}) on plants.

C. Other Organic Compounds

The potential hazardous effects on plants of organic chemicals other than PAHs and PCBs, e.g., halogenated benzenes, nitrobenzene, phenolic compounds, etc., which are released to the environment in substantial amounts, have hardly been investigated. The necessity of investigating has been realized by public administrations and agencies and as a consequence a number of plant species are now included and recommended as biological test species by the U.S. EPA, FDA, and OECD for toxicity testing and environmental assessment.

Based on recent studies it appears that these compounds have a toxic effect on some species in a concentration range which can be found in contaminated areas. Studies by Wang[23,57] reveal that the plant species tested (millet, radish, velvetleaf, lettuce) show an ascending order of toxicity for substituted phenols (phenol, chloro-, dichloro-, and trichlorophenol) using root-elongation tests. The sensitivity to these toxicants differed between the species (millet was the most sensitive). The NOEC values of 2,4,6-trichlorophenol for millet are of the same order of Cd, and these concentrations (5 mg/kg) have been observed in sewage sludge.[57] In a risk assessment study of xenobiotics such as 2,6-dichlorobenzo-nitrile, pentachlorophenol, and thiourea with *Brassica rapa* and *Avena sativa,* clear differences in toxicity were observed between species.[58]

It is evident that phytotoxicity tests of these compounds are necessary for a broad range of plant species in relation to environmental concentrations, and should be considered as a part of ecotoxicological risk assessment studies.

In a more physiological study concerning the phytotoxic effects of nitro-benzene on transpiration and photosynthesis, a difference in these activities was found between species[59] (Figure 3). Some species did not show any difference in transpiration and photosynthesis; in other species enhanced activities were detected, and one species (autumn olive) did not show any activity after 10 h exposure and seems to be very sensitive to nitrobenzene. This study also revealed a similarity in uptake mechanism, which was passive and proportional to the rate of water flux in each species. On the other hand, a high diversity of plant responses was observed to the toxicants. Some species (such as honeysuckle, *Lonicera catarica*) have a high capacity for chemically altering toxicants; other species (e.g., ash) have the potential for high rates of chemical volatilization. Therefore, plant species may have a much greater influence on the fate of waste chemicals than has ever been realized.

D. Conclusions

1. Organic substances (PCBs, PAHs, etc.) in contaminated soils, such as in sludges, are on average present in concentrations less than 10 mg kg^{-1} dry soil. Some sewage sludge contains unusually high concentrations (up to 1% d.w.), but in most cases the organics tested are below detection limits.
2. Major assimilative pathways for organic substances applied to the soil plant system include adsorption, volatilization, degradation, leaching, and plant uptake. Many organics (like PAHs and PCBs) are strongly adsorbed to soil organic matter and/or undergo degradation, reducing the potential for plant uptake.

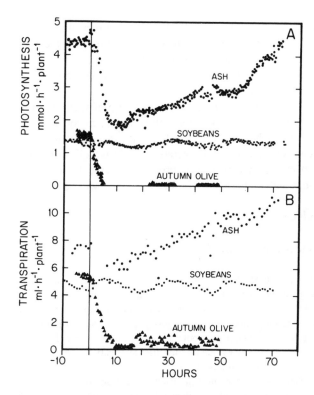

FIGURE 3. Photosynthesis (A) and transpiration (B) rate of ash, soybean, and autumn olive plants in controlled environments before and after exposure time to nitrobenzene (8.0 μg/ml) in the hydroponic solution. After McFarlane, C., Pfleeger, T., and Fletcher, J., *Environ. Toxicol. Chem.,* 9, 513, 1990. With permission.

3. PCBs and PAHs are not readily taken up by plants and will be mainly adsorbed to the other surfaces of the plant. It is not clear if, and to what amount, the plant is capable of metabolizing these compounds.
4. Other organic chemicals, such as substituted phenols, have a toxic effect on plants at concentrations which have been observed in sewage sludges. Plants take up and are capable of metabolizing these compounds. Due to the ability to chemically alter such compounds by metabolization, plants seem to play a more important role in chemically altering waste products. More studies are needed to examine these specific aspects.

IV. PESTICIDES

A. Introduction

The use of pesticides accounts for one of the major sources of xenobiotics in the environment. The most widely applied pesticides are herbicides (43%), insecticides (32%), fungicides (19%), and miscellaneous agrochemicals (5%), respectively.[60] Pesticides can be classified according to their use against pest organisms (as insecticides, nematicides, herbicides, etc.) or according to their chemical structure and chemically active components (thiocarbamates, organophosphates, triazines, etc.).

Herbicides belong to a heterogeneous group of chemical substances, which are primarily developed to kill (vascular) plants; many other pesticides (e.g., nematicides) show herbicidal side effects. As the use of herbicides is the main cause of negative effects on the vegetation inside as well as outside the application zone, I will restrict myself to this group of chemicals. The indirect effects of the herbicides and nonherbicides on the vegetation will be considered separately.

B. Herbicides

Herbicides can be classified according to:

- Selectivity: some of the most used herbicides are very selective. They cause damage on the crop and can be applied during the crop season (e.g., simazine). Other herbicides are not or are hardly selective, e.g., paraquat. Paraquat belongs to the contact herbicides group and acts almost immediately.
- Persistence: a number of herbicides are very persistent (i.e., simazine) and detectable in the soil even 9 months after application; it may retain phytotoxic action for up to 3 months.
- (Bio)chemical mechanism of toxification: the phytotoxic effect of herbicides may differ. The group of triazine compounds interferes directly with the photosynthesis process.[61] Most herbicides, however, deregulate simultaneously various metabolic processes inside the plant.

The net toxic effect of the applied herbicide depends on a lot of factors: the active substance, its dose, the time and the way it is used (via the foliage, on the soil), developmental stage of the plant, and weather conditions (temperature, relative humidity, wind, etc.). Higher plants can metabolize the herbicides,[60] and sometimes the intermediates during this metabolization process are more toxic than the original herbicides. Due to photosystem I, bipyridyl herbicides (paraquat, diquat, morfamquat, etc.) are reduced to bipyridyl radicals (very reactive intermediates) and further oxidized to less reactive products.[63] However, in most cases this metabolization process leads to detoxification of the herbicides. In general, herbicides are easily degradable, with the exception of some compounds such as paraquat and diquat, which are bound to clay minerals/particles and are therefore not biologically available (half-time about 20 years). Due to the changing use of the soil, these substances may become available at some future time as a chemical time-bomb.

The consequences of the multiple application of herbicides can vary from a temporary reduction of the vegetation, selection of herbicide-resistant species (particularly annual arable weeds), and niche variation in invasive species such as *Cyperus esculentus* and *Abutilon theophrasti*. However, species reduction in agricultural flora is not only due to the intensive use of herbicides, which started in about 1950. A number of characteristic weeds disappeared or became very rare due to other practices, such as purifying the sowing seed, mechanical control, fertilization, crop rotation, crop density, and reduction in the agricultural area. A strong decrease in the occurrence of *Agrostemma githago* (corn cockle), a very general and noxious weed in winter cereal fields in former years, is mainly due to the cleaning of seed.

The evolution of resistance, especially to the group of triazine herbicides, has been demonstrated for a number of annual weeds such as *Senecio vulgaris, Chenopodium album, Stellaria media, Poa annua,* and *Amaranthus retroflexus* and is due to these herbicides being applied repeatedly and regularly.[64-67]

Recently, resistance to chlorosulfuron and triazolo pyrimidine was discovered in *S. media* and *Arabidopsis thaliana*.[68] Apart from this, it is remarkable that the long-term application of herbicides has only sporadically led to the evolution of resistance in plants in contrast to the often very fast and frequent evolution of resistance to heavy metals, insecticides, and antibiotics in organisms upon exposure to these substances (for further discussion see Gressel and Segel).[69]

It has been recognized for many years that herbicide drift from agricultural land can affect the flora of field margins and adjacent vegetation. In this buffer zone (6 to 10 m) that separates treated areas from sensitive vegetation, the effects of herbicides are less deleterious in terms of decrease in diversity and species loss. Relatively little is known about the impact of such herbicide drift on individual species and species composition. Recent studies by Marrs and co-workers,[70] using experimental microcosms, have demonstrated that the drift of mecoprop (a test herbicide widely used in the U.K.) affected the aesthetic appearance of the buffer zone (reduction in either performance or yield of some attractive species) and possibly affected the fecundity of other species and the balance between species.

In order to prevent a further impoverishment of the flora in the buffer zone, more effective herbicides have to be applied in a more selective way (time of application dose). Moreover, management of this zone has to be changed. German and English researchers, based on long-term field experiments for many years, have shown[71,72] that in unsprayed zones more and sometimes rare arable weed species occur compared to the sprayed zones (Table 2). Similar findings have been made in a recent study in The Netherlands over a 1-year period.[73]

C. Effects of Herbicides/Nonherbicides on Plants Outside the Treated Areas

Little is known about the effects of pesticides on the vegetation outside the target area. Due to surface runoff, leaching losses, and evaporation, concentrations of pesticides in surface water may reach values of 1 μg l^{-1}. On the basis

**Table 2. Total Yield (g) of Vegetation per Microcosm and the %
Contribution of Each Component Species in Unsprayed
Microcosms in 1988 and 1989**

	Total yield	
Year	− *Lolium perenne*	+ *Lolium perenne*
1988	48.2 ± 1.1	65.1 ± 3.6
1989	11.8 ± 0.9	16.5 ± 0.7

	Percentage contribution of individual species			
	− *Lolium perenne*		+ *Lolium perenne*	
Species	1988	1989	1988	1989
Digitalis purpurea	8.0 ± 0.2	0.7 ± 0.7	11.9 ± 1.3	0.2 ± 0.2
Filipendula ulmaria	2.9 ± 0.8	5.0 ± 1.0	1.4 ± 0.5	2.4 ± 0.7
Galium mollugo	18.2 ± 2.4	16.1 ± 2.4	14.5 ± 1.6	11.1 ± 2.5
Hypericum hirsutum	18.8 ± 2.9	0.4 ± 0.2	13.4 ± 0.8	1.4 ± 0.5
Lolium perenne	—	—	12.0 ± 2.7	37.4 ± 4.4
Lychnis flos-cuculi	20.5 ± 1.7	59.3 ± 6.2	11.6 ± 3.4	36.7 ± 2.9
Primula veris	1.8 ± 0.5	4.5 ± 2.1	0.7 ± 0.2	1.9 ± 0.3
Ranunculus acris	10.8 ± 1.1	8.3 ± 1.1	4.5 ± 2.7	5.0 ± 1.7
Stachys sylvatica	18.8 ± 5.3	5.3 ± 1.7	29.9 ± 2.4	3.7 ± 0.9

Note: Mean values ± standard errors (*n* = 5) are presented.

From Marrs, R.H., Frost, A.J., and Plant, R.A., *Environ. Pollut.*, 73, 25, 1991. With permission.

of an ecotoxicological risk assessment study in an agricultural area of the polder Haarlemmermeer (The Netherlands), it was concluded that some concentrations exceeded the calculated critical concentrations for the ecosystem[74] according to an extrapolation procedure, in which the NOEC is used as parameter to calculate the hazardous concentration for p% of the species;[75] experimental data on the potential effects on waterplants are, however, lacking.

In addition to herbicides, a number of nonherbicides are used for agricultural purposes to control diseases and pests. These compounds also exert a weak phytotoxic effect. Particular fungicides may interfere negatively with mycorrhiza host relationships. Some evidence is given by Fitter,[76] who showed that application of benomyl (a fungicide) diminished the rate of infection of VAM in plants.

Furthermore, there are indications of indirect effects of nonherbicides on the vegetation via complex plant/animal interactions, but quantitative data are not available.

D. Conclusions

- With regular application of herbicides on agricultural land, an impoverishment of the flora will locally occur (including the buffer zone); nonresistant species will disappear and resistant plant species (particularly annual agricultural weeds) will become predominant.

- There is insufficient knowledge of the potential toxic effects of herbicides outside the application area. Due to leaching losses or surface runoff of herbicides and nonherbicides to surface water, a negative effect might be expected on water plants.
- Although herbicides have the greatest impact on the vegetation, a toxic side effect for the flora of other pesticides cannot be excluded. A shift in the population size of species may occur due to the application of pesticides which could cause significant changes at the plant community level.
- In order to prevent or diminish these negative effects the application of pesticides has to be more efficient and selective. This would increase species abundance in the buffer zone adjacent to the treated areas.

REFERENCES

1. Ernst, W.H.O. and Joosse, E.N.G., *Umweltbelastung durch Mineralstoffe; Biologische Effecte,* VEB Gustav Fischer Verlag, Jena, 1983.
2. Ernst, W.H.O., *Schwermetallvegetation der Erde,* G. Fischer Verlag, Stuttgart, 1974.
3. Ernst, W.H.O., Mine vegetation in Europe, in *Heavy Metal Tolerance in Plants: Evolutionary Aspects,* Shaw, A.J., Ed., CRC Press, Boca Raton, FL, 1990, 21.
4. Woolhouse, H.W., Toxicity and tolerance in the response of plants to metals, in *Encyclopedia of Plant Physiology, Vol. 12, Physiological Plant Ecology III,* Lange, O.L., Nobel, P.S., Osmond, C.B., and Ziegler, H., Eds., Springer-Verlag, Berlin, 1983, 245.
5. Antonovics, J., Bradshaw, A.D., and Turner, R.G., Heavy metal tolerance in plants, *Adv. Ecol. Res.,* 7, 1, 1971.
6. Ernst, W.H.O., Ökologisch-soziologische Untersuchungen der Schwermetall-Pflanzen-gesellschaften Mitteleuropas unter Einschluss der Alpen, in *Abh. Landesmus. Naturkd. Müster Westfalen,* 27(1), 1, 1965.
7. Ernst, W.H.O., Zink- und Cadmium-Immisionen auf Böden und Pflanzen in der Umgebung einer Zinkhütte, *Ber. Dtsch. Bot. Ges.,* 85, 295, 1972.
8. Dueck, Th.A., Impact of Heavy Metals and Air Pollutants on Plants, Dissertation, Vrije Universiteit, Amsterdam, 1986.
9. Dueck, Th.A., Ernst, W.H.O., Faber, J., and Pasman, F., Heavy metal emission and genetic constitution of plant populations in the vicinity of two metal emission sources, *Angew. Bot.,* 58, 47, 1984.
10. Harmsen, K., *Behaviour of Heavy Metals in Soils,* Pudoc, Wageningen, 1977.
11. Ietswaart, J.H., Griffioen, W.A.J., and Ernst, W.H.O., Seasonality of VAM infection in three populations of *Agrostis capillaris* (Gramineae) on soil with or without heavy metal enrichment, *Plant Soil,* 139, 67, 1992.
12. Baker, A.J.M., Grant, C.J., Martins, M.H., Shaw, S.C., and Whiterrook, J., Induction and loss of cadmium tolerance in *Holcus lanatus* L. and other grasses, *New Phytol.,* 102, 575, 1986.
13. Al-Hiyaly, S.A.K., McNeilly, T., and Bradshaw, A.D., The effects of zinc contamination from electricity pylons — evolution in a replicated situation, *New Phytol.,* 110, 571, 1988.

14. Al-Hiyaly, S.A.K., McNeilly, T., and Bradshaw, A.D., The effects of zinc contamination from electricity pylons — contrasting patterns of evolution in five grass species, *New Phytol.*, 114, 183, 1990.

15. Kraal, H. and Ernst, W.H.O., Influence of copper high tension lines on plants and soils, *Environ. Pollut.*, 11, 131, 1976.

16. Salomons, W. and Förstner, U., Metals in the Hydrocycle, Springer-Verlag, Berlin, 1984.

17. Otte, M.L., Heavy Metals and Arsenic in Vegetation of Salt Marshes and Flood Plains, Dissertation, Vrije Universiteit, Amsterdam, 1991.

18. Otte, M.L., Rozema, J., Koster, L., Haarsma, M.S., and Broekman, R.A., Iron plaque on roots of *Aster tripolium* L.: interaction with zinc uptake, *New Phytol.*, 111, 309, 1989.

19. Lepp, N.W., *Effect of Heavy Metal Pollution on Plants*, Vol. 2, Applied Science Publishers, London, 1981.

20. Baker, A.J.M. and Walker, P.L., Ecophysiology of metal uptake by tolerant plants, in *Heavy Metal Tolerance in Plants: Evolutionary Aspects*, Shaw, A.J., Ed., CRC Press, Boca Raton, FL, 1990, 155.

21. Berry, W.L. and Wallace, A., Toxicity: the concept and relationship to the dose response curve, *J. Plant Nutr.*, 3, 13, 1981.

22. Wilkins, D.A., The measurement of tolerance to edaphic factors by means of root growth, *New Phytol.*, 80, 623, 1978.

23. Wang, W., Literature review on higher plants for toxicity testing, *Water, Air Soil Pollut.*, 50, 381, 1991.

24. Macnicol, R.D. and Beckett, P.H.J., Critical tissue concentrations of potentially toxic elements, *Plant Soil*, 85, 107, 1985.

25. Davis, R.D., Beckett, P.H.T., and Wollan, E., Critical levels of twenty potentially toxic elements in young spring barley, *Plant Soil*, 49, 395, 1978.

26. Miles, L.J. and Parker, G.R., The effect of soil-added cadmium on several plant species, *J. Environ. Qual.*, 8, 229, 1979.

27. Schat, H. and ten Bookum, W.M., Genetic control of copper tolerance in *Silene vulgaris*, *Heredity*, 68, 219, 1992.

28. Verkleij, J.A.C. and Prast, J.E., Cadmium tolerance and cotolerance in *Silene vulgaris* (Moench.) Garcke (=*S. cucubalus* L. Wib.), *New Phytol.*, 111, 637, 1989.

29. Miller, W.E., Peterson, S.A., Greene, J.C., and Callahan, C.A., Comparative toxicology of laboratory organisms for assessing hazardous waste sites, *J. Environ. Qual.*, 14, 569, 1985.

30. Godbold, D.L. and Hüttermann, A., Effect of zinc, cadmium and mercury on root elongation of *Picea abies* (Karst.) seedlings, and the significance of these metals to forest die-back, *Environ. Pollut.*, 38, 375, 1985.

31. De Vos, C.H.R., Schat, H., Vooijs, R., and Ernst, W.H.O., Copper-induced damage to the permeability barrier in roots of *Silene cucubalus*, *J. Plant Physiol.*, 135, 164, 1989.

32. Verkleij, J.A.C., Koevoets, P., Van 't Riet, J., Bank, R., Nydam, Y., and Ernst, W.H.O., Poly (gamma-glutamylcysteinyl) glycines or phytochelatins and their role in cadmium tolerance of *Silene vulgaris*, *Plant, Cell Environ.*, 13, 913, 1990.

33. Lolkema, P., Copper Resistance in Higher Plants, Dissertation, Vrije Universiteit, Amsterdam, 1985.

34. Dueck, Th.A., Wolting, H.G., Moet, D.R., and Pasman, F.J., Growth and reproduction of *Silene cucubalus* Wib. intermittently exposed to low levels of air pollutants, zinc and copper, *New Phytol.*, 105, 639, 1987.
35. Ernst, W.H.O. and Bast-Cramer, W.B., The effect of lead contamination of soils and air on its accumulation in pollen, *Plant Soil*, 57, 491, 1980.
36. Searcy, K.B. and Mulcahy, D.L., Pollen selection and the gametophytic expression of metal tolerance in *Silene dioica* (Caryophyllaceae) and *Mimulus guttatus* (Scrophularaceae), *Am. J. Bot.*, 72, 1700, 1985.
37. Verkleij, J.A.C. and Schat, H., Mechanisms of metal tolerance in higher plants, in *Heavy Metal Tolerance in Plants: Evolutionary Aspects*, Shaw, A.J., Ed., CRC Press, Boca Raton, FL, 1990, 179.
38. Grill, E., Winnacker, E.L., and Zenk, M.H., Phytochelatins, a class of heavy-metal binding peptides from plants, are functionally analogous to metallothioneins, *Proc. Natl. Acad. Sci. U.S.A.*, 84, 439, 1987.
39. Grill, E., Löffler, S., Winnacker, E.-L., and Zenk, M.H., Phytochelatins, the heavy-metal binding peptides of plants, are synthesized from glutathione by specific γ-glutamylcystein dipeptidyl transpeptidase (phytochelatine synthase), *Proc. Natl. Acad. Sci. U.S.A.*, 86, 6838, 1989.
40. Schat, H. and Kalff, M., Are phytochelatins involved in differential metal tolerance or do they merely reflect metal-imposed strain?, *Plant Physiol.*, 99, 1475, 1992.
41. De Vos, C.H.R., Vonk, M.J., Vooijs, R., and Schat, H., Glutathione depletion due to copper-induced phytochelatin synthesis causes oxidative stress in *Silene cucubalus*, *Plant Physiol.*, 98, 853, 1992.
42. Van Assche, F. and Clijsters, H., A biological test system for the evaluation of the phytotoxicity of metal-contaminated soils, *Environ. Pollut.*, 66, 157, 1990.
43. Verkleij, J.A.C., Lolkema, P.C., and Ernst, W.H.O., The effect of heavy metals on isozyme gene expression in *Silene cucubalus*, in *Isozymes, Current Topics in Biological and Medical Research*, Rattazzi, M.C., Scandalios, J.G., and Whitt, G.S., Eds., Alan R. Liss, New York, 1987, 209.
44. Rabe, R. and Kreebs, K.H., Enzyme activities and chlorophyll and protein content in plants as indicators of air pollution, *Environ. Pollut.*, 19, 119, 1979.
45. Scandalios, J.G., Isozymes in development and differentiation, *Annu. Rev. Plant Physiol.*, 25, 225, 1974.
46. Alia, P.S., Proline accumulation under heavy metal stress, *J. Plant Physiol.*, 138, 554, 1991.
47. Wild, S.P., McGrath, S.P., and Jones, K.C., The polynuclear aromatic hydrocarbon (PAH) content of archived sewage sludge, *Chemisphere*, 20, 703, 1990.
48. Weber, J.B., Donney, J.R., and Overcash, M.R., Crop plants growth and uptake of toxic organic pollutants found in sewage sludge: polynuclear aromatics, in Proc. Triangle Conf. of Environm. Tech., Duke University, Durham, NC, March 1984, 1.
49. Wild, S.R., Berrow, M.L., McGrath, S.P., and Jones, K.C., Polynuclear aromatic hydrocarbons in crops from long-term field experiments amended with sewage sludge, *Environ. Pollut.*, 76, 25, 1992.
50. Jones, K.C., Contaminant trends in soil and crops, *Environ. Pollut.*, 69, 311, 1991.
51. Wegener, J.W.M., van Schaik, M.J.M., and Aiking, H., Active biomonitoring of PAHs by means of mosses, *Environ. Pollut.*, 76, 15, 1992.
52. Hutzinger, O., Safe, S., and Zitko, V., *The Chemistry of PCBs*, CRC Press, Cleveland, OH, 1974.

53. Brouwer, A., Reijnders, P.J.H., and Koeman, J.H., Polychlorinated biphenyl (PCB)-contaminated fish induces vitamin A and thyroid hormone deficiency in the common seal *(Phoca vitulina)*, *Aquat. Toxicol.*, 15, 99, 1989.

54. Strek, H.J. and Weber, J.B., Behaviour of polychlorinated biphenyls (PCBs) in soils and plants, *Environ. Pollut.*, 28, 291, 1982.

55. Moza, P., Scheunert, J., Klein, W., and Korte, F., Studies with 2,4',5 trichlorobiphenyl-14C and 2,2',4,4',6-pentachlorobiphenyl-14C in carrots, sugar beets and soil, *J. Agric. Food Chem.*, 27, 1120, 1979.

56. Gaggi, C. and Bacci, E., Accumulation of chlorinated Lydrocarbon vapours in pine needles, *Chemosphere*, 14, 451, 1980.

57. Wang, W., Root elongation method for toxicity testing of organic and inorganic pollutants, *Environ. Toxicol. Chem.*, 6, 409, 1987.

58. Günther, P. and Pestemer, W., Risk assessment for selected xenobiotics by bioassay methods with higher plants, *Environ. Manage.*, 14, 381, 1990.

59. McFarlane, C., Pfleeger, T., and Fletcher, J., Effect, uptake and disposition of nitrobenzene in several terrestrial plants, *Environ. Toxicol. Chem.*, 9, 513, 1990.

60. Dix, H.M., *Environmental Pollution*, John Wiley & Sons, New York, 1981.

61. Holt, J.S., Stemler, A.J., and Radosevich, S.R., Differential light responses of photosynthesis by triazine resistant and triazine susceptible *Senecio vulgaris* biotypes, *Plant Physiol.*, 67, 744, 1981.

62. Ashton, F.M. and Crafts, A.S., *Mode of Action of Herbicides*, John Wiley & Sons, New York, 1973.

63. Kearney, D.D. and Kaufman, D.D., *Herbicides: Chemistry, Degradation and Mode of Action*, Vol. 3, Marcel Dekker, New York, 1988.

64. Conard, S.G. and Radosevich, S.R., Ecological fitness of *Senecio vulgaris* and *Amaranthus retroflexus* biotypes susceptible or resistant to atrazine, *J. Appl. Ecol.*, 16, 171, 1979.

65. Warwick, S.I. and Thomson, B.K., Comparative growth and atrazine response of resistant and susceptible populations of *Amaranthus* from Southern Ontario, *J. Appl. Ecol.*, 19, 611, 1982.

66. Darmency, H. and Gasquez, I., Inheritance of triazine resistance in *Poa annua*: consequences for population dynamics, *New Phytol.*, 89, 487, 1981.

67. Hall, L.M. and Devine, M.D., Cross-resistance of a chlorsulfuron-resistant biotype of *Stellaria media* to triazolopyrimidine herbicide, *Plant Physiol.*, 93, 962, 1990.

68. Häusler, R.E., Holtum, J.A.M., and Powles, S.B., Cross-resistance to herbicides in annual ryegrass *(Lolium rigidum)*, *Plant Physiol.*, 97, 1035, 1991.

69. Gressel, J. and Segel, L.D., The paucity of plants evolving genetic resistance to herbicides: possible reasons and implications, *J. Theor. Biol.*, 75, 349, 1978.

70. Marrs, R.H., Frost, A.J., and Plant, R.A., Effect of mecoprop drift on some plant species of conservation interest when grown in standardized mixtures in microcosms, *Environ. Pollut.*, 73, 25, 1991.

71. Kees, H., Einfluss zehnjähriger Unkrautbekämpfung mit 4 unterschiedlichen Intensitätsstufen unter Berucksichtigung der Wirtschaftlichen Schadensschwelle auf Unkrautflora und Unkrautsamenvorrat in Boden, Proc. EWRS Symposium on Economic Weed Control, Wageningen, 1986, 399.

72. Marrs, R.H., Williams, C.T., Frost, A.J., and Plant, R.A., Assessment of the effect of herbicide spray drift on a range of plant species of conservation interest, *Environ. Pollut.*, 59, 71, 1989.

73. Smeding, F.W. and Joenje, W., Onbespoten en onbemeste perceelsranden in graan-akkers, in *Zonderwijk en VPO,* van Groenendael, J.M., Joenje, W., and Sykora, K.V., Eds., Vakgroep VPO, Adviesgroep Vegetatiebeheer, 129 (in Dutch).
74. Greve, P.A., Klapwijk, S.P., Linders, J.B.H.J., and van der Plassche, E.J., Bestrijdingsmiddelen in oppervlaktewater uit het akkerbouwgebied in de Haarlem-mermeer, Rapportnr. 638812002 RIVM, Bilthoven.
75. Van Straalen, N.M. and Denneman, C.A.J., Ecotoxicological evaluation of soil quality criteria, *Ecotoxicol. Environ. Saf.,* 18, 241, 1989.
76. Fitter, A.H., Effect of benomyl on leaf phosphorous concentrations in alpine grasslands. A test of mycorrhizal benefit, *New Phytol.,* 103, 767, 1986.

CHAPTER 10

Phytotoxic Organic Compounds in Spruce Forest Soil: Chemical Analyses Combined With Seedling Bioassays

A.-B. Steen, H. Borén, and A. Grimvall

TABLE OF CONTENTS

ABSTRACT

This study showed that phytotoxic leachates could be obtained from a podzolic soil in a seemingly unpolluted spruce forest in Sweden. Using alkaline leaching followed by ether extraction, all extracts of soil from three different layers were found to be strongly phytotoxic in a cucumber root bioassay. Some litter extracts produced by Milli-Q-water leaching were also phytotoxic. Ultrafiltration experiments showed that the observed phytotoxic effects could be attributed to low-molecular-weight organic compounds (M_w < 1000). Several known phytotoxic compounds were identified in the extracts exhibiting phytotoxicity. However, a test solution of eleven identified compounds (aliphatic decarboxylic acids, phenolic acids, and aldehydes) was considerably less toxic than the original extract. Fractionation according to polarity indicated that the phytotoxicity of the soil extracts was primarily caused by unidentified compounds in the carboxylic acid fraction.

I. INTRODUCTION

A recent survey revealed that several surface waters used in greenhouses in southern Sweden inhibited the root growth of different types of seedlings cultured in rolled filter paper in water.[1] The geographical distribution of the surface waters causing such inhibition indicated that the observed effects did not originate from local industrial emissions or herbicide spills. This was confirmed by another study, in which several water samples from remote areas dominated by mires or coniferous forests produced visible root abnormalities, including reduced length.[2] The same authors also suggested that the observed growth inhibition was caused by hydrophilic organic substances with a molecular weight exceeding 1000 or by low-molecular weight compounds associated with organic macro-molecules.

Soil leaching is a presumptive origin of phytotoxic organic compounds in surface water. In the literature on allelopathy, several naturally produced acids and aldehydes in soil have been reported to exhibit toxic effects, especially on seedlings.[3] A group of compounds that have been studied extensively in this regard are the phenolic acids and aldehydes, which are important intermediates in the formation of humus.[4-8]

The aim of the present study was to examine to what extent phytotoxic organic compounds could be leached from surface litter and soil from a catchment area with strongly phytotoxic surface water. Extracts obtained from surface litter and soil exhibiting phytotoxic effects were also examined to determine the chemical character of the compounds responsible for the observed toxicity. Special attention was paid to the role of phenolic acids and other known phytotoxic organic compounds.

II. MATERIALS AND METHODS

This study was based on combined chemical analyses and bioassays. Soil extracts or fractions thereof were dissolved in Milli-Q water (Millipore®) or in a buffer solution and then tested in a cucumber root bioassay. Gas chromatographic analyses were performed in parallel to establish possible relationships between phytotoxicity and identified compounds (Figure 1).

A. Soil Samples

Soil samples consisted of surface litter and soil from a spruce forest that produced strongly phytotoxic surface water. The forest is located in southern Sweden (57°N 14°E); the mean age of its spruces is 64 years and its soil profile is podzolic. Surface litter (L) and two subsurface horizons, a dark brown organic layer (O) and a grey eluvial layer (E), were selected and collected in April. The collected material was sieved (mesh <4 mm) while still moist and stored at −20°C until analyzed. Water content was determined by drying overnight at 105°C, and total organic content was determined by dry ashing at 600°C. The pH of the samples was measured in a 2:1 (vw) water-soil mixture and in a 2:1 (vw) 0.2 M KCl-soil mixture.

B. Leaching and Extraction

Aliquots of surface litter and soil were shaken overnight (under N_2) with 2 M NaOH or Milli-Q water. Extracts for chemical analysis were produced by leaching 50 g of soil or surface litter with 100 ml of NaOH or Milli-Q water;

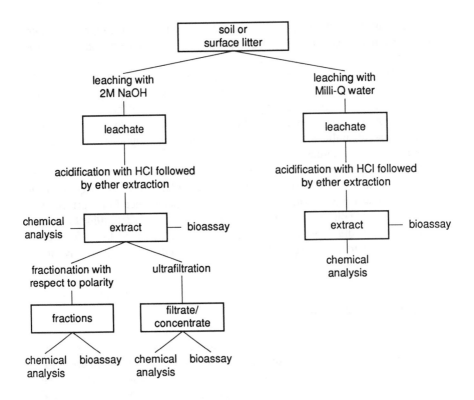

FIGURE 1. Schematic outline of the study.

bioassay extracts were obtained in the same manner, but using 150 g of soil or surface litter and 300 ml of NaOH or Milli-Q water. After centrifugation at 10,000 rpm for 15 min, the pH of the supernatant was adjusted to 2 with concentrated HCl, whereby a precipitate of humic acids was formed. After recentrifugation and removal of the new supernatant, the residue was washed with methanol. The residue-methanol mixture was again centrifuged, and the supernatant formed was combined with the supernatant removed in the previous step. The combined supernatants were extracted with diethyl ether (2 × 25 and 3 × 50 ml, respectively, for chemical analysis and bioassay extracts), and the ether extracts thus produced were dried over anhydrous $MgSO_4$, and finally the extracts were evaporated to dryness.

C. Cucumber Root Bioassay

Aqueous samples for cucumber root bioassay were produced by dissolving the evaporated ether extracts in 150 ml Milli-Q water or in 150 ml of a 5-mM buffer solution composed of 2-(N-morpholino) ethanesulfonic acid (MES) with the pH adjusted to 7.0. The cucumber root bioassays were performed according to the procedure described by Grimvall et al.[2] Seeds of field cucumber (Weibull Favör) were placed in a petri dish on a piece of filter paper (Munktell 3) moistened with the solution to be tested. After 3 d in darkness at 22°C and 70% humidity, ten seedlings of average size were selected and rolled into a new piece of wet filter paper. The paper roll was then put in a 300-ml beaker containing 50 ml of the test sample, and the seedlings were grown for 4 d at 24°C and 70% humidity. Photoperiod (10,000 lux) was 10 to 12 h/d. After 2 d in the climate room, the remaining part of the water sample (about 60 ml) was added. At the end of the bioassay, root length was measured for each seedling and visible root abnormalities were noted.

D. Chemical Analyses

Gas chromatographic (GC) analyses were performed on a Hewlett Packard® 5880 gas chromatograph. GC parameters: fused silica column DB-1, 60 m × 0.32 mm, 0.25 μm (J&W); carrier gas He, 40 cm/s; temperature program 140°C (2 min), 5°C/min, 230°C (2 min); flame ionization detector (FID); split injection. Confirming GC/mass spectroscopy (MS) analyses were performed on a Shimadzu® QP2000 system, using the GC parameters as described above.

The phenolic compounds and carboxylic acids were analyzed as TMS derivatives. TMS reagent (100 μl) and hexane (2 ml) were added to the evaporated ether extract before analysis. The TMS reagent was a mixture of 2.4 ml N,O-*bis*-(trimethylsilyl)-trifluoroacetamide, 0.24 ml trimethylchlorosilane, and 2.4 ml pyridine.

E. Fractionation and Filtration Methods

To determine which organic compounds were responsible for the observed phytotoxicity, soil extracts were fractionated into carboxylic acids, phenols, and neutral compounds (Figure 2). The extracts used in this procedure were obtained by leaching soil (from the organic layer, O) with 2 M NaOH. After fractionation, the ether extracts were evaporated to dryness. The evaporated extracts were then dissolved in MES buffer and used in the cucumber root bioassay (pH adjusted to 7.0). Chemical analysis was performed as described above. Molecular weight fractionation was performed using a stirred ultrafiltration cell (Amicon, model 8400) equipped with a YM-10 filter (Amicon, M_w cut-off 1000). Evaporated ether extracts, originating from NaOH-leached soil (O-horizon), were dissolved in MES buffer (pH 7.0) before filtration. After filtration, both the concentrate (diluted to the volume of the filtrate) and the filtrate were subjected to bioassay and chemical analysis.

F. Bioassay of Synthetic Test Solution

Cucumber seedlings were also grown in a synthetic test solution of 11 identified phytotoxic compounds. This solution was prepared to be identical

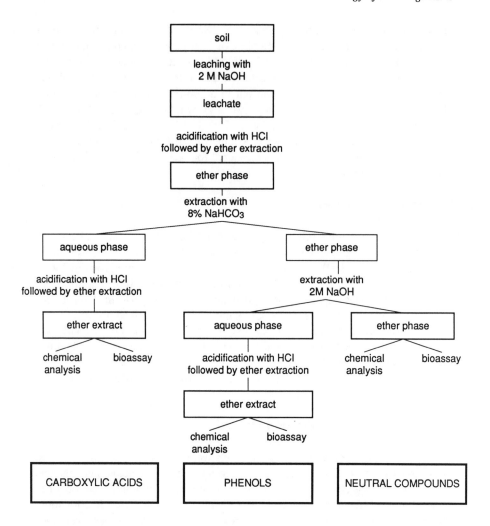

FIGURE 2. Fractionation of a soil extract into carboxylic acids, phenols, and neutral compounds.

regarding compounds included and their concentrations with one of the tested soil extract solutions. The compounds were dissolved in ethanol, and part of this solution was then mixed with MES buffer to produce solution samples for bioassay. The pH of the synthetic test solution was adjusted to 7.0 prior to use (Table 1).

III. RESULTS

The basic characteristics of the soil samples, including pH levels and the content of water and organic matter, are listed in Table 2.

Table 1. Concentrations of the Organic Compounds Included in the Synthetic Test Solution

	Concentration in test solution	
Organic compound	mg/l	mM
p-Hydroxybenzaldehyde	1.6	0.013
Vanillin	2.7	0.018
Heptanedioic acid	2.2	0.014
p-Hydroxybenzoic acid	3.3	0.024
Octanedioic acid	5.8	0.033
Vanillic acid	7.8	0.046
Nonanedioic acid	14.2	0.076
3,4-Dihydroxybenzoic acid	17.7	0.110
Decanedioic acid	0.8	0.004
p-Coumaric acid	6.0	0.037
Ferulic acid	2.2	0.011

A. Bioassay and Chemical Analysis of Soil Extracts Obtained by NaOH Leaching

Extracts obtained by NaOH leaching had a strong phytotoxic effect on cucumber seedlings: the roots of tested seedlings were spongy and brownish, and lateral roots and root hairs were lacking. Furthermore, root length was strongly reduced (Figure 3).

GC/FID analysis of NaOH-leached extracts revealed the presence of considerable amounts of aliphatic dicarboxylic acids, phenolic acids, and aldehydes, the identities of which were confirmed by GC/MS analysis. Table 3 presents a list of organic compounds found in the soil extracts, together with the concentrations of these compounds in the dissolved extracts used in the bioassays.

B. Bioassay and Chemical Analysis of Soil Extracts Obtained by Leaching With Milli-Q Water

Some water leachates of surface litter caused a pronounced inhibition of root growth (Figure 3). In general, however, extracts obtained by water leaching had no significant effect on root growth. Chemical analysis showed that none of the organic compounds identified in NaOH-leached extracts could be detected in the water extracts.

Table 2. Basic Characteristics of the Soil Samples

Soil horizon	pH (H$_2$O)	pH (0.2 M KCl)	Water content (%)	Organic content (% of d.w.)
L	3.9	3.2	63	93
O	3.6	2.6	66	93
E	3.8	2.9	25	6

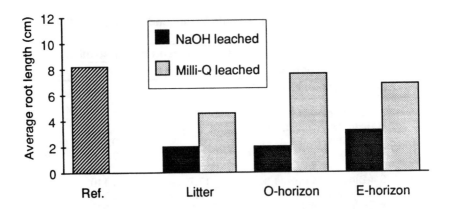

FIGURE 3. Average root length of ten cucumber seedlings grown in Milli-Q water solutions of soil extracts obtained by leaching with NaOH or Milli-Q water followed by ether extraction. Ref. denotes Milli-Q water. Standard error of the mean: 0.11 to 0.66.

C. Bioassay and Chemical Analysis of Fractions of a Soil Extract

To determine the main phytotoxic organic compounds in a particular soil extract, it was fractionated into carboxylic acids, phenols, and neutral compounds. Subsequent bioassays showed that the carboxylic acid fraction had a strong phytotoxic effect (Figure 4). When discussing the results of the bioassays it should be kept in mind that complete separation was not achieved with the fractionation method used. The GC/FID chromatograms did, however, confirm that a majority of the carboxylic acids were found in the carboxylic acid fraction, and a majority of the phenols were found in the phenolic fraction (Figure 5). GC analysis of the nonderivatized neutral fraction did not result in any peaks.

D. Bioassay and Chemical Analysis of Ultrafiltration Filtrate and Concentrate

Bioassays of samples obtained by ultrafiltration of soil extracts dissolved in MES buffer showed that the phytotoxic effects could be attributed to compounds with molecular weights that did not exceed 1000 (Figure 6). GC analysis of the ultrafiltration fractions showed that the major part of the organic acids and aldehydes appeared in the low-molecular weight fraction (Figure 7). It is noteworthy, however, that a considerable amount of the dicarboxylic acids were found in the high-molecular weight fraction ($M_w > 1000$).

Table 3. Organic Compounds Found in Soil Extracts Obtained by NaOH Leaching; Concentrations of These Compounds in the Soil Extract Solutions Used in the Cucumber Root Bioassay Are Also Shown

Organic compound	Soil horizon	Concentration in soil extract solution	
		mg/l	mM
p-Hydroxybenzaldehyde	L	1.5	0.012
	O	1.6	0.013
	E	0.5	0.004
Vanillin	L	3.6	0.024
	O	2.7	0.018
	E	1.5	0.010
Heptanedioic acid	L	1.6	0.010
	O	2.2	0.014
	E	1.3	0.008
p-Hydroxybenzoic acid	L	4.3	0.031
	O	3.3	0.024
	E	1.2	0.009
Octanedioic acid	L	5.1	0.029
	O	5.7	0.033
	E	5.5	0.032
Vanillic acid	L	8.8	0.052
	O	7.8	0.046
	E	3.4	0.020
Nonanedioic acid	L	14.2	0.076
	O	14.4	0.077
	E	15.4	0.082
3,4-Dihydroxybenzoic acid	L	14.4	0.094
	O	17.7	0.110
	E	0.4	0.003
Decanedioic acid	L	0.9	0.005
	O	0.8	0.004
	E	1.0	0.005
p-Coumaric acid	L	16.3	0.099
	O	6.0	0.037
	E	0.6	0.004
Ferulic acid	L	4.5	0.023
	O	2.2	0.011
	E	1.3	0.007

E. Bioassay of Synthetic Test Solution

Cucumber seedlings exposed to the synthetic test solution described in Section II exhibited a reduced average root length, but the growth inhibition was less pronounced than for seedlings exposed to the authentic soil extract solution (Figure 8). Furthermore, the synthetic test solution did not cause any visible root

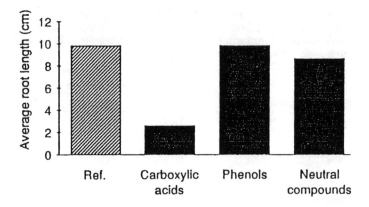

FIGURE 4. Average root length of ten cucumber seedlings grown in MES buffer solutions of different fractions of a soil extract. Ref. denotes MES buffer. Standard error of the mean: 0.16 to 1.04.

abnormalities. Consequently, the organic compounds used in the test solution could not entirely explain the negative effects of the soil extracts on seedling root growth and appearance.

IV. DISCUSSION

The present study showed that phytotoxic leachates could be produced from all investigated layers of a podzolic soil in a seemingly unpolluted spruce forest in southern Sweden. This confirms the results of a recently published Russian study[9] of phytotoxic substances in litter and soil from spruce forests. Furthermore, it has previously been shown that known phytotoxic compounds occur in ethyl acetate extracts of a spruce-dominated soil.[10,11] The large differences in phytotoxicity between NaOH leachates and water leachates found in the present study may be attributed to the pH dependence of the recovery of organic compounds from soil.[12] Model experiments in which phenolic compounds were added to soil have demonstrated that they are partly adsorbed to organic matter.[13] It has even been claimed that such substances can be so strongly associated to humic matter, that they are not released from soil upon nonalkaline aqueous leaching.[10,11]

In the previously mentioned study by Grimvall and co-workers,[2] it was remarked that phytotoxic effects of high-molecular weight fractions of organic compounds may, in fact, be caused by low-molecular weight compounds adsorbed to naturally occurring organic macromolecules. In the present study the separation of low- and high-molecular weight compounds was improved. As a result of the extraction procedure used, most of the high-molecular weight organic

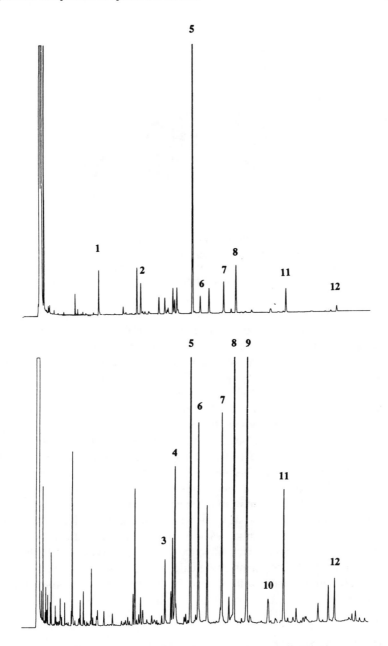

FIGURE 5. GC/FID chromatograms of the phenolic fraction (top) and the carboxylic acid fraction (bottom) of the same soil extract as in Figure 4. 1 = *p*-hydroxybenzaldehyde; 2 = vanillin; 3 = heptanedioic acid; 4 = *p*-hydroxybenzoic acid; 5 = internal standard; 6 = octanedioic acid; 7 = vanillic acid; 8 = nonanedioic acid; 9 = 3,4-dihydroxybenzoic acid; 10 = decanedioic acid; 11 = *p*-coumaric acid; 12 = ferulic acid.

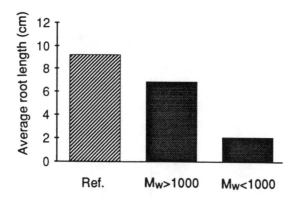

FIGURE 6. Average root length of ten cucumber seedlings grown in MES buffer solutions of different molecular weight fractions (M_w <1000 and M_w >1000, respectively) of organic compounds from a soil extract; fractionation by ultrafiltration. Standard error of the mean: 0.10 to 1.12.

matter was removed prior to the bioassay. The subsequent ultrafiltration experiments confirmed that the observed phytotoxic effects were caused by low-molecular weight organic compounds (M_w <1000).

Chemical analyses of phytotoxic soil extracts revealed several organic compounds which are known to be phytotoxic in rather low concentrations. However, the test solution composed of 11 synthetic compounds complying with the identified aliphatic dicarboxylic acids, phenolic acids, and aldehydes exhibited considerably lower phytotoxicity than the original extract did. This discrepancy is not unique. On the contrary, many other investigators[14-18] have also found specific organic acids in phytotoxic extracts, but at concentrations too low to explain the effects of the entire extract. This might be due to the low recovery of hydrophilic compounds with the analytical procedures used. In the present study, fractionation experiments showed that despite the generally low recovery of hydrophilic acids in the ether extraction step, the carboxylic acid fraction had a strong phytotoxic effect in the cucumber assay. This indicates that there are very potent and still unidentified phytotoxic compounds in the carboxylic acid fraction, and further studies of the origin of phytotoxic organic compounds found in soil leachates and surface water will therefore be concentrated on the role of hydrophilic acids.

ACKNOWLEDGMENT

Financial support was obtained from the Swedish Council for Forestry and Agricultural Research.

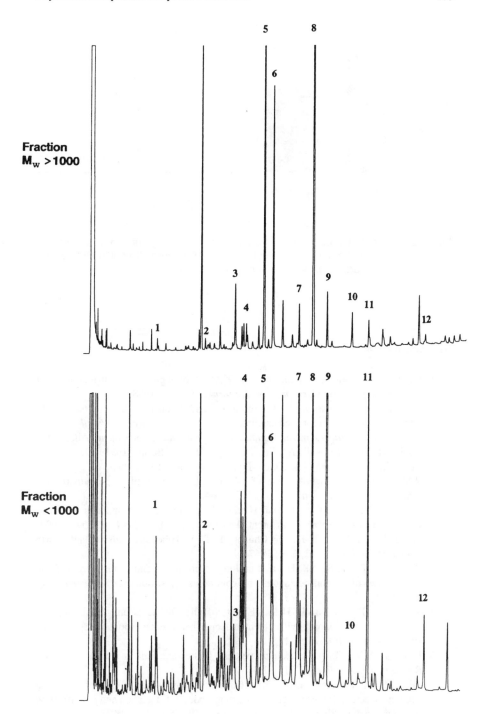

FIGURE 7. GC chromatograms of the ultrafiltration fractions of the same soil extract as in Figure 6. Peak numbers are as in Figure 5.

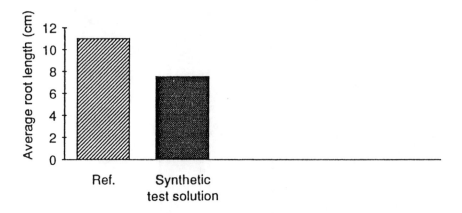

FIGURE 8. Average root length of 20 cucumber seedlings grown in a synthetic test solution (see text). Ref. denotes 1000 ppm ethanol in MES buffer. Standard error of the mean: 0.49 to 0.79.

REFERENCES

1. Lundin, P., Wibrandt, B., and Jönsson, K.O., A biological test of irrigation water (in Swedish), *Växtskyddsrapporter, Jordbruk,* 49, 191, 1988.
2. Grimvall, A., Bengtsson, M.-B., Borén, H., and Wahlström, D., Phytotoxic substances in runoff from forested catchment areas, in *Lecture Notes in Earth Sciences 33. Humic Substances in the Aquatic and Terrestrial Environment,* Allard, B., Borén, H., and Grimvall, A., Eds., Springer-Verlag, Berlin, 1991, 397.
3. Rice, E.L., *Allelopathy,* Academic Press, New York, 1984, chap. 10.
4. Whitehead, D.C., Identification of p-hydroxybenzoic, vanillic, *p*-coumaric and ferulic acids in soils, *Nature,* 202, 417, 1964.
5. Whitehead, D.C., Dibb, H., and Hartley, R.D., Phenolic compounds in soil as influenced by the growth of different plant species, *J. Appl. Ecol.,* 19, 579, 1982.
6. Wang, T.S.C., Yang, T.-K., and Chuang, T.-T., Soil phenolic acids as plant growth inhibitors, *Soil Sci.,* 103, 239, 1967.
7. Kuiters, A.T. and Sarink, H.M., Leaching of phenolic compounds from leaf and needle litter of several deciduous and coniferous trees, *Soil Biol. Biochem.,* 18, 475, 1986.
8. Kuiters, A.T. and Sarink, H.M., Effects of phenolic acids on growth, mineral composition and chlorophyll content of some herbaceous woodland species, *Z. Pflanzenernähr. Bodenkd.,* 150, 94, 1987.
9. Lysikov, A.V., On phytotoxic properties of litter and soil in spruce forests (in Russian), *Lesovedenie,* 3, 31, 1989.
10. Jalal, M.A.F. and Read, D.J., The organic acid composition of Calluna heathland soil with special reference to phyto- and fungitoxicity. I. Isolation and identification of organic acids, *Plant Soil,* 70, 257, 1983.

11. Jalal, M.A.F. and Read, D.J., The organic acid composition of Calluna heathland soil with special reference to phyto- and fungitoxicity. II. Monthly quantitative determination of the organic acid content of Calluna and spruce dominated soils, *Plant Soil,* 70, 273, 1983.
12. Whitehead, D.C., Dibb, H., and Hartley, R.D., Extractant pH and the release of phenolic compounds from soils, plant roots and leaf litter, *Soil Biol. Biochem.,* 13, 343, 1981.
13. Shindo, H. and Kuwatsuka, S., Behavior of phenolic substances in the decaying process of plants. IV. Adsorption and movement of phenolic acids in soils, *Soil Sci. Plant Nutr.,* 22, 23, 1976.
14. Jain, R., Singh, M., and Dezman, D.J., Qualitative and quantitative characterization of phenolic compounds from Lantana (Lantana camara) leaves, *Weed Sci.,* 37, 302, 1989.
15. Krogmeier, M.J. and Bremner, J.M., Effects of aliphatic acids on seed germination and seedling growth in soil, *Commun. Soil Sci. Plant Anal.,* 21, 547, 1990.
16. Kaminsky, R., The microbial origin of the allelopathic potential of Adenostoma fasciculatum, *Ecol. Monogr.,* 51, 365, 1981.
17. Singh, M., Tamma, R.V., and Nigg, H.N., HPLC identification of allelopathic compounds from Lantana camara, *J. Chem. Ecol.,* 15, 81, 1989.
18. Wójcik-Wojtkowiak, D., Politycka, B., Schneider, M., and Perkowski, J., Phenolic substances as allelopathic agents arising during the degradation of rye (Secale cereale) tissues, *Plant Soil,* 124, 143, 1990.

CHAPTER 11

The Yorkshire Water Standard Plant Growth Trial for Toxicity Testing of Soils, Sludge, and Sediments

P.D. Hiley and B. Metcalfe

TABLE OF CONTENTS

0-87371-530-6/94/$0.00 + $.50

179

ABSTRACT

Plants are grown to maturity in mixtures of a fertile synthetic soil and the test material in order to detect and monitor their potential for harming agricultural crops consequent on recycling to farmland. The synthetic soil retains its structure when mixed with 30 times the normal annual quantity of liquid sewage sludge for farmland (150t dry solids ha^{-1}). Tomato (sensitive to herbicides), lettuce, (sensitive to metals), barley (normal arable crop), and ryegrass (normal forage crop) are grown in environmentally controlled greenhouses. The plants are harvested at several growth stages and at maturity. Crop weights, chemical content, photographs and notes of the plants condition combine to give a comprehensive assessment of potential toxicity. Between 1983 and 1990 about 200 sludges and soils were tested.

I. INTRODUCTION

The Plant Growth Trials Unit was begun in 1980 with the objective of maximizing the recycling of sewage sludges to agriculture and land reclamation by identifying those sludges which do not harm growing plants. Sludges from sewage works in industrial catchments may meet the European Community guidelines for metals content, yet contain harmful residues, e.g., from herbicide manufacture. Such residues may have damaged bulb fields in 1972, resulting in restriction of the cheap land disposal option at a major sewage works. Biological tests identify effects[2] and can detect potentially harmful conditions no matter what their source, reducing the risk of crop damage.

In order to be sure that the amounts of sludge applied to farmland will have no effect it is necessary to test at much higher concentrations. However, most sewage sludges have little structure, do not normally support crops on their own, and damage soil structure when mixed with natural soils at high concentrations. In addition, the variability of soil mixtures and plants requires many replicates for accurate results. The Yorkshire Water trial uses a synthetic soil which is mixed with the sludge in varying proportions up to 30 times the normal application onto farmland, i.e., 150 t dry solids per hectare (t ds ha^{-1}). The history of the test and its current protocol are described in this paper.

II. METHODS

A. Reference Soil Development

An unchanging reference soil is essential to ensure that test interpretations do not vary with time. In studies of sludge applications heavy metal concentrations are of particular concern, so the reference soil should contain as little heavy metal as possible. "Uncontaminated" natural soils go to mud if more than about 50 t ds ha^{-1} of sludge is added to them, and at lower applications partial soil structure damage may result in poor root aeration. These natural soils were found to have variable contents of heavy metals, trace elements, etc., sometimes due to contamination from ancient metal mining. In no location could we guarantee our ability to obtain a consistent reference soil over a 20-year period. A synthetic soil to fulfill the requirements of the test was therefore created in 1984 from the reliable components peat, sand, and expanded rock. All three components are very low in content of the heavy metals of particular concern in sludge disposal (Table 1). The small increases in metals content of the soil after the trial are mainly due to the impurities in the two fertilizers used.

Table 1. Average Analyses of Synthetic Soil (mg kg⁻¹)

	Cr	Pb	Zn	Cd	Cu	Ni
Before trial	<1.0	<1.0	<1.0	<0.1	<1.0	<1.0
After trial	<1.0	<1.0	1.8	0.1	0.7	1.35

B. Test Species

Four plant species were selected, two to represent likely crops and two sensitive to contaminants of particular concern.

Ryegrass — relevant to recycling to grassland for forage and grazing (CV Hercules)

Barley — relevant to recycling to arable land; herbage and grain are analyzed separately (CV Hart)

Lettuce — sensitive to metal contamination (CV All Year Round or Arctic King); this is a particularly difficult crop subject to fungal attack, even on healthy plants, but alternatives such as beet have other drawbacks

Tomato — sensitive to herbicide contamination which produces morphological symptoms; herbage and fruit are monitored separately (CV Moneymaker)

C. Trial Design

The trial employs dual controls: fertilized synthetic soil alone, and mixtures of fertilized synthetic soil with uncontaminated test material (e.g., domestic sewage sludge of similar nutrient content). Above-ground production of the four species in the tests and controls is monitored at intermediate stages and at maturity. Six replicates of each test per species are grown in randomized blocks, each of which contains an example of every test and control. Therefore, 120 (31, 18-cm standard shape) pots are used for testing each sample. The greenhouse environment is optimized for each species through the use of computer control of temperature, illumination, humidity, individual watering, and ventilation.

D. Synthetic Soil Preparation

An animal feed mixer can be used to ensure homogeneity, which is checked by adding some lithium-labeled soil to a mix and analyzing subsamples. After mixing the components, 20 g are dried to constant weight at 120°C to obtain the %ds (A% in subsequent calculations).

E. Test Material Preparation

The test material should be analyzed for content of potential contaminants, for instance, the metals Cr, Ni, Cu, Zn, Cd, and Pb would be commonly expected in a sewage sludge.[5] The plant nutrients N, P, and K should also be analyzed

Table 2. Weights of Materials Per Pot For Sludge Tests (Thickened Sludge in 10*FAR and 30*FAR)

Application rate	Dry weight of synthetic soil (g)	Dry weight of sludge (g)	Wet weight of synthetic soil (g)	Wet weight of sludge (g)
FAR	1996	4.46	(1996*100)/A%	(4.46*100)/B%
10*FAR	1955	44.6	(1955*100)/A%	(44.6*100)/C%
30*FAR	1866	134	(1866*100)/A%	(134*100)/C%

to aid interpretation of growth results. The %ds of the test material (B%) is required in the mixture calculations. It is better to dewater dilute materials like untreated sewage sludges for concentrations in excess of the equivalent of 5 t ds ha^{-1} year^{-1}, rather than make mixtures and then subject them to a long period of drying (see Section II.I). They may be dewatered up to 15 to 20 %ds (C% in the calculations) by freezing and thawing, centrifugation, or similar technique.

F. Test Substance Application Rates

The normal sludge concentrations used in the trials are field application rate (FAR) = 5 t ds ha^{-1}, 10*FAR, and 30*FAR. On average, 1 ha of land contains 2240 t of topsoil to a 15-cm normal rooting depth.[10] Therefore, 5 t ds ha^{-1} are equivalent to an addition of (5/2240)*100 = 0.223% ds of sludge and pro rata, evenly throughout the soil. For sludgelike materials and fine powders, 7% ds is a practical maximum concentration to avoid growth restriction due to the blinding of pore spaces at such concentrations. The second control identifies these effects separately from those due to toxicity.

G. Mixture Calculations, Sludge Example

Dry soils and sludges are very difficult to rewet. Therefore, mixtures are made with wet materials of known %ds. Each pot is filled with 3 l of medium (2 kg ds approx.) composed as in Table 2. Immediately prior to filling the pots, 10 g of slow-release fertilizer (14:14:14 of N:P:K) and lime (Ca[OH]$_2$) to pH 6 to 7 are mixed in.

H. Mixing

The required synthetic soil and test substance are weighed to an accuracy of ±1% and mixed until homogeneous. Wet mixes are dried until friable in a low humidity at 10 to 20°C, turning daily. The fertilizer is then mixed in, crumbling the mixture prior to filling the pots. The pots are rested for 10 d before planting to allow settlement of the mixtures, maintaining normal water content as in horticultural practice.

I. Supernatants

The supernatants retained from sludge thickening are tested for soluble toxicants by the root-elongation test[1] and Microtox.[3] If the supernatant is nontoxic it may be discarded. If toxic it should be returned to the mixtures in their pots in order that its potential effects can be assessed. The supernatant is added in 200- to 300-ml stages, allowing drying after each, so that the medium does not lose its structure or undergo denitrification. The amount of supernatant to be added to each pot is back-calculated using B% and C%.

J. Planting

1. Ryegrass

Dry seed (0.48 g $\pm$ 1%) is sown evenly per pot (equivalent to the standard field rate of 20 g m^{-2}) at 5 mm depth. The pots are covered to retain moisture in the surface layer until the grass has germinated, which is day 1 of the trial.

2. Barley

Five plants are required and, since germination is irregular, seed is pregerminated under water in the dark at 20°C for 36 to 48 h. Five seeds with 2-mm roots are sown evenly round the rim of each pot at day 1 of the trial.

3. Lettuce and Tomato

The seeds are grown in 1-cm^3 peat plugs which minimize handling damage and maximize contact with the medium. Five seedlings with the first true leaf are planted evenly round the rim of each pot at day 1 of the trial.

K. Watering and Feeding

Each pot remains on its own saucer so that leachate does not escape. Water to the demand of each pot is given by adjustable drip feeds, since capillary matting may allow lateral migration of toxicants into uncontaminated pots. Feeding with complete fertilizer (10:10:27 of N:P:K and containing micronutrients) is carried out as follows: ryegrass, total 90 kg ha^{-1} of N (and pro rata for P and K) in 9 equal doses at day 14 and at 7-d intervals thereafter. Barley and lettuce, 140 kg ha^{-1} total with the first dose of 20 kg on day 14 and 30 kg at 14-d intervals. Tomato, 240 kg ha^{-1} total with 20 kg at days 14 and 23, 40 kg at days 30 and 37, then 50 kg at 7-d intervals.

L. Observation

The plants are checked daily, and since disease and pest susceptibility may be higher in plants stressed by toxicants, pesticide treatment is given if damage to the control plants is anticipated. Preparations containing persistent chemicals or heavy metals may affect the final analyses of crops and soils. Subjective records of significant adverse symptoms compared to the controls, e.g., leaf curl in tomato, premature yellowing in barley, above average deaths, etc., are made.

M. Harvesting

Four harvests are taken from each replicate during the trials, avoiding contamination from the growth mixtures. The first harvest for barley, lettuce, and tomato is at day 29 (two plants), then one plant at days 43 and 57. The last tomato is taken when all healthy plants have a ripe fruit, and the last barley when the ears begin to yellow. Randomly chosen plants are cut level with the soil. For the final harvest, records are made of the tomato fruit dry weight, longest barley tiller, number of tillers, and ear dry weight. Tomato fruit is wet weighed, pulped, and a 100-g subsample dry weighed to permit calculation of the total dry weight. Ryegrass is harvested at day 22, then at 21-d intervals by turning the pot at an angle and cutting level with the rim. All harvested material is dried in a moisture-extraction oven at 100°C to constant weight.

N. Sample Preparation For Metals Analysis

The dried herbage, barley ears, tomato pulp, and soil mixtures from each set of six replicates are pooled, mixed, and milled to pass a 1-mm-mesh screen, then analyzed for suspect contaminants, e.g., the six metals commonly found in sludge. Many types of mill and blender can contaminate samples through abrasion of metal liners, and silicon carbide linings are strongly recommended. Milling of synthetic soil followed by metals analysis is used as a regular check (Table 1).

III. TREATMENT OF RESULTS

Test samples are failed if any of the determinations is significantly (95%) outside the norm. Subsequent interpretation of the reasons for the failure may indicate the possibility for remedial measures, such as aerobic composting, to reduce content of degradable toxicants.

A. Crop Production

The production of all crop materials must be $\geq$ controls in all concentrations of test substance, and the subjective visual symptoms must not be greatly different to the controls. Table 3 is a typical set of sludge results in which, due perhaps to a better balance of nutrients, the crop weights increase with test substance

Table 3. Crop Weights (g ds per Pot) For Barley With a Nontoxic Sewage Sludge

Test	Crop 1	Crop 2	Crop 3	Crop 4	Grain	Tiller no	Leaf mm
Control	0.7	3.0	6.8	12.8	7.7	24	834
FAR	0.9	3.5	8.7	13.8	45.3	29	767
10*FAR	1.7	4.9	8.7	15.8	53	35	770
30*FAR	1.8	4.9	8.8	17.9	64	27	778

concentration. Crop weights may decrease compared to control due to lockup of nutrients by degradable organics (e.g., cellulose) in the test substance, but still increase with sludge concentration. In contrast, a toxicant may cause decreases in crop weights with increased sludge concentration as in Table 4, crops 1 and 2. This effect reduced with time, but the relatively low fruit weight compared to the second control is another symptom of toxicity.

B. Metals Content

Test substances, field soils, and herbages may have to comply with regulations on metals content, as in Table 5. (The values for Zn, Cu, and Ni are for pH 6 to 7; for Cd, Pb, Cr pH >5). The threshold concentration of Cu, at which herbage yield begins to reduce, is lower than that for livestock, i.e., "passes" must be purpose related. In Table 6, Cu levels are over the threshold limit for ryegrass, due to enhanced uptake from an apparently satisfactory sludge.

C. Remaining Mixtures

In Table 7 the metals Cr, Pb, and Cu remaining in the mixtures (as a proportion of the calculated content) were lower compared to those from a toxic sludge in Table 8, probably due to increased uptake into the greater plant biomass in the former. Allowances are made for the normal contamination (see Section II.A and Table 1).

Table 4. Crop Weights (g ds per Pot) For Tomato With a Toxic Sewage Sludge

Test	Crop 1	Crop 2	Crop 3	Crop 4	Fruit
Control	13.4	7.6	12.1	13.5	1.4
FAR	15.5	8.5	11.5	13.2	1.2
10*FAR	12.6	7.0	16.5	26.7	1.9
30*FAR	9.2	4.7	18.3	28.3	1.9
2nd control (30*FAR)	14.1	12.4	20.8	34.3	2.4

Table 5. Concentrations of Metals, Norms, and Standards, mg kg^{-1} (threshold values for grass and livestock)

	Cr	Pb	Zn	Cd	Cu	Ni	Ref.
Normal range in plants		0.1–10	15–200	0.2–0.8	4–15	1–10	6
Ryegrass norm	1	3	50	0.5	11	2	7
Ryegrass threshold	10	35	221	10	21	14	7
Livestock feed threshold	50	3[a] 10[b]	900	3[a] 10[b]	50	50	10
Sludge	1500	1200	4000	40	1750	400	5, 11
Soil for agriculture	400	300	300	3	135	75	5, 11
Soil for grassland	600	300	500	3.5	225	125	11

[a] Concentration for sheep.
[b] Concentration for cattle.

Table 6. Metal Content of Ryegrass Grown on a Toxic Sludge (mg kg^{-1})

Test	Cr	Pb	Zn	Cd	Cu	Ni
Control	1.3	<2.0	36.0	<0.5	8.5	1.3
FAR	2.0	<2.0	38.0	<1.0	12.0	3.5
10*FAR	2.0	<2.0	78.0	<1.0	22.0	9.0
30*FAR	1.8	<2.0	99.0	<1.0	27.0	8.0
Sludge mg kg^{-1}	181	354	1226	6.4	385	156

Table 7. Metal Content of Remaining Soils From Barley Grown on a Nontoxic Sludge (mg kg^{-1})

Test	Cr	Pb	Zn	Cd	Cu	Ni
Control	<1.0	<1.0	1.5	0.1	<1.0	1.4
FAR	<1.0	<1.0	1.9	0.24	<1.0	2.6
10*FAR	2.8	3.8	8.2	0.1	3.5	2.0
30*FAR	6.0	8.2	15.1	0.2	8.7	1.4
Calculated 30*FAR	10.7	14.7	36.3	0.13	30.8	1.1

Table 8. Metal Content of Remaining Soils From Tomato Grown on a Toxic Sludge (mg kg^{-1})

Test	Cr	Pb	Zn	Cd	Cu	Ni
Control	2.2	1.0	2.8	0.28	1.6	1.2
FAR	3.0	1.4	5.8	0.32	1.8	2.0
10*FAR	5.0	4.2	14.0	0.4	5.0	3.2
30*FAR	13.6	18.0	47.2	0.64	17.0	9.2
Calculated 30*FAR	12.1	23.7	82	0.42	25.8	10.4

Table 9. Typical Results, Grass Control (g ds per Pot)

Pot no.	Crop 1	Crop 2	Crop 3	Crop 4	Crop 5
1	0.89	2.16	3.02	3.69	7.2
2	1.0	1.97	3.25	4.0	5.77
3	0.57	1.81	3.7	4.25	6.24
4	1.01	2.73	6.36	5.0	4.41
5	0.9	2.77	4.81	5.69	8.07
6	0.99	2.19	3.64	4.29	7.83
Mean	0.89	2.34	4.13	4.5	6.58
Standard deviation	0.15	0.36	1.14	0.67	1.28
Coefficient of variance (cv)	17%	16%	28%	15%	19%

D. Reproducibility

Table 9 is typical of the irregular coefficients of variance obtained with all crop species in the uneven environmental conditions of the old greenhouse facilities. However, some chemical tests, such as BOD, can have similar coefficients of variance (cv) (up to 20%). Programs are in hand to reduce cv to 10 to 15% in the new (1991) greenhouses. Given that 30 times the annual application rate of sludge is tested, changes in crop production of around 1% are assumed to be detected with the current protocol.

IV. DISCUSSION

Reductions of farm crop yields by a few percent are serious, but may not be predicted by short-term assays like the root test[1] or the *Photobacterium phosphoreum* test.[3] The standard trial is more sensitive, but involves a delay of 4 months, which usually means that some fields will have received sludge before the results are known. Short-term tests may therefore be used to screen out the obviously toxic[2] and reduce this risk.

A. Recycling of Sewage Sludges to Land

There is increasing pressure from our government to maximize the recycling of sewage sludge, a potentially valuable resource for creating or improving the fertility of agricultural and amenity land. However, the regulatory standards are increasingly stringent. The standard trial helps to ensure that the presence of substances not specified in the regulations, but likely to damage crops, can be detected. In respect of metals content (e.g., Pb, Cu, Cd, Zn, Cr, and Ni), many sewage sludges are similar to farmyard manure; the problem of relatively high contents of heavy metals being a feature of certain types of industrial waste. Sludges not meeting the regulatory standards are not recycled to land, but disposed by other means, e.g., by incineration. Ever tighter control of industrial effluents, including their toxicity, is helping to reduce the quantity of unsatisfactory sewage sludge.

B. Soil Creation

The standard trial has helped the development of the technique of soil creation for the restoration of despoiled and unproductive land, including acidic mine waste.[4] The soils created can meet the relevant regulatory limits for metals content in themselves and in the crops. Further information on the various uses of sewage sludge on land is given in Reference 9.

ACKNOWLEDGMENTS

We are most grateful to John Oldroyd, our greenhouse technician, for his continuing assistance in the development and running of these trials, and to Yorkshire Water for permission to publish.

REFERENCES

1. Hiley, P.D., The use of barley root elongation in the toxicity testing of sediments, sludges and sewages, this volume, Ch. 12.
2. Hiley, P.D. and Fearnside, D., The use of multiple toxicity assays to improve sewage treatment performance, *Toxicity Assessment,* in press.
3. Fearnside, D. and Hiley, P.D., The use of toxicity testing methods to identify and trace treatability problems on sewage treatment works, *Toxicity Assessment,* in press.
4. Metcalfe, B. and Lavin, J.C., Consolidated sewage sludge as soil substitute in colliery spoil reclamation, in *Alternative Uses for Sewage Sludge,* Hall, J.E., Ed., Pergamon Press, Oxford, 1991 (see also Discussion pp 97–110).
5. E.E.C., Council directive on the protection of the environment, and in particular of the soil, when sewage sludge is used in agriculture, *Official Journal of the European Communities,* 86/278/EEC., 1986. And Amendment, ibid, March 1990.
6. Alloway, W.H., Agronomic controls over environmental cycling of trace elements. *Adv. Agron.,* 20, 235, 1968.
7. Davis, R.D. and Beckett, P.H., Upper critical levels of toxic elements in plants. II. Critical levels of copper in young barley, wheat, rape, lettuce, and ryegrass, and of nickel and zinc in young barley and ryegrass, *New Phytol.,* 80, 23, 1978.
8. Underwood, E.J., *Trace Elements in Human and Animal Nutrition,* Academic Press, 1977.
9. Hall, J.E., Ed., *Alternative Uses for Sewage Sludge,* Pergamon Press, Oxford, 1991.
10. Chumbley, C.G., Permissible Levels of Toxic Metals in Sewages Used on Agricultural Land, Agricultural Development and Advisory Service, July 1971.
11. Her Majesty's Stationery Office, The sludge (use in Agriculture) (amendment) regulations 1990. Statutory instrument 1990 no 880, ISBN 0 11 113880 0.

CHAPTER 12

The Use of Barley Root Elongation in the Toxicity Testing of Sediments, Sludges, and Sewages

P.D. Hiley

TABLE OF CONTENTS

0-87371-530-6/94/$0.00 + $.50

© 1994 by CRC Press, Inc.

ABSTRACT

This is a simple and highly reproducible screening test for the phytotoxicity of entire sludges, sediments, and soils. It allows the early identification of problems, reducing the need for expensive growth to maturity crop tests. A relatively inert, contaminant free, expanded rock medium is used which can adsorb half its volume of liquid, yet retain an aerobic root environment. Pregerminated seed (e.g., barley) is grown for 5 days at 20°C, 100% humidity, in mixtures of test materials and medium in 4'' plastic flower pots. Ammonia in excess of 100 mg 1^{-1} has a toxic effect and can be removed (by raising the mixture to pH 10, then drying for 5 days before rewetting), or allowed for by spiking the second control. The test is sensitive to organic and inorganic contaminants such as herbicides, heavy metals, and dichlorophenol. Results are expressed as statistical differences from the controls, or as EC_{50} concentrations calculated from tests on a series of dilutions.

I. INTRODUCTION

Yorkshire Water Services required a screening test for use in connection with the recycling of materials (e.g., sludges, liquors, sediments) to agricultural or amenity land, of 1 to 7 d duration, to follow the fast screens of relatively low relevance (e.g., the 5-min *Photobacterium phosphoreum* test)[1] which are used both for emergencies and to reject any obviously toxic materials.

In general, it is not possible to store these materials for the duration of the 3-month growth to maturity tests on crop species.[2] The screening test would have to be simple, precise, reproducible, and easy to run. It should be able to deal with complex mixtures, the composition of which was unlikely to be known, predicting their effects in the soil.

Biological tests identify effects and can thus detect the presence of potentially toxic substances, synergisms, etc., without the need for detailed chemical analysis.[2] The 28-d growth test using *Brassica parachinensis*[10] in soil mixtures does not take the plants to maturity, neither is it fast enough. Ingredients of the media used for algae and *Lemna*[3] tests may interact with the components of the test substances, rendering them more or less toxic than will in fact occur on the land. Dense suspensions and colored solutions may not allow sufficient light for algae growth. Tests on aqueous extracts of the sediments may give falsely low toxicity predictions, since insoluble contaminants can be active in soil mixtures,[5] and the root-elongation test could permit intimate contact with sediments.

Part of the toxicity observed[3,4] in phytotoxicity tests may be due to ammonia if removal or compensation is not carried out. Ammonia is of little concern as a toxicant in sludges, since it is degraded in the soil environment, so its influence on test results should be minimized. Root-elongation tests have been widely used for examining the toxicity of isolated components of sludges,[4,5,7-10] e.g., heavy metals, either in sand media or on absorbent paper. For a 7-d trial, seeds of grasses, cabbages, cereals, etc., have sufficient food and nutrient reserves[9] to allow 7 d growth in distilled water. Smaller seeds may be relatively more sensitive to toxicants.[8] The sand medium used in previous tests allowed a relatively small volume of test substance to contact the roots, and evaporation could cause concentration of toxicants. Removal of the sand from the seedlings was difficult, and special pots were made to retain the sand during the trial.

The modifications to the root-elongation test described in this paper enable a rapid estimation of likely toxicity of sludges, sediments, etc., when applied to agricultural land. Barley was chosen as it permitted direct comparison with the data from the growth to maturity trial,[2] and the results of these comparisons will be reported in a future paper. Although root growth[8] is more sensitive to heavy metals, due presumably to the close proximity of the meristematic tissue to the toxicants, the measurement of shoots and root numbers was retained to check for possible effects of other substances.

II. METHODS

The extent of root elongation (growth), shoot growth, and other indicators of health of seedlings (e.g., barley) in the sample under test are compared to two controls: distilled water, and uncontaminated test substance with a similar chemical and physical content to the sample, in 5 d growth of pregerminated seed at 20°C. The double control method ensures, firstly, that the response to all the standard ingredients remains within the established limits (e.g., Table 1), and secondly that any conditions of the sample which influence the test, but are not required to be measured, are eliminated. By using dilutions of the sample in distilled water an EC_{50} estimate can be obtained. The expanded siliceous mineral test medium is widely available as a horticultural product. It can adsorb

Table 1. Root Test Reproducibility Data

Test number	Control mean root length (cm)	3-5 DCP EC_{50} mg l^{-1}	Real sample EC_{50} dilution
1	9.95	21	* 57 (1.8%)
2	8.23	18	* 50 (2%)
3	8.9	19	* 48 (2.1%)
4	7.55	16	* 52 (1.9%)
5	8.4	17	* 55 (1.8%)
6	8.4		* 53 (1.9%)
Mean	8.57	18.2	* 52.5
Standard deviation (n)	0.73	1.72	2.99
Coefficient of variance (%)	8.5	9.5	5.7

up to half its'own volume of water while retaining an aerobic root environment, and is easy to wash from the roots of the test seedlings.

A. Normal Method For Liquids

In a new, 10-cm-diameter standard shape plastic plant pot which has been rinsed in distilled water, 300 ml of liquid or suspension are added to 650 ml (79.5 ± 5 g, excluding pot) of test medium. The pot has five barley seeds with 2- to 5-mm roots (pregerminated for 48 h)[8] inserted root down into holes 1 cm deep regularly spaced, and is loosely enclosed in a clear polythene bag to prevent evaporation and consequent localized concentration of solutes. Incubation is at 20 ± 2°C with a day:night cycle of 18:6, the first 48 h of which are continuously dark. All tests are performed in duplicate; i.e., two pots, ten seeds per concentration. After gently washing all medium from the seedlings with tap water, the longest root of each is measured together with the number of roots and the length of the shoot. Aberrant seedlings (see Section II.E) are ignored and the results calculated accordingly. Photography can be used to record other details of the seedlings. The EC_{50} is defined as that concentration giving 50% of the root length obtained in the control. The pH of the test mixtures can be measured with a soil pH probe and adjusted to the range of the soil mixture resulting on site after the disposal of the test material (usually pH 6 to 8) by addition of KOH or HCl. All plastic materials are disposed after use, since they can adsorb and then subsequently release contaminants.

B. Modification For Sediments

A known quantity of sediment dried to constant weight at 100°C is first mixed with the test medium and 300 ml of distilled water are then added to the pot. If rewetting is difficult the amounts of wet sediment to be added can be calculated using the wet to dry weight ratio (obtained by drying a small pre-weighed sample), and adding distilled water to give a total water content of 300 ml. The amount of sediment that can be added before physical conditions become unsuitable for root growth will vary widely with the nature of the sediment, but normally a maximum of 200 ml dry sediment per pot is recommended. The physical nature (e.g., particle size composition) of the uncontaminated (control) sediment must match that of the test as closely as possible. The results are expressed as g ds^{-1} (dry solids) per liter liquid added.

C. Modification For Ammonia Compensation

This method is recommended for all samples with >25 mg l^{-1} ammonia (as N) content. The ammonia content of the second control is raised with ammonium sulfate to that of the contaminated material, and parallel dilutions are prepared for comparison with the test material dilutions.

D. Modification For Ammonia Removal

If ammonia (as N) of the test material exceeds 200 mg l^{-1} at the lowest dilution used, the root growth in the second control (see Section II.C) would be too small to detect extra toxicity. Ammonia should be removed only if the suspect toxins will not volatilize during the process. Raise the pH of the test material to 10 with KOH, mix with the test medium, and dry below 50% relative humidity in a dust-free environment at 15 to 20°C for 7 d, agitating daily. The mixture is then treated as in Section II.B. KOH is used, as the Na content of sewages is often high and the use of NaOH might raise it to toxic levels.

E. Seeds

The seeds should germinate evenly, with fewer than 5% showing aberrations after 7 d growth, to maintain the precision of the test. Undressed seed, of consistent genotype, is stored according to the recommendations of the supplier and replaced when germination[6] falls below its initial rate. For barley, individual packs containing just enough seed at 8% moisture content for a run of tests (there are approximately 150 seeds in 8 g of dry seed), are stored at −15°C. Dormancy, if present, is broken before the seeds are stored. All batches of seed are tested against standard toxicants (Section III.D) prior to use in tests.

III. REPRODUCIBILITY

Sources of error were identified and dealt with progressively, then reproducibility trials, on which the credibility of the test is based, were conducted.

A. Ammonia

In five separate trials, each employing Organization for Economic Cooperation and Development (OECD) synthetic sewage[12] spiked with a range of ammonia concentrations, no effects were found at 25 mg l^{-1} ammonia (as N), small effects at 50 mg l^{-1}, 100 mg l^{-1}, then roots were progressively shortened in concentrations of 200 mg l^{-1} and higher. Compensation for the potential effects of ammonia should therefore be made at concentrations above 25 mg l^{-1}. (Settled domestic sewage has an ammonia content of approximately 30 mg l^{-1}.) Ammonia removal with nitrifying bacteria at 20°C and 14 d aeration failed at concentrations above 100 mg l^{-1}.

B. Evaporation

The problems of evaporation were not mentioned by Cheng et al.[8] (12-d test), but even at 80% relative humidity in our 5-d tests efflorescence was noticed on the tops of some pots, indicating evaporation and the potential for toxicants to become locally concentrated. Watering would create a dilute layer, worsening the heterogeneity. Tests in which half the pots were enclosed in clear polythene bags recorded higher toxicity (i.e., less root growth) in the uncovered pots. The loss of liquid after 5 d was 33% (70 ml in 300 ml) in the uncovered pots and 1% (4 ml in 300 ml) in the covered pots. No adverse effects on seedlings were noticed, but the effects of volatile toxicants were enhanced.

C. Light

The test medium allows more light transmission than soil, and root/shoot emergence varies according to how the seed is positioned, as well as its depth. Keeping the pots in the dark for the first 48 h allows the roots to develop to a depth where light inhibition is no longer a problem.

D. Reproducibility Tests

A series of tests was run using dichlorophenol (3,5-DCP), an internationally recognized standard toxicant used, for example, in the verification of the *P. phosphoreum* and respiration inhibition tests,[13] and a real sample of a nonde-gradable industrial waste with a low ammonia concentration. The EC_{50} results on these are given in Table 1, along with the control data and coefficients of variance (cv). The reproducibility of <10% cv compares well with chemical tests such as COD.

IV. DISCUSSION

The root test cannot be expected to predict 100% of potential toxicities, because it involves different circumstances from those of actual test material use. Interactions with the soil can only be properly anticipated if the soil from the field is used,[11] but then relatively high test material concentrations cannot be used to increase sensitivity.[2] The test medium helps to ensure an aerobic root environment at high test material concentrations, but it has a small buffering capacity and may adsorb certain toxicants, reducing their apparent effect. Such adsorption is likely to occur in normal soils, but perhaps at a different rate. The testing of each potentially toxic component is impractical because of the complexity of the mixtures. Therefore, it is recommended that at least one other toxicity test[1] be used to provide a more robust screening estimation of potential toxicity.

The retrieval of the entire root mass is simple and could allow estimation of the dry weight.[4] However, inhibited roots tend to develop a mass of laterals which may weigh as much as normal roots, thus reducing the sensitivity of the test. Enhanced growth at low toxicant doses[9] is only a problem if it occurs at the maximum concentration of the test material, since toxicity is otherwise revealed at higher concentrations.

ACKNOWLEDGMENTS

I am most grateful for the assistance of David Fearnside, Barbara Metcalfe, and John Oldroyd in the development of this test, for the advice from Bob Davies (Water Research Centre, England) in the setting of the original sand-medium test, and for permission from Yorkshire Water Services Ltd. to publish this paper.

REFERENCES

1. Hiley, P.D. and Fearnside, D., The use of multiple toxicity tests in the testing of sewages and trade wastes, *Toxicity Assessment,* in press.
2. Hiley, P.D. and Metcalfe, B., The Yorkshire Water standard plant growth trial for toxicity testing of soils, sludges and sediments, this volume, Ch. 11.
3. Urquhart, C. and Chalk, E.C., The use of Lemna minor in the testing of sewages and trade wastes, Internal Reports of Yorkshire Water Authority 1979–82.
4. Wang, W. and Williams, J., The use of phytotoxicity tests (common duckweed, cabbage and millet) for determining effluent toxicity, *Environ. Monitoring Assessment,* 14, 45, 1990.
5. Dutka, B.J., Tuominen, T., Churchland, L., and Kwan, K.K., Fraser river sediments and waters evaluated by the battery of screening tests technique, *Hydrobiologia,* 188/189, 301, 1989.
6. International Seed Testing Association, International rules for seed testing, *Seed Sci. Technol.,* 13(2), 299, 1985.
7. Schat, H. and Ten-Bookum, W.M., Genetic control of copper tolerance in *Silene vulgaris, Heredity,* 68(3), 219, 1992.

8. Cheung, Y.H., Wong, M.H., and Tam, N.F.Y., Root and shoot elongation as an assessment of heavy metal toxicity and "Zn equivalent value" of edible crops, *Hydrobiologia*, 188/189, 377, 1989.

9. Wilkins, D.A., The measurement of tolerance to edaphic factors by means of root growth, *New Phytol.*, 80, 623, 1978.

10. Wong, M.H., Phytotoxicity of refuse compost during the process of maturation, *Environ. Pollut.*, A37, 159, 1985.

11. Metcalfe, B. and Lavin, J.C., Consolidated sewage sludge as soil substitute in colliery spoil reclamation, in *Alternative Uses for Sewage Sludge*, Hall, J.E., Ed., Pergamon Press, Oxford, 1991.

12. E.E.C., Biodegradation — activated sludge simulation tests, Official Journal of the European Communities, L133 Journal 31: 106–117, 30 May 1988.

13. Fearnside, D. and Hiley, P.D., The use of toxicity testing methods to identify and trace treatability problems on sewage treatment works, *Toxicity Assessment*, in press.

CHAPTER 13

Accumulation of Putrescine in Chromium-Exposed Barley and Rape: A Potential Biomarker in Higher Plant Tests

M.Z. Hauschild

TABLE OF CONTENTS

0-87371-530-6/94/$0.00 + $.50
© 1994 by CRC Press, Inc.

ABSTRACT

In an investigation of the capabilities of putrescine (1,4-diaminobutane) for revealing low levels of chromium stress of higher plants grown in contaminated soils, barley and rape plants were exposed in hydroponic culture to Cr(III) and Cr(VI) in concentrations up to 100 mg*1^{-1} for 1–14 days. Concomitant development of other stress symptoms such as reductions in leaf water content or in growth of seedling or root was recorded. Putrescine accumulation occurred earlier than any other stress symptom recorded implying that this compound may have potentials as an early warning biomarker in a higher plant test.

I. INTRODUCTION

Most certified plant tests apply reduced growth as a measure of the stress experienced by the test plants.[1,2] Due to the inherent variability of plant growth, the sensitivity of these plant tests is generally too low to reveal subtle chronic stress experienced by plants growing in contaminated soil. Biomarkers[3] in the form of early changes in the plant metabolism in response to the applied stress might be used as stress indicators in a more sensitive type of plant test.

Putrescine (1,4-diaminobutane) is known to accumulate in plants following exposure to a range of different biotic or abiotic stresses,[4] and particularly to various heavy metals.[5] During exposure of barley (a monocotyledonous plant species) or rape (a dicotyledonous species) to chromium in the form of Cr^{3+} (Cr[III]) or CrO_4^{2-} (Cr[VI]), putrescine accumulation was followed together with various known stress symptoms to evaluate the possible use of putrescine as a biomarker in a higher plant test.

The two chromium forms were chosen to represent different types of inorganic pollutant stresses. Cr^{3+} is a typical toxic trace metal cation of low solubility and mobility in the plant, while CrO_4^{2-} as an anion presumably will follow other routes into the plant and within the plant, and through its potential oxidative capacity probably exert its toxicity by other mechanisms than Cr(III).

II. EXPERIMENTAL

Plants were exposed in hydroponic culture for between 1 and 14 d to either Cr(III) or Cr(VI). Chromium concentrations applied were 0, 10, 30, 50, or 100 ppm. After 1, 2, 3, 4, 6, 8, 10, and 14 d (Cr[III]) or 1, 2, 3, 4, 7, 8, and 10 d (Cr[VI]), four plants were harvested from all chromium concentrations, and a measure of growth was performed together with an evaluation of visible stress symptoms (different degrees of chlorosis and coloring) and determination of leaf water content. Samples were taken for determination of putrescine. Experimental details are given in Reference 6.

III. RESULTS AND DISCUSSION

Figure 1 shows the concentration of putrescine in barley exposed to Cr(III) (A) or to Cr(VI) (B). Both forms of chromium clearly lead to a significant accumulation of putrescine in the leaves of the exposed plants. There are, however, important differences in the pattern of putrescine accumulation during exposure to the two chromium forms. Putrescine levels gradually increase after 1 week of exposure to Cr(III) and continue increasing with Cr(III) concentration or exposure time for the rest of the experiment, reaching levels that are more than ten times higher in barley exposed to 100 ppm Cr(III) than in unexposed control plants.

Exposure to Cr(VI) gives a quicker putrescine accumulation than Cr(III) exposure. Exposure to 100 ppm Cr(VI) thus causes an immediate surge in leaf concentrations of putrescine, followed by a gradual decline, but also the lower Cr(VI) concentrations lead to a quicker putrescine response than Cr(III) exposure at the same nominal level.

A similar development is found for rape,[6] pea,[7] and tomato[7] exposed to Cr(VI). Though the accumulation of putrescine is initiated sooner by Cr(VI) exposure, it reaches higher levels by exposure to Cr(III), which also causes significant putrescine accumulation at lower chromium concentrations. The putrescine figures alone, therefore, do not give a conclusive answer as to which of the chromium species is more phytotoxic.

Figure 2 shows the relative growth of the barley plants exposed to either of the two chromium species. The growth pattern of the chromium-exposed plants shows less consistency than the putrescine accumulation. As seen from Figure 2A, significant growth reduction only occurs after 10 d of exposure to 100 ppm Cr(III). Lower Cr(III) concentrations or shorter exposure times do not lead to consistent growth reductions at all.

Figure 2B reveals that Cr(VI) is indeed more phytotoxic than Cr(III), causing significant growth reductions by exposure to Cr(VI) concentrations as low as 30 ppm Cr(VI).

Comparison of Figures 1 and 2 clearly demonstrates that for exposure to both forms of chromium, putrescine accumulation is a quicker and more sensitive stress indicator than growth reductions, revealing the stressed condition of the plant up to 1 week sooner and responding to exposure to lower chromium concentrations.

Together with putrescine accumulation and growth, various other possible stress indicators were measured, including leaf content of other polyamines, root length, visible damage symptoms of leaves, water content of leaves, and chitinase activity of leaves. Of all these symptoms, putrescine accumulation was always the first to develop to statistical significance.[6]

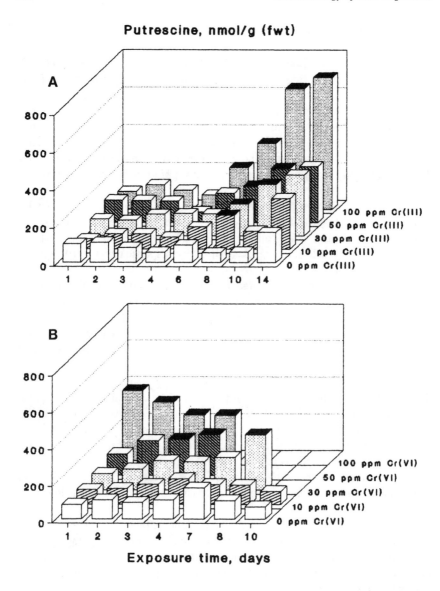

FIGURE 1. Concentration of putrescine in third-generation leaves of barley exposed to chromium in the form of Cr(III) (A) or Cr(VI) (B). Each bar represents the mean putrescine concentration of leaves from four plants. Chromium treatment means and control means are compared in a *t*-test at a 5% level, and significant difference is marked by shading the top of the treatment bar.

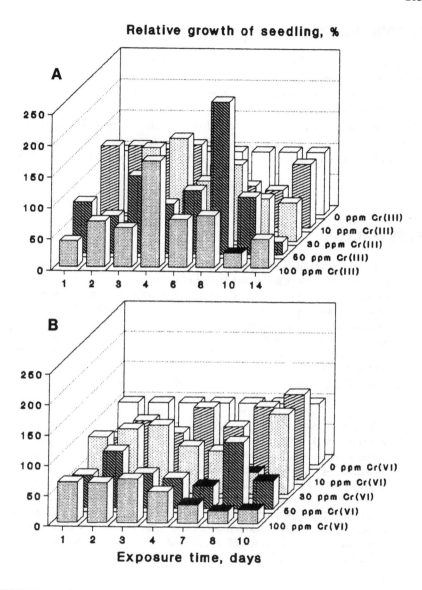

FIGURE 2. Relative growth of barley exposed to chromium in the form of Cr(III) (A) or Cr(VI) (B). The growth is measured from the start of the experiment to the end of the chromium exposure. It is expressed in percent of the growth of the control plants. Each bar represents the mean relative growth of four plants. Chromium treatment means and control means are compared in a *t*-test at a 5% level, and significant difference is marked by shading the top of the treatment bar.

IV. CONCLUSION

Putrescine is accumulated in both barley and rape plants following exposure to either of the chromium species, Cr(III) and Cr(VI), and the accumulation reaches statistical significance prior to any other stress symptom recorded (and especially as much as a week before growth reductions of the exposed plants). The accumulation of putrescine reveals lower levels of stress than most of the other stress symptoms recorded. Accumulation of putrescine thus seems to be a potentially functional biomarker for use in a more sensitive higher plant test.

REFERENCES

1. Organisation for Economic Cooperation and Development (OECD), Second Addendum to OECD Guidelines for Testing of Chemicals, OECD, Paris, 1984.
2. Swanson, S. and Peterson, H., Development of Guidelines for Testing Pesticide Toxicity to Non-target Plants, SRC Publication No. E-901-20-E-88, draft version, Environment Canada, Canada, 1988.
3. Sanders, B., Stress proteins: potentials as multitiered biomarkers, in *Biomarkers of Environmental Contamination,* McCarthy, J.F. and Shugart, L.R., Eds., Lewis Publishers, 1990, 165.
4. Flores, H.E., Protacio, C.M., and Signs, M.W., Primary and secondary metabolism of polyamines in plants, in *Recent Advances in Phytochemistry, Vol. 23, Plant Nitrogen Metabolism,* Poulton, J.E., Romeo, J.T., and Conn, E.E., Eds., Plenum Press, New York, 1989, 329.
5. Wettlaufer, S.H., Osmeloski, J., and Weinstein, L.H., Response of polyamines to heavy metal stress in oat seedlings, *Environ. Toxicol. Chem.,* 10, 1083, 1991.
6. Hauschild, M.Z., Putrescine (1,4-diaminobutane) as an indicator of pollution induced stress in higher plants. Barley and rape stressed with Cr(III) or Cr(VI), *Ecotoxicol. Environ. Saf.,* in press.
7. Hauschild, M. and Jacobsen, S., Putrescine as a marker of pollution induced stress in higher plants, in *Polyamines and Ethylene: Biochemistry, Physiology and Interactions,* Flores, H.E., Arteca, R.N., and Shannon, J.C., Eds., American Society of Plant Physiologists, 1990, 405.

CHAPTER 14

Ecotoxicological Test Systems for Terrestrial Invertebrates

C.A.M. Van Gestel and N.M. Van Straalen

TABLE OF CONTENTS

0-87371-530-6/94/$0.00 + $.50

ABSTRACT

This chapter gives a brief overview of the present status of the development of laboratory tests in terrestrial ecotoxicology. In the introduction, a number of criteria is given for the selection of test species. Representativity of test species with respect to taxonomic groups, ecological function, and route of exposure are considered to be the most important selection criteria. In toxicity tests with terrestrial organisms, different exposure systems are applied. To allow for a comparison of the sensitivity of different species, it is recommended to use the same substrate for all species. For reasons of standardization and representativity for soil ecosystems, for tests with soil dwelling organisms the use of the artificial soil prescribed by OECD for acute toxicity tests with earthworms should be recommended. No standardized test methods are available yet for protozoans, nematodes, isopods, millipedes, oribatid mites, enchytraeids and molluscs, although a number of test systems using these organisms have been described in the literature. For collembola and earthworms standardized test methods are available, all using the OECD artificial soil. A number of tests are available using beneficial arthropods; these tests are mainly used to assess the side effects of pesticides in integrated crop protection. Available tests include hymenopteran parasitoids and parasites, beetles, predatory mites, spiders, and honey bees.

Exposure routes in these tests are residual uptake from contaminated surfaces, oral uptake, or direct exposure. Microcosm and field tests may give more information on the possible ecological effects of chemicals than the single-species tests described before. Tests using microcosms have not been standardized yet. Some standardized protocols are available for performing field tests on beneficial arthropods. The final section of this chapter evaluates the tests described with respect to the criteria mentioned in the introduction and two additional criteria: the potential for standardization and the usefulness of test results for the deriviation of environmental criteria.

I. INTRODUCTION

This chapter gives a short review of laboratory test methods used currently for assessing the potential ecotoxicological risk of chemical substances for terrestrial invertebrates. For a more extensive review of test methods and an overview of the literature used, see Van Straalen and Van Gestel;[1,2] these reviews formed the basis of this chapter.

A. Diversity of Terrestrial Invertebrates and Their Life Histories

There is a tremendous variety of multicellular animals in the soil. Protozoans and nematodes are the most abundant, with populations of up to 10^9 individuals per square meter. Nematodes and other small organisms such as protozoans and tardigrades move through the soil pore water with no influence on the soil particles. Other, larger soil animals disturb the soil by ingestion or by burrowing. These include mites (Acarina), springtails (Collembola), woodlice (Isopoda), snails and slugs (Mollusca), millipedes (Diplopoda) and earthworms (Lumbricidae).

A large group of the terrestrial invertebrates live in the litter layer covering the soil surface and in the vegetation. Among these organisms, arthropods are the most abundant. These organisms do not have direct contact with soil, and are therefore only affected indirectly by soil contamination. They play, however, an important role in the terrestrial ecosystem and are the first to be influenced by, e.g., pesticide spraying and pollutants deposited onto soil and vegetation through the air. For these reasons, tests on these organisms are also discussed in this chapter.

Soil animals are often classified according to their size, that is, microfauna, mesofauna, and macrofauna. However, within most groups there is little correlation between body size and ecological function. For most soil organisms, physiological attributes and life histories are often correlated better with the average positions of animals in the soil profile, rather than with body size. This may be because the trend for higher metabolic rate with decreasing size is counteracted by the fact that smaller animals (like arthropods) often live deeper

Table 1. Comparison of Criteria for the Toxicity of Some Chemicals For the Earthworms *Eisenia andrei,* in Reproduction Toxicity Tests in OECD Artificial Soil

	Cadmium	Chromium	Benomyl	Parathion
LC_{50} (3 weeks) (μg g^{-1})	>1000	>1000	5.7	>180
NOEC body growth (μg g^{-1})	100	320	1.0	<10
NOEC reproduction (μg g^{-1})	<10	32	1.0	10

From Van Gestel, C.A.M. et al., *Ecotoxicol. Environ. Saf.,* 23, 206, 1992. With permission.

in the soil, avoiding such factors as fluctuating temperature and predatory pressure. The well-known allometric relationship between physiological variables and body size, therefore, does not hold strictly for soil organisms.

For two groups, earthworms and springtails, classification systems have been proposed in which life history strategies are related to vertical distribution.[3-5] In both groups, there is a trend towards higher fertility, sexual dimorphism, and increased seasonality in species living close to the surface, compared to species living in the soil interstices. This is correlated with physiological variables such as drought resistance, metabolic rate, and temperature dependence, which tend to be higher in species with surface living habits.

Life history patterns are relevant to ecotoxicological testing. It has been shown that the population consequences of toxic action critically depend on the trade-offs between life history components.[6] Species used for toxicity tests often have a high reproductive output, as this makes them manageable in cultures. Therefore, effects demonstrated in test species do not cover the broad range of possible responses in the field. Depending on energy-budget priorities under stress, there will be a variable level of response, between species, to the same level of intoxication.

To illustrate the importance of life histories, Table 1 provides an example of the toxicity of a number of chemicals for the earthworm *Eisenia andrei*. Effects on population growth follow the effects on reproduction, so it can be concluded from the table that, for the chemicals tested, the LC_{50} is a poor predictor of population response.

Life history differences are also relevant in the context of recovery after a disturbance. For example, the fungicide benomyl is extremely toxic to earthworms, and populations were severely affected in grassland plots following spraying.[8] While the species *Lumbricus festivus* recovered within a month, populations of *L. terrestris* were unrestored after a year.

B. Interactions Between Terrestrial Organisms

Ecological processes in soil are often the result of the interactions between different species. Soil pollution studies must not only consider the effects of contaminants on isolated species, but also changes in community structure due to interactions. Some of the interactions in soil take a key position, that is, one may expect major effects for the stability and the functioning of the soil as an ecosystem when they are disturbed. The following groups of critical interactions can be recognized.

The microfloral community carries out the majority of decomposition processes in soil, but it is stimulated in this function by the activities of soil animals, specifically saprotrophs feeding on decaying organic matter, and microbivores grazing the microflora. The effects of soil animals can be direct contributions to elemental cycles (production of carbon dioxide, excretion of mineral nitrogen) or indirect (aeration of soil as a result of burrowing activities, comminution of leaves, grazing of fungal mycelium, spreading of propagules). In a review of nitrogen dynamics in natural and agro-ecosystems, Verhoef and Brussaard[9] estimated that soil fauna accounted for about 30% of the nitrogen mobilization. Effects of contamination on soil animals may thus affect indirectly decomposition processes. This is illustrated in a study by Killham and Wainwright,[10] where coke plant emissions decreased litter decomposition, not through a direct effect on microflora, but through decreased animal activity.

Another group of critical interactions are those between predators and their prey. By means of density-dependent and size-selective predation, arthropod predators such as gamasid mites, carabid beetles, and linyphiid spiders have a strong structuring effect on their prey communities. Ecotoxicological studies on pesticide sprays in the field provide classic examples of these relationships. Some predators are more sensitive than their prey to insecticides and as a consequence, prey populations (Collembola, oribatid mites) often increase following treatment.[11]

Finally, interactions between parasite and host, plant and herbivore, and between detritus and detritivore can be found in terrestrial ecosystems.

Chemical ecological research has demonstrated the role of various allelochemicals in trophic interactions with terrestrial arthropods. Classical examples are found in the chemical defenses of plants against herbivores, using feeding deterrents and toxicants, which in several cases has led to the coevolution of plant defense systems and herbivore detoxification capacity. More recently it has been demonstrated that the behavior of the predator is also influenced greatly by chemicals from the plant; for example, the predatory mite *Amblyseius potentillae* is able to locate host plants infested by spider mites, and even to discriminate between plants infested by different herbivore species.[12] Plants may benefit from producing volatile chemicals when damaged by herbivores attracting predators.

C. Criteria for Selecting Test Species

When considering the use of invertebrates for ecotoxicological tests, the species should be selected on the basis of their "representativeness" to the community of organisms that is to be protected from the adverse effects of the chemicals tested. A set of test species should give an indication of the ecological responses that are expected to follow from the introduction of the test chemical into the environment. There are at least three criteria for selecting species.

First, there may be toxicological similarities between taxonomically related species. In contrast to aquatic toxicology, there is little comparative data for soil organisms. A study using four earthworm species, by Neuhauser et al.,[13] suggests that correlations for LC_{50} values among earthworm species are very high. Whether this also holds for chronic toxicity data or for other soil animal groups is not known. Furthermore, Neuhauser et al.[13] results show that in spite of the good correlation, the four earthworm species were by no means equally sensitive to the chemicals tested. Still, it may be better to cover a wide taxonomic range of species, rather than to concentrate on earthworms alone: the probability of finding a more sensitive group is increased when the taxonomic range is extended.

Secondly, the ecological role of a species can be taken into account; that is, the set of test species should be such that the various functions exerted by terrestrial invertebrates are covered. From the preceding text it may be concluded that on the basis of this argument, one would like to include predators, parasitoids and parasites (especially those of pests), microbivores, saprotrophs, and other groups associated with organic matter decomposition, as well as pollinators.

Thirdly, the set of test species should include the various ways in which soil organisms may be exposed to contamination, that is, through contact with contaminated surfaces, through soil pore water, by soil ingestion, by ingestion of organic matter, by feeding on fungi, through soil air, etc. This calls for a set of species with diverse feeding habits.

These arguments are supplemented with various practical reasons. The most obvious ones are the suitability of a species for culture in the laboratory and the availability of the same life stages throughout the year. These are often the main factors restricting the choice of species, due to insufficient ecological background knowledge on food requirements and responses to temperature and photoperiod.

II. EXPOSURE SYSTEMS

Invertebrate toxicity tests often use a species-specific exposure method in relation to the expected uptake route in the field. Bioavailability of the chemical tested is different for each exposure system; also, the various methods differ with respect to the suitability for standardization. Therefore, it is expedient to review the various exposure systems used before discussing the tests proper.

Organisms can be exposed to pollutants present in soil via two major routes: (1) oral uptake or (2) dermal uptake of pollutants from the soil solution. Another possible route of exposure is the inhalation of pollutants present in the soil air phase. So far, no literature data are available on the importance of this route compared to other routes of exposure. For organisms with a firm cuticle or exoskeleton, such as many arthropods, direct dermal uptake of pollutants from the soil may be less important when compared to oral uptake. The results of Everts et al.[14] suggest, however, that dermal exposure to contaminated soil may be of importance for spiders. No data are available on the contribution of this route relative to oral uptake for other soil arthropods. For soft-bodied organisms living in close contact with the soil such as earthworms and enchytraeids, exposure via soil pore water will be most important. This can be demonstrated for earthworms by Van Gestel and Ma;[15] from their results it can be concluded that, for earthworms, the equilibrium partitioning concept developed in aquatic ecotoxicology may be valid. This means that, for earthworms and probably also for other soft-bodied soil organisms, such as protozoa, tardigrades, nematodes, and enchytraeids, the toxicity of chemicals in the soil is determined mainly by the pore water concentration, which can be derived from the total concentration using sorption data.

A. Direct and Indirect Application

Topical dosing is generally applied to arthropods. A small droplet of the toxic solution is applied directly to a predetermined area of the arthropod body surface (usually the thorax). The animal is more or less immobilized, either mechanically or with cold or carbon dioxide anesthetic. This method of topical application allows the dose to be expressed as an absolute amount per animal. From a toxicological point of view, this is a preferred exposure method, since any effect can directly be related to the dose; disturbing factors such as consumption, walking behavior, and activity can be eliminated as sources of variation.

Arthropods may also be dosed by whole-body contact with a chemical through immersion in a solution. The time of immersion is standardized and the animals are allowed to dry before noting the effect. Wingless arthropods such as predatory mites can be glued conveniently to a surface by the dorsal shield, to allow for ease of dipping and manipulation. The dipping technique is easy to carry out, but it has the disadvantage that the dose received is unknown. Effects are expressed in terms of the concentration of the chemical in the dipping solution. Soil-living invertebrates, such as earthworms, nematodes, and protozoa, may also be tested in aqueous solutions of the test chemical. In these tests, exposure via the aqueous phase is considered to be the most important route of exposure, and the interaction of the organisms with the soil solid phase is neglected.

The relevance of both topical and whole-body exposure techniques is mainly restricted to laboratory research methods. In the field, other routes of exposure may be important as well.

An important exposure route for arthropods in agricultural fields treated with sprayed chemicals is residual uptake. Surfaces coated with films of pesticides will act as a source for uptake by arthropods as they walk over the surface, especially high-surface activity species, such as predatory mites, spiders, beetles, and springtails.

In experiments, arthropods may be present during application of the toxicant, in which case the effect is a mixture of both direct and residual exposure. More frequently, however, the treated surface is allowed to dry and arthropods are placed on the treated surface for the test. Surfaces used in such tests include plant leaves, sand, natural soils, or artificial substrates such as filter paper or glass. The bioavailability of the residue depends very much on the topography of the substrate, its tendency to adsorb the chemical, and its moisture content. For the purposes of standardization the use of an inert material that will neither absorb nor react with the chemical, i.e., glass or sand, is recommended. Effective doses established in this way are, however, very difficult to translate to field situations, as the inert surfaces do not resemble the natural situation.

B. Chemicals Mixed in With the Soil

For soil-dwelling animals, such as earthworms, enchytraeids, nematodes, and protozoa, tests in soil seem to best simulate the natural exposure routes. Although arthropods normally do not ingest mineral soil, many live in close contact with it and take up chemicals from the soil/air interface. This is thought to be mediated via a water film. In this manner, the actual uptake mechanism is similar to the exposure systems considered in the previous section, but the way the substrate is prepared allows the dose to be expressed as a concentration per mass unit of dry soil, instead of a concentration per unit of surface area.

The substrates used in soil tests vary from natural material taken from the field to soils composed artificially out of commercially available materials. The type of soil has a large influence on bioavailability of the chemical; the soil factors that determine the distribution of chemicals over solid phase and pore water will have a large effect on toxicity.

For organic chemicals, toxicity can often be related directly to soil organic matter content. Van Gestel and Ma[15] demonstrated this for the toxicity of trichlorobenzene and dichloroaniline for earthworms. For chlorophenols, however, pH as well as organic matter affects the toxicity for earthworms. Van Gestel et al.[16] demonstrated that after correcting for the pH effect the toxicity of all chemicals tested on earthworms (chlorophenols, chlorobenzenes, dichloroaniline) could almost entirely be related to the organic matter content of the soil. Similar relationships between toxicity and soil organic matter content are found in the literature for the effect of pesticides on the cricket *Gryllus pennsylvanicus*[17] (see Figure 1), the wireworm *Melanotus communis*,[18] and the nematode *Heterodera rostochiensis*.[19]

For heavy metals, different factors appear to play a role. From data obtained by Jäggy and Streit[20] it can be concluded that the acute toxicity of copper to the earthworm *Octolasium cyaneum* is solely correlated with organic matter content. No such data are available for other metals. From data on bioaccumulation in earthworms, however, it can be concluded that the bioavailability of metals may depend on soil pH, organic matter content, cation exchange capacity (CEC),

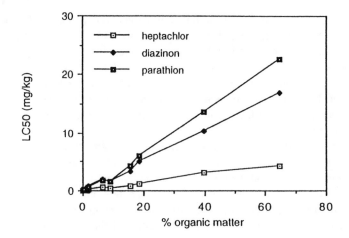

FIGURE 1. Acute toxicity of some pesticides for *Gryllus pennsylvanicus* in relation to soil organic matter content. After Harris, C.R., *J. Econ. Entomol.*, 59, 1221, 1966. With permission.

and clay content. Quantification of the contribution of each factor to the toxicity and bioavailability of metals for earthworms appears to be almost impossible.[21] Wohlgemuth et al.[22] found a significant correlation between the toxicity of cadmium for the springtail *Folsomia candida* and the calculated CEC of several artificial and natural soils. Cadmium toxicity did not correlate with soil pH and only slightly with soil organic matter content.

For the purposes of standardization the use of an artificial soil is recommended, not only for earthworms, but also for soil arthropods. This artificial soil is prepared according to guideline 207 of the Organization for Economic Cooperation and Development (OECD)[23] and is a mixture of 10% *Sphagnum* peat, 20% kaoline clay, and 70% industrial quartz sand. The ingredients are mixed dry, and an amount of pulverized calcium carbonate is added to adjust the pH to 6.0 ± 0.5. Finally, deionized water is added to bring the water content of the soil to 40 to 60% of the water holding capacity. This deionized water may contain the test substance, or the latter may be applied separately using a volatile solvent.

C. Chemicals Added to the Food

Dietary uptake is a direct route for chemicals sprayed on leaf surfaces, acting as stomach poisons in phytophagous invertebrates, for chemicals associated with dead organic matter exerting an effect on saprotrophs. Dietary exposure may also occur via the food chain, e.g., microbivorous arthropods feeding on fungi that concentrate chemicals from the soil, or predatory arthropods feeding on contaminated prey.

In toxicity tests, chemicals are mixed homogeneously through the food, and the effective dose is expressed per dry mass of food. Alternatively, the food (e.g., leaves) may be treated with a spray, or soaked into a solution of the test chemical. Other dietary routes are via drinking water, or via sucrose solutions. If the test animals take in the amount fed completely, or if consumption can be determined by weighing the food left, the dose can be expressed in absolute units per animal. This allows comparison with doses taken up via other routes, e.g., topical application.

The food used in feeding experiments depends greatly on the species. For example, at our laboratory isopods are fed with ground poplar leaf amended with commercial pet food. Collembola may be fed with algae, with fungi grown on contaminated agar, or with yeast. Beetles, spiders, and predatory mites may be fed with arthropods previously exposed to the chemical.

The uptake efficiency of chemicals added to the diet is rather variable; effective concentrations are difficult to compare between species, as they will be influenced by the type of food used and the physiological condition of the animal. In some arthropods, avoidance of contaminated food is a common response. In these cases, the effect is a function of both the dose absorbed and a reduced energy intake. Many saprotrophic arthropods have rather specific feeding preferences, which may be changed by contamination. Tranvik and Eijsackers[24] showed that springtails of the species *F. fimetarioides* were able to avoid metal-contaminated food, while the species *Isotomiella minor* could not.

D. Exposure via the Air

When arthropods are tested for their susceptibility to gaseous air pollutants, exposure units might be air-tight and are then flushed with a known concentration of the chemical in air. The effective concentration is expressed in $nl\ l^{-1}$ (ppb-vol), or in $\mu g\ m^{-3}$. Alternatively, a static system is used. This will not, however, provide a constant exposure concentration, as air pollutants may be deposited on surfaces present in the test units.

Several pesticides, especially those applied as fumigants, exert their effect through aerial exposure and are tested on flies, fleas, ticks, etc., in this way. Even for some pesticides sprayed on a surface, actual exposure may be via the air, as the chemical evaporates from the surface film to reach toxic concentrations above the surface. In contact toxicity tests, this effect is usually not included, and air flow ventilation of test units is recommended.[25]

III. TESTS ON SOIL FAUNA

The current status of soil animals as test organisms is poor. Many species do, however, show promise as they are easy to culture and their size allows for a small-scale experimental setup and ease of replication. In this section, tests on saprotrophic species only are described; tests on predatory species will be discussed in the next section.

A. Protozoans and Nematodes

Protozoans and nematodes live in the soil pore water, and the most suitable test methods are adapted from aquatic toxicology. Among the protozoans the ciliates *Tetrahymena pyriformis, Colpoda cucullus,* and *Paramecium aurelia* have been considered for use as test animals,[26-28] as have the nematode species *Caenorhabditis elegans, Panagrellus silusiae,* and *Plectus acuminatus.*[29-32] However, an accepted test procedure is not available.

The problem with these types of test is not their implementation, but the extrapolation of test results to the field. Usually, tests are done in an artificial medium (nutrient solution, agar, etc.), the composition of which affects toxicity greatly. If concentrations in test solutions can be equated with pore water concentrations, sorption data may be used to express the toxicity per unit of soil. This extrapolation, however, still has to be validated.

B. Isopods and Millipedes

Isopods are an interesting group of animals for heavy metal research, because of their unique ability to concentrate extreme amounts of metals in their bodies.[33] Their use as a test animal for soil toxicity studies, is, however, restricted to a few cases and no attempts have yet been made to arrive at standardization.

Usually, isopods are kept on plaster of Paris and are fed with partly decomposed leaves, either intact or ground, to which chemicals may be added. Growth increase over several weeks is observed, but is rather variable, even for one individual. Reproduction is also difficult to assess, due to the fact that, after mating, females may retain the sperm for a long period before producing eggs, which are carried in a brood pouch.

Porcellio scaber, Oniscus asellus, Philoscia muscorum, and *Trichoniscus pusillus* have been used by various authors[34-39] in toxicity experiments. Among these, *T. pusillus* seems to be the most suitable as a test species, as it has a somewhat shorter life cycle compared to the other three, and is parthenogenetic. Most isopods are very easy to culture and do not require special conditions, although their relatively long life cycle (1 to 3 years) does not allow a rapid buildup of cultures.

Millipedes (Diplopoda) are another important group of saprotrophic soil invertebrates, but they have never been considered seriously as test animals. The most widely investigated species is *Glomeris marginata.*[40] The species *Cylindroiulus britannicus* is also well suited to be a test animal.[39] Test conditions for millipedes are similar to those for isopods.

C. Oribatid Mites

The parthenogenetic oribatid mite *Platynothrus peltifer* seems to be the only oribatid used so far in soil toxicity experiments,[41] although oribatids comprise hundreds of species and are usually the most numerous group of arthropods in forest soils.

The test with *P. peltifer,* in which the animals are exposed to contaminated algae and the number of eggs are counted, is very laborious and rather lengthy (9 to 12 weeks) due to the low rate of egg production in this species and its long life cycle (1 year).

In experiments using cadmium, copper, and lead, *P. peltifer* proved to be rather resistant in terms of lethality, but very susceptible in terms of egg production. Due to their peculiar habits, species such as *P. peltifer* tend to be forgotten in the development of toxicity tests; it is, however, the most sensitive soil invertebrate tested so far for cadmium, while it is more sensitive than springtails for copper and lead.[41,42]

D. Collembola

The Collembola are a relatively well-investigated group of soil animals. Several species have been used frequently in toxicity experiments: *Onychiurus armatus, O. quadriocellatus, F. candida, Tullbergia granulata,* and *Orchesella cincta.*[42-47] The first three species are parthenogenetic (thelytokous), *O. cincta* is sexual, sperm being transferred indirectly through spermatophores deposited on the substrate by a male. Three exposure systems have been described: feeding on fungi grown on contaminated agar, feeding on directly contaminated food, and residual exposure (treated substrate, e.g., sand, leaves, soil, etc.).

When testing Collembola with contaminated fungi, the animals are kept on glass fiber, filter paper, or a plaster of Paris substrate, and fed on a piece of agar, overgrown with hyphae, cut from a malt agar gel inoculated with a fungus (e.g., *Verticillium bulbillosum*).[43] Egg production, growth, and survival are recorded regularly throughout a period of several weeks. The advantage of this system is that substances are offered in a natural way, i.e., after being taken up and possibly transformed to naturally occurring complexes.

Collembola can be tested using directly contaminated food (algae, yeast, ground leaf material), made up with a specific concentration of the test chemical in water or acetone. Food can be offered as droplets on filter paper discs, with the animals kept on a plaster or sand. In this manner, concentrations can be manipulated easily. Growth, egg production, and survival are then monitored over a period of several weeks.[42]

The third system for testing soil-dwelling Collembola is the *F. candida* test,[22,48] using the artificial soil medium developed for earthworm toxicity tests (see Section II.B). Juvenile Collembola (10 to 12 d old) are placed on artificial soil with dry yeast provided for food. The number of remaining animals and the number of offspring produced are used as test parameters. The *Folsomia* test is easy to carry out, it requires little attention during the test, and it gives reproducible results. Another advantage is the use of the Organization for Economic Cooperation and Development (OECD) artificial soil used also in earthworm tests; thus, experimental results can be compared between earthworms and springtails. The only disadvantage of the test is that reproduction cannot be observed directly, and cannot be separated from juvenile mortality and hatching success. Reproduction depends largely on the body size of the adult, so any chemical that inhibits growth will also decrease the number of offspring recovered after 28 d. This was the main effect of cadmium on *F. candida* in experiments reported by Crommentuijn et al.:[49] population increase was retarded due to reduced growth rate in adults.

E. Enchytraeids

Enchytraeids can be cultured easily on agar and on (artificial) soil substrates, when fed rolled oats.[50] The species used in toxicity tests are of the genus *Enchytraeus*. The well-known species *Cognettia sphagnetorum* can also be bred easily in the laboratory, but its tendency to fragment upon handling makes it less suitable for toxicity tests.

Westheide et al.[51] described a test in which the test chemical is incorporated in 1.5% nutrient agar and two species are used, *Enchytraeus crypticus* and *E. minutus*. Adult enchytraeids are placed on an agar surface and after exposure for 25 d the number of cocoons produced and the number of hatched cocoons are counted. This method seems to provide an easy and reproducible test; however, as agar does not resemble a real soil, it should be concluded that the usefulness of this test for risk assessment of chemicals in soil is limited.

Römbke[50] described a test with *E. albidus,* using the OECD artificial soil (see Section II.B), in which adult animals are exposed to different concentrations of a test chemical mixed, homogeneously, through the artificial soil. After 4 weeks surviving adult animals are removed and the test substrate is incubated for another 4 weeks to allow for the hatching of cocoons. The test is ended with a determination of the total number of juveniles produced. This test combines both acute and sublethal effects. An important positive aspect of the test is the use of the same OECD artificial soil substrate as is used in the internationally accepted earthworm toxicity tests.[23,52]

F. Lumbricids

In the OECD[23] and EEC[52] guidelines on acute toxicity testing using earthworms, the species *Eisenia fetida* and *E. andrei* are recommended. These are not real soil-dwelling species, but are commonly found in compost and dung heaps, and can be cultured easily in the laboratory on a substrate of horse or cow dung. According to these guidelines, other soil-dwelling species may also be used; such species are, however, hard to culture in the laboratory because they have long generation times and need large volumes of soil. Thus the two *Eisenia* species are recommended for practical reasons.

Three acute toxicity tests exist; in the filter paper contact test,[23] adult earthworms of the species *Eisenia* spp. are exposed for 48 h to filter paper wetted with a solution of the test substance. Mortality is assessed, and the LC_{50} value is expressed as $\mu g \ cm^{-2}$. The method has been shown to be easy, fast, and highly reproducible, but it has no predictive value for the effect of chemicals in the soil.

In the artificial soil test,[23,52] adult earthworms of the species *Eisenia* spp. are exposed to the test chemical, mixed in with the OECD artificial soil (see Section II.B), for 14 d. Mortality is the only test parameter, and LC_{50} values are expressed as mg kg^{-1} dry soil. The test has been shown to be reproducible. The results obtained in the artificial soil can easily be translated to natural soils by using sorption data.[15] For this reason, it is acceptable to use the artificial soil, as it has sufficient predictive value with respect to effects that occur in the field.

In the Artisol test,[53] adult earthworms of *Eisenia* spp. are exposed to chemicals mixed through a substrate of glass balls covered with a slurry of amorphous silica gel (Artisol) for 14 d. Survival is the only test parameter, and results are expressed in terms of LC_{50} values. It should be noted that the silica gel substrate does not bear any resemblance with natural soil, thus for reasons of ecological realism and extrapolation towards natural soil, this test cannot be recommended.

Two sublethal toxicity tests with earthworms have been described.[54,55] In both tests, the OECD artificial soil substrate (see Section II.B) and the earthworm species *Eisenia* spp. are used. In the first test,[54] chemicals are mixed homogeneously in with the artificial soil, and after 3 weeks of exposure the effects on the growth and cocoon production by adult earthworms are determined. The worms are fed by supplying a small amount of (untreated) cow dung in a small hole made in the middle of the soil. By incubating the cocoons produced for 5 weeks in untreated artificial soil, effects on hatchability (% fertile cocoons, number of juveniles per cocoon) and the total number of offspring per adult worm can be determined.

The second test[55] was developed for determining sublethal effects of pesticides on earthworms. The pesticide is sprayed onto the soil surface, and earthworms are fed by applying about 0.5 g cow dung per animal to the soil surface once a week. Pesticides are applied in two treatment levels, corresponding to the recommended dose and a fivefold dose. After 6 weeks of incubation, adult worms are removed from the substrate and weighed. The test substrate containing cocoons and juveniles is incubated for another 4 weeks. After 10 weeks, the juveniles are extracted from the substrate by hand sorting or heat extraction and counted. Effects on earthworm growth and on the total number of offspring produced per box are determined.

G. Molluscs

In the limited number of toxicity tests using terrestrial molluscs, exposure was via the food. In a toxicity experiment with *Helix aspersa*,[56] snails were kept on a substrate of moist quartz sand covered by a piece of woven glass towel and fed a diet of ground Purina® Lab-Chow (formulation for rats, mice, and hamsters) supplemented with 10% CaCO₃. The toxicant is introduced into the diet by mixing with the CaCO₃ and adding to the Purina® Lab-Chow. Effect parameters are survival, reproductive behavior, dormant state, new shell growth, and food consumption. Similar test methods using the snail *H. pomatia* or the slug *Arion ater* have been described by other authors.[57-59]

IV. TESTS ON BENEFICIAL ARTHROPODS

Arthropods that may improve the production of agricultural products are designated as ''beneficials'', and there is commercial interest in designing and applying pesticides in such a way that beneficials are least affected. The working group on ''Pesticides and Beneficial Organisms'' of the International Organization for Biological and Integrated Control of Noxious Animals and Plants (IOBC) has significantly contributed to designing ecotoxicological test methods and decision schemes for evaluating the hazard of pesticides. These methods have been reviewed by various authors[60-62] and are summarized in Table 2.

The hymenopteran groups Ichneumonidae, Braconidae, and Chalcidoidea contain a large number of parasitoid species. The female insect deposits an egg in or on a host (usually an insect egg or larva), which is then gradually eaten as the offspring develop. The host selection process and the life cycle of the parasitoid are finely tuned to the host, and many species will attack only a single or a few host species. As well as the species given in Table 2, other hymenopteran species used in toxicity tests include *Diaeretiella rapae*, an internal parasite of aphids such as *Myzus persicae*, *Phygadeuon trichops*, a parasite of *Delia* species (bulb flies), *Coccygomimus* (= *Pimpla*) *turionellae*, a polyphagous parasite of Lepidoptera (Tortricidae, Geometridae, Noctuidae), and *Opius* sp., a parasite of leaf mining insects. The methods used for these species are similar to those described for *Trichogramma* and *Encarsia* in Table 2.

Within the order of the Coleoptera, the families Carabidae (ground beetles), Staphylinidae (rove beetles), and Coccinellidae (lady birds) contain representatives that are common in agricultural fields and are recognized for their predation of pests.

Among the various arthropod groups, spiders seem to be a particularly sensitive group. This often appears in field tests with pesticides, where catches of surface-active spiders are reduced in a manner similar to that of predatory mites following pesticide application. The families Erigonidae and Linyphiidae (money spiders) are important groups with a great species richness.

Table 2. Some Tests on Beneficial Arthropods Described by IOBC[60] and OEPP/EPPO[63]

Group/species	Life stage	Type of test[a]	Substrate	Test parameters
1. Hymenopteran parasitoids and parasites				
Trichogramma cacoeciae	Adult	Res. cont.	Glass	Parasitism of host eggs
	Pupae	Dir. cont.	Dip. sol.	Parasitism of host eggs
Encarsia formosa	Adult	Res. cont.	Glass	Survival/parasitism of host eggs
	Pupae	Dir. cont.	Dir. spray	Emergence
2. Beetles				
Bembidion lampros	Adult	Dir. cont.	Dir. spray	Survival
Aleochara bilineata	Adult	Res. cont.	Sand	Survival/egg production/hatching rate
Coccinella septempunctata *Semiadalia undecimnotata*	Larvae	Res. cont.	Glass	Survival/development Time/egg production/hatching rate
3. Predatory mites				
Phytoseiulus persimilis	Juvenile	Res. cont.	Bean leaves/glass	Survival/egg production
4. Spiders				
Lepthyphantes tenuis *Oedothorax apicatus*	Adult	Res. cont.	Glass	Survival/predatory capacity/behavior
5. Honey bees				
Apis mellifera	Adult	Dir. cont.	Thorax appl.	Survival (LD$_{50}$)
		Feeding	Sucrose sol.	Survival (LD$_{50}$)

a Dir. cont. = direct contact; res. cont. = residue contact.

The recommendations made by the International Commission for Plant Bee Relations have been included in a guideline of the European and Mediterranean Plant Protection Organization (EPPO) for evaluating the hazards of pesticides to the honey bee *Apis mellifera*.[63] Several countries have slightly different national guidelines for testing pesticides on honey bees.

Several other beneficial arthropods have been proposed as test species:[25,60] *Chrysoperla carnea* (Neuroptera, Chrysopidae), *Anthocoris nemorum* (Heteroptera, Anthocoridae), *Syrphus corollae* and *S. vitripennis* (Diptera, Syrphida), and *Drino inconspicua* (Diptera, Tachinidae).

V. MICROCOSM TESTS

Single species tests are carried out under rather artificial conditions and disregard ecological interactions between different species. To evaluate effects of chemicals under more natural conditions, model ecosystems, microcosms, or microecosystems have been designed that simulate certain aspects of real ecosystems, and are yet simple enough for experimental use. Saprotrophic invertebrates have been considered for such systems, because their activities can be assessed conveniently in terms of system functions such as leaf litter fragmentation and nutrient conversions.[64]

Several terrestrial model ecosystems have been described, without attempt to arrive at standardization. Usually, the system comprises a soil column, sometimes with a plant or small tree. The system may either be closed or it may be open to the ambient air, and contain intact core samples from a natural habitat or a more or less standardized soil. The effects of pretreatments, such as drying, sterilizing, inoculation, litter-type, age of the litter, etc., have a significant impact on the behavior of the system and need to be investigated thoroughly.

Natural rainfall may be stimulated and leachate can be collected. Various chemical analyses of the leachate solution may indicate aspects of the saprotrophic activity: dissolved organic carbon, NH_4, NO_3, pH, Ca, etc. One disadvantage of the leaching procedure is that the humidity of soil and litter is not stable and is difficult to standardize. Moreover, toxicants added to the system may be displaced through the column or leached out. In a closed system, air samples can be taken; carbon dioxide concentrations in the air above the soil column are often determined to estimate the rate of decomposition processes in microcosms.

Van Wensem et al.[65,66] added chemicals to poplar leaf litter, which is incubated for 4 weeks after which some replicates are terminated to determine DOC, NH_4, NO_3 and pH. The remaining replicates are incubated for another 4 weeks, with eight isopods (*Porcellio scaber*) added to each system. Survival and growth of the isopods can be assessed, as well as particle size distribution and concentrations of minerals of the remaining litter. The organotin fungicide triphenyltin hydroxide increased the concentration of soluble ammonium in the litter, partly due to excretion by isopods and partly due to stimulating effects on the microflora. In systems with isopods, the organotin decreased ammonification in treatment levels higher than 10 μg g^{-1}, but in systems without isopods the organotin had no significant effect. The addition of isopods in this case, therefore, made the system quite sensitive, which was not expected (triphenyltin is a fungicide) and would not have been noticed in a single species test using isopods.

VI. FIELD TESTS

The tests described in the preceding sections will only provide a rough estimate of the possible hazard imposed by a chemical in the environment. In many cases this will be sufficient. Usually the laboratory test is considered to be a "worst case" situation, since test animals are exposed to a constant concentration which is relatively available because the test medium is prepared freshly. Under field conditions exposure may be lower, since the chemical is not distributed uniformly over the habitat and bioavailability will often (not always) be lower due to various sorption processes. On the other hand, the laboratory test considers the test organism under optimal conditions, without secondary stresses, such as those of food shortage, drought, and cold. The uncertainties attached to the laboratory-field extrapolation can be avoided by considering experiments under semifield or field conditions.

Various organizations have issued recommendations for test protocols in the field.[60,63] In several cases guidelines for field tests are part of the national registration procedures for pesticides. In addition, an extensive amount of scientific research has been done in which side effects of pesticides on nontarget arthropods have been documented, and several reviews on this aspect have been published.[11,67-69] Some attempts have also been made to develop standardized procedures for field tests to assess the effects of pesticides on earthworms.[70] No attempt will be made to recapitulate this information.

VII. CONCLUSIONS

It may be concluded that the array of methods used in testing terrestrial invertebrates is diverse, mainly because different tests have been developed with different aims. Many methods are still poorly described, especially in relation to the medium to which the chemical is applied, and the consequences for bioavailability. Section II of this chapter has illustrated the wide variety of exposure systems.

The three criteria for selecting test species on the basis of their representativeness to the community of organisms to be protected can be supplemented with arguments derived from suitability for culture, potential for standardization (based on the extent of interlaboratory calibration already achieved), and the possibility of using the test results in deriving environmental quality criteria. Table 3 provides a rough indication of the results of such an evaluation. In this table, species are included for which test systems are still under development or which seem to be promising for test development.

In Table 3 the ecological relevance of the test species and their occurrence in ecosystems to be protected is used as an evaluation criterion. For some species, such as the earthworm *E. fetida*, it is obvious that they are not occurring in natural ecosystems; they are used as surrogate species because they are more easy to culture and handle in the laboratory than natural species.

The potential for standardization of a test system is important when one strives for international use of test methods. For some species, it is still uncertain whether or not it is possible to arrive at a standardized method; this holds, e.g., for the oribatid mite *Platynothrus peltifer,* which has a very long life cycle and a low reproduction rate not allowing for a proper determination of effects on reproduction within a reasonable test duration. For others, such as the earthworms and the collembola *F. candida,* tests have already been standardized at an international level.

The usefulness of a test system for deriving environmental quality criteria, mentioned in Table 3, also comprises the substrate used. When real or artificial soils are used as a test substrate, test results may be used directly for deriving soil quality criteria. This is not the case for tests on soil organisms using other substrates or exposure routes, such as water, agar, or nutrient solutions used in tests with protozoans and nematodes. Results of such tests cannot be translated directly to soil quality criteria. This holds also for exposure routes used in tests on beneficial arthropods; such tests can only be useful for the risk assessment of pesticides when test results can be related to natural exposure routes.

It may be argued that each species should be tested under its optimal conditions and that interspecies harmonization of conditions, e.g., by using the same test substrate, is not to be recommended. The usefulness of more than one test on the same chemical is, however, very limited when the results for two species cannot be compared to each other. Standardization of the test substrate should therefore be considered in the further development of test methods.

Ecotoxicological test methods for soil are relatively underdeveloped; standardization has been achieved in only a few cases. Ecotoxicological risk assessment calls not only for better tests, it also requires the use of sublethal endpoints instead of lethality, as well as tests on more species representing a wider range of soil organisms. Instead of putting too much effort into standardization, it may be useful to consider testing other (rare, difficult) species or to test sublethal criteria such as reproduction.

Table 3. Overview of Selected Laboratory Tests Using Terrestrial Invertebrates, Evaluated According to Three Criteria

Invertebrate species used in the test		Ecological relevance; occurrence in ecosystems to be protected	Potential for standardization culture and use in different labs	Test results may be used to derive environmental quality criteria
Protozoans	Colpoda cuculus	+	+	−
Nematodes	Plectus acuminatus	+	−	−
Isopods	Porcellio scaber	+	−	±
	Trichoniscus pusillus	+	−	±
Mites	Platynothrus peltifer	+	+	±
Collembola	Folsomia candida	+	±	+
	Orchesella cincta	±	+	±
Enchytraeidae	Enchytraeus albidus	−	+	+
Lumbricidae	Eisenia fetida	+	+	+
Molluscs	Helix aspersa		±	±
Hymenopteran parasitoids and parasites	Encarsia formosa	±	+	−
	Trichogramma cacoeciae	+	±	−
Beetles	Bembidion lampros	+	+	±
	Aleochara bilineata	+	±	±
Predatory mites	Phytoseiulus persimilis	±	+	−
Spiders	Oedothorax apicatus	+	−	±
Honey bees	Apis mellifera	+	+	±

REFERENCES

1. Van Straalen, N.M. and Van Gestel, C.A.M., Soil, in *Handbook of Ecotoxicology,* Calow, P., Ed., Blackwell Scientific, Oxford, 1993, chap. 15.
2. Van Straalen, N.M. and Van Gestel, C.A.M., Ecotoxicological Test Methods Using Terrestrial Arthropods, Detailed Review Paper for the OECD Test Guidelines Programme, 1992.
3. Satchell, J.E., r Worms and K worms: a basis for classifying lumbricid earthworm strategies, in *Soil Biology as Related to Land Use Practices,* Dindal, D.L., Ed., Office of Pesticides and Toxic Substances, U.S. EPA, Washington, D.C., EPA-560/13-80-038, 1980, 848.
4. Petersen, H., Population dynamic and metabolic characterization of Collembola species in a beech forest ecosystem, in *Soil Biology as Related to Land Use Practices,* Dindal, D.L., Ed., Office of Pesticides and Toxic Substances, U.S. EPA, Washington, D.C., EPA-560/13-80-038, 1980, 806.
5. Van Straalen, N.M., Comparative demography of forest floor Collembola populations, *Oikos,* 45, 253, 1985.
6. Sibly, R.M. and Calow, P., A life-cycle theory of responses to stress, *Biol. J. Linn. Soc.,* 37, 101, 1989.
7. Van Gestel, C.A.M., Dirven-van Breemen, E.M., Baerselman, R., Emans, H.J.B., Janssen, J.A.M., Postuma, R., and Van Vliet, P.J.M., Comparison of sublethal and lethal criteria for nine different chemicals in standardized toxicity tests using the earthworm *Eisenia andrei, Ecotoxicol. Environ. Saf.,* 23, 206, 1992.
8. Edwards, P.J. and Brown, S.M., Use of grassland plots to study the effects of pesticides on earthworms, *Pedobiologia,* 24, 145, 1982.
9. Verhoef, H.A. and Brussaard, L., Decomposition and nitrogen mineralization in natural and agro-ecosystems: the contribution of soil animals, *Biogeochemistry,* 11, 175, 1990.
10. Killham, K. and Wainwright, M., Deciduous leaf litter and cellulose decomposition in soil exposed to heavy atmospheric pollution, *Environ. Pollut. (Ser. A),* 26, 79, 1981.
11. Edwards, C.A. and Thompson, A.R., Pesticides and the soil fauna, *Res. Rev.,* 45, 1, 1973.
12. Dicke, M., Sabelis, M.W., Takabayashi, J., Bruin, J., and Posthumus, M.A., Plant strategies of manipulating predator-prey interactions through allelochemicals: prospects for application in pest control, *J. Chem. Ecol.,* 16, 3091, 1990.
13. Neuhauser, E.F., Durkin, P.R., Malecki, M.R., and Anatra, M., Comparative toxicity of ten organic chemicals to four earthworm species, *Comp. Biochem. Physiol.,* 83C, 197, 1986.
14. Everts, J.W., Aukema, B., Mullié, W.C., Van Gemerden, A., Rottier, A., Van Katz, R., and Van Gestel, C.A.M., Exposure of the ground dwelling spider *Oedothorax apicatus* (Blackwall) (Erigonidae) to spray and residues of deltamethrin, *Arch. Environ. Contam. Toxicol.,* 20, 13, 1991.
15. Van Gestel, C.A.M. and Ma, W.-C., An approach to quantitative structure-activity relationships in terrestrial ecotoxicology: earthworm toxicity studies, *Chemosphere,* 21, 1023, 1990.
16. Van Gestel, C.A.M., Ma, W.-C., and Smit, C.E., Development of QSAR's in terrestrial ecotoxicology: earthworm toxicity and soil sorption of chlorophenols, chlorobenzenes and dichloroaniline, *Sci. Total Environ.,* 109/110, 589, 1991.

17. Harris, C.R., Influence of soil type on the activity of insecticides in soil, *J. Econ. Entomol.*, 59, 1221, 1966.
18. Campbell, W.V., Mount, D.A., and Heming, B.S., Influence of organic matter of soils on insecticidal control of the wireworm *Melanotus communis, J. Econ. Entomol.*, 64, 41, 1971.
19. Hoestra, H., Effect of benomyl on the potato nematode, *Heterodera rostochiensis, Neth. J. Plant Pathol.*, 82, 17, 1976.
20. Jäggy, A. and Streit, B., Toxic effects of soluble copper on *Octolasium cyaneum* Sav. (Lumbricidae), *Rev. Suisse Zool.*, 89, 881, 1982.
21. Van Gestel, C.A.M., The influence of soil characteristics on the toxicity of chemicals for earthworms; a review, in *Ecotoxicology of Earthworms,* Greig-Smith, P.W., Becker, H., Edwards, P.J., and Heimbach, F., Eds., Intercept Press, Andover, Hants, 1992, 44.
22. Wohlgemuth, D., Kratz, W., and Weigmann, G., The influence of soil characteristics on the toxicity of an environmental chemical (cadmium) on the newly developed mono-species test with the springtail *Folsomia candida* (Willem), in *Environmental Contamination,* Barcelo, J., Ed., 4th International Conference, Barcelona, October 1990, CEP Consultants Ltd., Edinburgh, 1990, 260.
23. OECD, *Guideline for Testing of Chemicals. No. 207. Earthworm Acute Toxicity Tests,* 1984.
24. Tranvik, L. and Eijsackers, H., On the advantage of *Folsomia fimetarioides* over *Isotomiella minor* (Collembola) in metal polluted soils, *Oecologia,* 80, 195, 1989.
25. Hassan, S.A., Standard methods to test the side-effects of pesticides on natural enemies of insects and mites developed by the IOBC/WRPS Working Group "Pesticides and Beneficial Organisms", *Bull. OEPP/EPPO Bull.*, 15, 214, 1985.
26. Berhin, F., Houba, C., and Remada, J., Cadmium toxicity and accumulation by *Tetrahymena pyriformis* in contaminated river waters, *Environ. Pollut. (Ser. A)*, 35, 315, 1984.
27. Nyberg, D. and Bishop, P., High levels of phenotypic variability of metal and temperature tolerance in *Paramecium, Evolution,* 37, 341, 1983.
28. Janssen, M.P.M., unpublished data, 1992.
29. Sturhan, D., Influence of heavy metals and other elements on soil nematodes, *Rev. Nématol.*, 9, 311, 1986.
30. Haight, M., Mudry, T., and Pasternak, J., Toxicity of seven heavy metals on *Panagrellus silusiae*: the efficacy of the free-living nematode as an in vivo toxicological bioassay, *Nematologica,* 28, 1, 1982.
31. Van Kessel, W.H.M., Brocades Zaalberg, R.W., and Seinen, W., Testing environmental pollutants on soil organisms: a simple assay to investigate the toxicity of environmental pollutants on soil organisms using cadmium chloride and nematodes, *Ecotoxicol. Environ. Saf.*, 18, 181, 1989.
32. Kammenga, J.E., unpublished data, 1992.
33. Hopkin, S.P., *Ecophysiology of Metals in Terrestrial Invertebrates,* Elsevier Applied Science, London, 1989.
34. Dallinger, R. and Wieser, W., The flow of copper through a terrestrial food chain. I. Copper and nutrition in isopods, *Oecologia,* 30, 253, 1977.
35. Van Capelleveen, H.E., Ecotoxicity of Heavy Metals for Terrestrial Isopods, Ph.D. thesis, Vrije Universiteit, Amsterdam, 1987.

36. Hopkin, S.P., Species-specific differences in the net assimilation of zinc, cadmium, lead, copper and iron by the terrestrial isopods *Oniscus asellus* and *Porcellio scaber, J. Appl. Ecol.,* 27, 460, 1990.

37. Van Straalen, N.M. and Verweij, R.A., Effects of benzo(a)pyrene on food assimilation and growth efficiency in *Porcellio scaber* (Isopoda), *Bull. Environ. Contam. Toxicol.,* 46, 134, 1991.

38. Eijsackers, H., Side effects of the herbicide 2,4,5-T affecting the isopod *Philoscia muscorum* Scopoli, *Z. Angew. Entomol.,* 87, 28, 1978.

39. Crommentuijn, T., unpublished data, 1992.

40. Hopkins, S.P., Watson, K., Martin, M.H., and Mould, M.L., The assimilation of metals by *Lithobius variegatus* and *Glomeris marginata* (Chilipoda; Diplopoda), *Bijdr. Dierk.,* 55, 88, 1985.

41. Denneman, C.A.J. and Van Straalen, N.M., The toxicity of lead and copper in reproduction toxicity tests using the oribatid mite *Platynothrus peltifer, Pedobiologia,* 35, 305, 1991.

42. Van Straalen, N.M., Schobben, J.H.M., and De Goede, R.G., Population consequences of cadmium toxicity in soil microarthropods, *Ecotoxicol. Environ. Saf.,* 17, 190, 1989.

43. Bengtsson, G., Gunnarson, T., and Rundgren, S., Influence of metals on reproduction, mortality and population growth in *Onychiurus armatus* (Collembola), *J. Appl. Ecol.,* 22, 967, 1985.

44. Thompson, A.R. and Gore, F.L., Toxicity of twenty-nine insecticides to *Folsomia candida*: laboratory studies, *J. Econ. Entomol.,* 65, 1255, 1972.

45. Subagja, J. and Snider, R.J., The side effects of the herbicides atrazine and paraquat upon *Folsomia candida* and *Tullbergia granulata* (Insecta, Collembola), *Pedobiologia,* 22, 141, 1981.

46. Tomlin, A.D., Toxicity of soil applications of the fungicide benomyl, and two analogues to three species of Collembola, *Can. Entomol.,* 109, 1619, 1977.

47. Eijsackers, H., Side effects of the herbicide 2,4,5-T affecting mobility and mortality of the springtail *Onychiurus quadriocellatus* Gisin (Collembola), *Z. Angew. Entomol.,* 86, 349, 1978.

48. BBA, *Reproductive Toxicity for Folsomia candida (Willem), Collembola, in Artificial Soil,* Biologische Bundesanstalt, Berlin, BBA-CP-411-Ri, 1990.

49. Crommentuijn, T., Brils, J., and Van Straalen, N.M., Influence of cadmium on life-history characteristics of Folsomia candida (Willem) in an artificial soil substrate, *Ecotoxicol. Environ. Saf.,* in press.

50. Römbke, J., *Enchytraeus albidus* (Enchytraeidae, Oligochaeta) as a test organism in terrestrial laboratory systems, *Arch. Toxicol.,* 13, 402, 1989.

51. Westheide, W., Bethke-Beilfuss, D., and Gebbe, J., Effects of benomyl on reproduction and population structure of enchytraeid oligochaetes (Annelida) — sublethal tests on agar and soil, *Comp. Biochem. Physiol.,* 100C, 221, 1991.

52. EEC, *EEC Directive 79/831. Annex V. Part C. Methods for determination of ecotoxicity. Level I. C(II)4: Toxicity for earthworms. Artificial soil test,* DG XI/128/82, 1985.

53. Ferrière, G., Fayolle, L., and Bouché, M.B., Un nouvel outil, essentiel pour l'écophysiologie et l'écotoxicologie: l'élevage des lombriciens en sol artificiel, *Pedobiologia,* 22, 196, 1981.

54. Van Gestel, C.A.M., Van Dis, W.A., Van Breemen, E.M., and Sparenburg, P.M., Development of a standardized reproduction toxicity test with the earthworm species *Eisenia fetida andrei* using copper, pentachlorophenol, and 2,4-dichloroaniline, *Ecotoxicol. Environ. Saf.,* 18, 305, 1989.

55. Kokta, C., A laboratory test on sublethal effects of pesticides on *Eisenia fetida,* in *Ecotoxicology of Earthworms,* Greig-Smith, P.W., Becker, H., Edwards, P.J., and Heimbach, F., Eds., Intercept Press, Andover, Hants, 1992, 213.

56. Russell, L.K., DeHaven, J.I., and Botts, R.P., Toxic effects of cadmium on the garden snail (*Helix aspersa), Bull. Environ. Contam. Toxicol.,* 26, 634, 1981.

57. Marigomez, J.A., Angulo, E., and Saez, V., Feeding and growth responses to copper, zinc, mercury and lead in the terrestrial gastropod *Arion ater* (Linné), *J. Moll. Stud.,* 52, 68, 1986.

58. Meincke, K.-F. and Schaller, K.-H., Uber die Brauchbarkeit der Weinbergschnecke (*Helix pomatia* L.) im Freiland als Indikator für die Belastung der Umwelt durch die Elemente Eisen, Zink und Blei, *Oecologia (Berlin),* 15, 393, 1974.

59. Moser, H. and Wieser, W., Copper and nutrition in *Helix pomatia* (L.), *Oecologia (Berlin),* 42, 241, 1979.

60. IOBC, Guidelines for testing the effects of pesticides on beneficials: short description of test methods, *WPRS Bull.,* 11, 1, 1988.

61. Croft, B.A., *Arthropod Biological Control Agents and Pesticides,* John Wiley & Sons, New York, 1990.

62. Samsøe-Petersen, L., Sequences of standard methods to test effects of chemicals to terrestrial arthropods, *Ecotoxicol. Environ. Saf.,* 19, 310, 1990.

63. OEPP/EPPO, *Draft Guideline for Evaluating the Hazards of Pesticides to Honey Bees, Apis mellifera L.,* 1991.

64. Eijsackers, H., Litter fragmentation by isopods as affected by herbicide application, *Neth. J. Zool.,* 41, 277, 1991.

65. Van Wensem, J., A terrestrial micro-ecosystem for measuring effects of pollutants on isopod-mediated litter decomposition, *Hydrobiologia,* 188/189, 507, 1989.

66. Van Wensem, J., Jagers op Akkerhuis, G.A.J.M., and Van Straalen, N.M., Effects of the fungicide triphenyltin hydroxide on soil fauna mediated litter decomposition, *Pestic. Sci.,* 32, 307, 1991.

67. Eijsackers, H. and Van de Bund, C.F., Effects on soil fauna, in *Interactions Between Herbicides and the Soil,* Hance, R.J., Ed., Academic Press, London, 1980, 255.

68. Inglesfield, C., Pyrethroids and terrestrial non-target organisms, *Pestic. Sci.,* 27, 387, 1989.

69. Jepson, P.C., The temporal and spatial dynamics of pesticide side-effects on non-target invertebrates, in *Pesticides and Non-target Invertebrates,* Jepson, P.C., Ed., Intercept Ltd., Wimborne, 1989, 95.

70. Greig-Smith, P.W., Becker, H., Edwards, P.J., and Heimbach, F., Eds., *Ecotoxicology of Earthworms,* Intercept Press, Andover, Hants, 1992.

CHAPTER 15

Comparison of Effects of Two Pesticides on Soil Organisms in Laboratory Tests, Microcosms, and in the Field

J. Römbke, Th. Knacker, B. Förster, and A. Marcinkowski

TABLE OF CONTENTS

ABSTRACT

The objective of the study presented here was to compare ecotoxicological data obtained from the field with data obtained from laboratory test systems and microcosms. The intention of the comparison was to find out whether microcosms can be considered as a useful tool to obtain ecologically relevant data when investigating the ecotoxicological effects of chemical stressors. Two pesticides, the insecticide E 605 (active ingredient: Parathion) and the herbicide Ustinex (active ingredients: Amitrole and Diuron) were used as chemical stressors. Each pesticide was applied in two concentration levels on a grassland field site and on soil-core microcosms extracted from the grassland. In addition, the pesticides were tested by using laboratory test systems. Test parameters were the abundance and biomass of earthworms (Lumbricidae) and Enchytraeidae. With regard to the parameter observed, the chemical applied and the organizational levels investigated, each oligachaete family reacted to each of the pesticides in a specific manner. However, the ability to mimic effects of pesticides observed under field conditions was much better when using microcosms compared to results obtained from laboratory test systems.

I. INTRODUCTION

National and international legislations on plant protection require data to evaluate the effects of agrochemicals to terrestrial ecosystems. Usually, this evaluation is based on results derived from laboratory test systems, though these tests are not representative for ecosystems.[1] However, field tests which can represent ecosystems are difficult to reproduce, costly, and time consuming. An alternative approach is the use of microcosms or model ecosystems.[2]

The objective of the study presented here was to compare ecotoxicological data obtained from the field with data obtained from laboratory test systems and microcosms which have been developed at Battelle.[3] The intention of the comparison was to find out whether or not microcosms can be considered as a useful tool to obtain ecologically relevant data when investigating the ecotoxicological effects of chemical stressors. As such stressors the herbicide Ustinex, containing the active ingredients Amitrole and Diuron, and the insecticide E 605, containing the active ingredient Parathion, were selected. The abundance and biomass of enchytraeids and earthworms were chosen to quantify the possible ecotoxicological effects on all three differently complex organizational levels: laboratory, microcosms, and field. The Oligochaeta were chosen as representative and important members of the soil fauna which are typical for many soils of the temperate zones.[4]

II. MATERIALS AND METHODS

The laboratory tests with earthworms were performed according to the OECD Guideline No. 207 using the compost worm *Eisenia fetida*.[5] For enchytraeids, a laboratory test system developed at Battelle was performed using the species *Enchytraeus albidus*.[6] In both tests the parameter mortality and the effects on biomass, reproduction, and changes of behavior were determined.

The soil cores (= microcosms) with a depth of 60 cm and a diameter of 18 cm were extracted from the field without affecting the layering of the soil. The soil cores, encased in high-density polyethylene tubes, were transported to the greenhouse and placed in temperature-controlled wooden containers. A detailed description of the extraction and storage of these microcosms is given by Knacker et al.[3]

The field study was performed on the same grassland site from where the soil cores were extracted. The grassland was a *Arrhenathereum elatioris* association with rows of apple trees at the boundaries located approximately 10 km north of Frankfurt a. M. The mean annual temperature was 9.8°C, the average precipitation was 648 mm. The site was divided into four treated plots and a control plot, each covering an area of 25 m^2. The site was within a distance of approximately 10 m to the plot where the soil cores were extracted. The soil was a brown soil on loess. The control plot and the soil core plot showed a pH value of 5.8 and an organic matter content of 2.4%, whereas the plots treated with chemicals revealed higher pH values of 6.5 to 7.7 and organic matter contents of 2.7 to 3.1%.

This variation of the physicochemical properties of the soil was accompanied by differences in the enchytraeid species composition which were not observed in the microcosms. With regard to abundance, biomass, and species composition, the oligochaete populations were within the normal range found for other European grasslands.[7] The dominant enchytraeid species, *Achaeta microcosmi*, was described from this site for the first time.[8]

Each month and for each treatment and the control, one earthworm and one enchytraeid sample was collected from a single microcosm. The destructive sampling was conducted by using a split-core extractor for the enchytraeids samples. The remaining soil of the microcosm was then hand sorted to collect the earthworms. For microcosms there were no replicate samples per sampling date.

In the field, three replicate earthworm and enchytraeid samples were taken per plot. For earthworms the method of hand sorting, followed by a Formol treatment (0.3% formalin solution applied three times after intervals of 10 min) was used. For enchytraeids, a split-core extractor was used to collect the samples. The enchytraeids were extracted from the soil by using the wet-funnel method.

Table 1. Amount of Test Chemicals Applied on the Microcosms and the Field (A) milliliters of E 605 and grams of Ustinex per Hectare (B) milligrams of E 605 and milligrams of Ustinex per kg Dry Weight of Soil

	E 605		Ustinex	
Plot marking	E1	E5	U1	U5
A	610	3000	1000	5000
B	0.6	3.0	10	50

Samples were taken starting in April 1989, i.e., 3 months prior to application, until September 1989, and additionally in May 1990 1 year after application.

The insecticide E 605 forte (active ingredient: Parathion) and the herbicide Ustinex (active ingredients: Amitrole and Diuron), which are commonly used in apple orchards, were selected as test chemicals. In a laboratory study using sandy loam comparable to the soil of the grassland described above, Parathion showed a half-life of about 4 weeks.[9] Amitrole persists in soil for 2 to 4 weeks and diuron for 4 to 8 months.[10]

E 605 and Ustinex were applied to the field plots and microcosms in the highest recommended application rate and in a fivefold higher concentration once in June 1989 (Table 1). These concentrations were transferred to mg/kg using the equation: 1000 g/ha equals 1 mg/kg. In the field the tests chemicals were applied using a parcel sprayer. In the greenhouse the chemicals were applied on top of the microcosms using micropipettes. For each of the four treatments and the control, one field plot and eight microcosms, respectively, were used. E 605 caused no visible effect on the vegetation. Ustinex, however, eliminated nearly all plants and caused an increased soil moisture both in the field and in the microcosms (Figure 1).[11]

III. RESULTS AND DISCUSSION

A. Population Dynamics in the Microcosms and the Field

One purpose of this study was to find out whether or not the type of microcosm described in this report could be a useful tool to predict the ecotoxicological effects of chemicals on Oligochaeta living in natural environments. To answer this question it was necessary to demonstrate the similarity of the population dynamics in the field and in the microcosms.

Using the biomass as an example, it is shown in Figure 2 that in fact the population dynamics of enchytraeids in the field and microcosms were comparable, although in the microcosms the abundance and biomass were higher than in the field. It is assumed that this was caused by the high soil moisture of the soil cores in the greenhouse where additional watering was required to avoid harmful stress for the plants. Despite the fact that *F. galba* was always

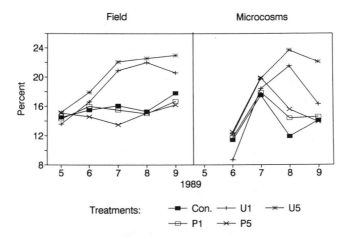

FIGURE 1. Soil moisture in the field and in the microcosms.

the dominant species when considering biomass, changes in the species composition occurred. Especially, the amount of *A. microcosmi* was much higher in the microcosms compared to the field. However, the ratio of juvenile to adult worms as well as the vertical distribution of the enchytraeids were not modified in the microcosms.[12]

In the soil cores the number of earthworms was slightly increased compared to the field, while the biomass decreased (Figure 3). The typical annual fluctuation of the population with a minimum in summer and a maximum in autumn

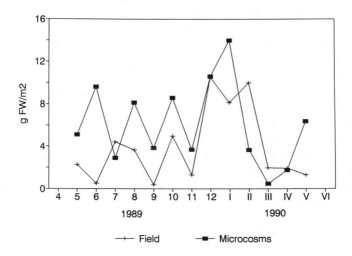

FIGURE 2. Biomass of the Enchytraeidae in the control plot of the field and in the untreated microcosms.

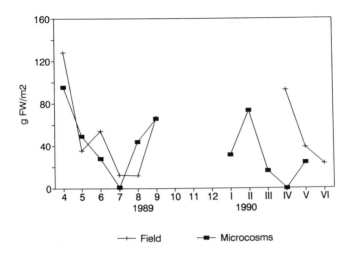

FIGURE 3. Biomass of the earthworms in the control plot of the field and in the untreated microcosms.

and spring was measured in the field as well as in microcosms. However, in comparison to the field the number of earthworm species decreased in the microcosms from 12 to 10. Especially, the large species *Lumbricus terrestris* and *Aporrectodea longa,* dominant in the field plots, were hardly found in the soil cores. Moreover, the individual weight of the worms was reduced in the microcosms by approximately a factor of two. The difference between microcosm and field was thought to be caused by the size of the soil cores and the limited amount of food for earthworms in the soil cores.

The standard variation of the replicate samples of enchytraeids in the field was on an average ±66%, which can be expected to be similar in the microcosms. Considering the variability of these data, which are in the same range as in other investigations,[13] the population dynamics of the worms in the microcosms showed excellent correlation with the development of the oligochaete cenosis in the field. As a result it can be stated that the specific microcosms design maintained abiotic and biotic conditions which allowed us to investigate members of the soil fauna for at least one vegetation period.

Based on these findings, it was expected that the ecotoxicological effects caused by the pesticides in the field and the microcosms were similar. First, however, the results obtained when applying the test materials on laboratory test systems are shown.

B. Laboratory Tests With Pesticides

In the laboratory tests, Ustinex caused neither effects on earthworms nor on enchytraeids up to a concentration of 1000 mg/kg. In the literature, however,

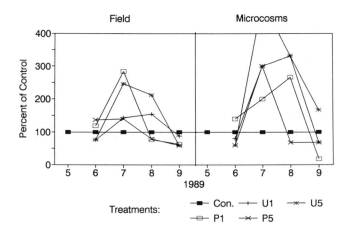

FIGURE 4. Effects of the test chemicals on the abundance of earthworms in the field and in the microcosms.

effects of the active ingredient Amitrole were reported: the biomass of the earthworm *A. caliginosa* was reduced under laboratory conditions at a concentration of 100 mg/kg.[14] The effects of Ustinex on Enchytraeidae were investigated in this study for the first time.

When applying E 605, LC_{50} values of 135 mg/kg for the earthworm *Eisenia fetida* and 124 mg/kg for the enchytraeid *Enchytraeus albidus* were determined. Moreover, effects on the biomass at approximately 100 mg/kg, on the reproduction and on the behavior at 10 mg/kg were detected. These findings were in agreement with those reported from other authors who measured an LC_{50} value of 34 mg/kg for *L. terrestris* and an LC_{50} value ranging between 100 and 1000 mg/kg for *Eisenia fetida*.[15] For the latter species, effects on biomass and reproduction were found at concentrations of 10 to 32 mg/kg.

Bethge-Beilfuss and Westheide[16] investigated the effects of Parathion on enchytraeids by using agar as a test substrate, which can be considered as an aqueous medium for enchytraeids. Dependent on the species selected and the test conditions chosen, the effects of Parathion were found to range between 0.21 and 21 mg/kg in soil.

C. Microcosm and Field Studies With Pesticides

Since all microcosms were extracted from within a small area of the grassland, the fluctuations of the soil characteristics in the microcosms were narrow. Consequently, the variation of the abundance for both the earthworms and enchytraeids (Figures 4 and 5) in the microcosms was small when comparing the data obtained from all treatments at the beginning of the investigation. In all treated soil cores, increased earthworm abundance and biomass were measured 1 month after application of the chemicals. This increase was lower in the

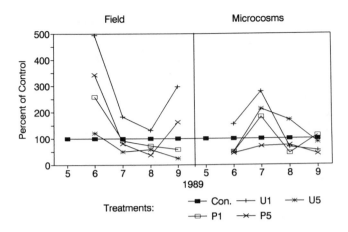

FIGURE 5. Effects of the test chemicals on the abundance of Enchytraeidae in the field and in the microcosms.

microcosms treated with E 605 than in those treated with Ustinex. In all soil cores a return to control levels or below was observed after 3 months (Figure 4).

Despite the variability of the soil properties in the field, Figure 4 shows that there were no obvious differences between the earthworm cenoses of the grassland plots before applying the chemicals. Immediately after application the number and biomass of the Lumbricidae increased for 2 months and returned to the level determined for the control plot. On the plot treated with the high E 605 concentration, the timely limited increase of the abundance was less pronounced than on the other plots.

The effects of the chemicals on the enchytraeids in the microcosms were comparable to those measured for the earthworms. Only in the microcosms treated with the high concentration of E 605, the number and biomass decreased (Figure 5).

When comparing the grassland plots before applying the pesticides the enchytraeid cenoses were rather different (Figure 5). In particular, their abundance on all treated plots was higher than on the control plot. Despite the difficulty of comparing the treated plots with the control, it seemed that there were no significant changes in number and biomass during the study period.

For Table 2, the single values derived from the microcosms on a monthly basis during the investigation period from June to September 1989 were averaged to calculate the mean abundance and biomass as well as the total species number for enchytraeids and earthworms. In Table 3, the triplicate values derived from the field were exhibited as in Table 2.

The abundance of earthworms was enhanced when microcosms were treated with Ustinex. The biomass increased when the recommended application rate was used for Ustinex and E 605, but decreased at the fivefold higher E 605

Table 2. Effects of the Test Chemicals Ustinex (U1 and U5) and E 605 (E1 and E5) on Earthworms and Enchytraeidae in the Microcosms

	U1	U5	E1	E5	Control
Earthworms					
Number of species	4	4	5	5	5
Mean abundance/m²	247	388	211	141	176
Mean biomass g/m²	47	39	54	22	34
Enchytraeidae					
Number of species	14	17	15	14	15
Mean abundance/m²	44,200	36,400	25,300	18,800	34,100
Mean biomass g/m²	7.7	11.7	7.1	2.8	6.1

Note: Average values based on four single samples taken between June and September 1989.

concentration (Table 2). In soil cores treated with Ustinex, juveniles of the genus *Octolasion* were most abundant while in all other microcosms juveniles of the genus *Aporrectodea* were dominant.[12]

Compared to the control the abundance of the Enchytraeidae was increased in the soil cores treated with the low Ustinex concentration, while the biomass was enhanced only in those microcosms treated with the high Ustinex concentration. The high E 605 concentration reduced the number and biomass of the enchytraeids (Table 2). Although *Achaeta microcosmi* was the dominant species in all microcosms concerning number and *F. galba* concerning biomass, changes in the dominance spectrum occurred in all treated soil cores except those treated with the low Ustinex concentration.[12]

In Table 3 it is shown that on average the biomass of the earthworms on the field plot treated with the low Ustinex concentration was increased, whereas no changes occurred on the other plots. The individual weight of most lumbricid species was enhanced on the Ustinex plots and reduced on the E 605 plots compared to the control. The weight increase of individual organisms was at least partly due to the enhanced amount of dead organic matter, which was available as food on the soil surface after application of the herbicide Ustinex,

Table 3. Effects of the Test Chemicals Ustinex (U1 and U5) and E 605 (E1 and E5) on Earthworms and Enchytraeidae in the Field

	U1	U5	E1	E5	Control
Earthworms					
Number of species	10	10	9	10	10
Mean abundance/m²	101	115	108	86	94
Mean biomass g/m²	48	40	32	29	36
Enchytraeidae					
Number of species	18	15	15	17	13
Mean abundance/m²	22,300	5,900	10,400	11,400	9,700
Mean biomass g/m²	5.2	0.9	2.6	2.6	2.2

Note: Average values based on four triplicate samples taken between June and September 1989.

whereas the reduction of biomass observed at the end of the investigation period (Figure 4) was probably caused by food shortages, since the herbicidal effect lasted for approximately half a year and did not allow any growth of plants. The population dynamics observed on the field plots treated with E 605 (Table 3, Figure 4) could be due to the reduction of epigean earthworm predators like carabid or staphylinid beetles, which were certainly affected by the insecticide.[17]

For Enchytraeidae, the abundance and biomass were increased on the plots treated with the low Ustinex and reduced on the plots treated with the high Ustinex concentration (Table 3). Based on average values covering the entire study period, the treatment with E 605 caused hardly any effect on the abundance and biomass of enchytraeids in the field (Table 3). However, on all treated plots the number of species was slightly higher than on the control plot. Additionally, both the high and low concentration of E 605 altered the ratio of juvenile to adult individuals.[12]

Other field investigations with the two pesticides are scarce (data from microcosms are not available). For example, Van Rhee[18] found small alterations of the behavior and reproduction of the earthworm *A. caliginosa* at much lower concentrations of Ustinex (1 to 5 kg/ha, corresponding to 1 to 5 mg/kg) than in laboratory tests.

A review of Haque and Pflugmacher[19] on the effects of Parathion to earthworms in the field revealed that some authors found effects,[20] while others, even at high application rates of 4.5 to 9 l/ha, could not detect any effect on the worms.[21] Unfortunately, the data basis of these studies is limited and in some cases appropriate methods have not been used.

A reduction of the number of enchytraeids, particularly juveniles, was measured in a loamy and sandy soil treated with Parathion under field conditions.[20] However, due to an inappropriate extraction method used by Weber,[20] which leads to underestimated numbers of enchytraeids, the results are difficult to compare with the results obtained in this study.

IV. CONCLUSIONS

In general, the Enchytraeidae were easier to collect and seemed to react slightly more sensitive to the agrochemicals than the earthworms. On the other hand, the determination to the species level and the evaluation of the biomass was more time consuming for enchytraeids than for earthworms.

To characterize ecotoxicological effects it seems that the number and biomass of the worms are sensitive parameters. The evaluation of the ecotoxicological effects on single species has not been completed yet. However, there are several indications that in both oligochaete families single species react specifically towards chemical stress.

Summarizing the results, it is important to remember that in comparison with the results of the laboratory tests the concentrations of the test chemicals applied in the microcosms and in the field were within the range where acute effects were not expected to occur.

In the microcosms, the high concentration of E 605 caused negative effects on the biomass of earthworms and enchytraeids, whereas the treatment with Ustinex enhanced the number and biomass of the Oligochaeta. In the field, the low Ustinex concentration produced effects corresponding to the results obtained from microcosms, while at the high concentration the number and biomass of enchytraeids were decreased and the number and biomass of earthworms remained unchanged.

The ability to mimic effects of pesticides observed under field conditions was much better when using microcosms compared to results obtained from laboratory test systems. However, the decrease of the enchytraeid population on the field plot treated with the high Ustinex concentration was not reflected by results derived from microcosms.

REFERENCES

1. Kimball, K.D. and Levin, S.A., Limitations of laboratory bioassays: the need for ecosystem-level testing, *BioScience,* 35, 165, 1985.
2. Gillett, J.W., The role of terrestrial microcosms and mesocosms in ecotoxicological research, in *Ecotoxicology: Problems and Approaches,* Levin, S.A. et al., Eds., Springer-Verlag, Berlin, 1989, 367.
3. Knacker, Th., Schallnass, H.-J., Marcinkowski, A., Förster, B., and Vincena, R., *Einsetzbarkeit seminatürlicher (terrestrischer) Systeme für die Bewertung der Umweltgefährlichkeit nach dem ChemG,* Report for the Umweltbundesamt (UBA), Berlin, FE 106 03 069, Battelle Institut, 1990.
4. Petersen, H. and Luxton, M., A comparative analysis of soil fauna populations and their role in decomposition processes, *Oikos,* 39, 287, 1982.
5. OECD, Guidelines for Testing of Chemicals, Paris, 1984.
6. Römbke, J., *Fortentwicklung eines Reproduktionstests an Bodenorganismen-En-chytraeen,* Report for the Umweltbundesamt (UBA), Berlin, FE 106 03 051/01, Battelle Institut, Teil A 36 S., Teil B 90 S., 1989.
7. Standen, V., Production and diversity of enchytraeids, earthworms and plants in fertilized hay meadow plots, *J. Appl. Ecol.,* 21, 293, 1984.
8. Heck, M. and Römbke, J., Two new species of Achaeta (Enchytraeidae, Oligochaeta) from meadow and pasture soils of Germany, *Zool. Scr.,* 20, 215, 1991.
9. *Agrochemicals Handbook,* The Royal Society of Chemistry, Thomas Graham House, Cambridge, 1987.
10. Harris, C.R., Laboratory studies on the persistence and biological activity of some insecticides in soil, *J. Econ. Entomol.,* 62, 1437, 1969.
11. Chalupsky, J., The influence of Zeazin 50 on Enchytraeidae (Oligochaeta) in an apple orchard soil, *Pedobiologia,* 33, 361, 1989.

12. Knacker, Th., Römbke, J., Förster, B., Marcinkowski, A., and Schallnass, H.-J., *Die Wirkung von Pestiziden auf die Mesofauna und den CO_2-Gas-Austausch eines terrestrischen Ökosystems,* Report for the Umweltbundesamt (UBA), Berlin, FE 106 03 069/01, Battelle Institut, 1992.

13. Nielsen, C.O., Studies on Enchytraeidae. 2. Field studies, *Nat. Jutl.,* 4, 1, 1955.

14. Martin, N.A., The effects of herbicides used on asparagus on the growth rate of the earthworm Allolobophora caliginosa, Proc. 35th N.Z. Weed and Pest Control Conf., 328, 1982.

15. Kokta, C., personal communication, 1990.

16. Bethge-Beilfuss, D. and Westheide, W., Subletale Schädigungen terrestrischer Enchytraeiden (Oligochaeta, Annelida) durch Pflanzenbehandlungsmittel: Veränderungen der Kokonproduktion und der Eifertilität, *Verh. Ges. Ökol.,* XVI, 417, 1987.

17. Boller, E., Bigler, F., Bieri, M., Häni, F., and Stäubli, A., Nebenwirkungen von Pestiziden auf die Nützlingsfauna landwirtschaftlicher Kulturen, *Schweiz. Landwirtsch. Forsch.,* 28, 3, 1989.

18. Van Rhee, J.A., Effects of soil pollution on earthworms, *Pedobiologia,* 17, 201, 1977.

19. Haque, A. and Pflugmacher, J., Einflüsse von Pflanzenschutzmitteln auf Regenwürmer, *Ber. Landwirtsch.,* 198, 176, 1985.

20. Weber, G., Die Makrofauna leichter und schwerer Ackerböden und ihre Beeinflussung durch Pflanzenschutzmittel, *Z. Pflanzenernähr. Düng.,* 61, 107, 1953.

21. Edwards, C.A., Thompson, A.R., and Lofty, J.R., Changes in soil invertebrate populations caused by some organophosphate insecticides, Proc. 4th Brit. Insecticide Fungicide Conf., 48, 1967.

CHAPTER 16

Species-Specific Sensitivity Differences of Earthworms to Pesticides in Laboratory Tests

H. Kula

TABLE OF CONTENTS

ABSTRACT

Acute and sublethal effects of the two insecticides parathion and propoxur on earthworms were studied in an artificial soil substrate in the laboratory. The epigeic species *Eisenia fetida* was compared with the endogeic mineral soil species *Aporrectodea caliginosa* and *Allolobophora chlorotica* and the anecic species *Aporrectodea longa*. In the acute toxicity tests the anecic and endogeic earthworm species were more susceptible to the test substances than *Eisenia fetida*. Factors of sensitivity differences ranged up to 80. In reproduction toxicity tests the sublethal parameters body weight development, sexual maturation, cocoon production and survival of juveniles were investigated in *Eisenia fetida*. Significant effects on reproductional success were observed at pesticide concentrations corresponding to normal field application rates. In reproduction toxicity tests with *Aporrectodea caliginosa* and *Allolobophora chlorotica* sexual maturation and survival of juveniles could not be investigated, but a significant reduction of cocoon production was observed also. Because of its short generation time and high reproductive potential, *Eisenia fetida* was more appropriate as a test organism than the other earthworm species, since the whole life-cycle could be studied in laboratory tests. In hazard assessment the observed sensitivity differences have to be taken into account when establishing adequate uncertainty factors.

I. INTRODUCTION

An acute toxicity test with *Eisenia fetida* (Savigny, 1826) according to the Organization for Economic Cooperation and Development (OECD) guideline No. 207[1] is generally used to determine the toxicity of pesticides to earthworms in the laboratory. *E. fetida* is well suited for this purpose as it is easy to handle in the laboratory. Its short generation time and high reproductive potential[2,3] ensure rearing of sufficient numbers of test animals of defined age in synchronized cultures under standardized laboratory conditions. However, in Germany the distribution of *E. fetida* is restricted to places rich in organic matter, such as compost and dung heaps.

Haque and Ebing[4] and Heimbach[5] demonstrated different sensitivity towards pesticides between *E. fetida* and *Lumbricus terrestris* L., 1758. In these studies, however, different soil substrates were used for the two species. As the biological activity of a pesticide among others depends on chemical properties of the soil,[6] such comparisons are difficult to interpret. For this reason, in the present study artificial soil according to OECD guideline No. 207 was used as the test substrate for all species.

In order to investigate possible sensitivity differences between the epigeic species *E. fetida* and indigenous earthworm species with different ecological strategies,[7] the endogeic mineral soil species *Aporrectodea caliginosa* (Savigny,

1826) and *Allolobophora chlorotica* (Savigny, 1826) and the anecic species *Aporrectodea longa* (Ude, 1885) were tested in acute toxicity and reproduction toxicity tests.

II. MATERIALS AND METHODS

Two pesticides were tested using two different experimental approaches; the organophosphorous insecticide E 605 forte (active ingredient: parathion) and the carbamate insecticide Unden flüssig (active ingredient: propoxur). Test substances (formulated products) were applied as aqueous solutions. The test substrate was an artificial soil,[1] consisting of 10% sphagnum peat, 69% fine sand, 20% kaolin clay, and about 1% calcium carbonate with a moisture content of 35% of dry weight of the soil.

A. Acute Toxicity Test

Acute toxicity tests according to OECD guideline No. 207[1] were conducted with adult, laboratory-bred *E. fetida* and adult, field-captured *A. caliginosa, A. longa,* and *Allolobophora chlorotica.* Laboratory-bred juveniles (age 4 to 6 weeks) of *Aporrectodea caliginosa* and *Allolobophora chlorotica* were also tested, because field studies with parathion and propoxur showed higher decreases in juvenile than in adult abundance. This might be explained by a higher susceptibility of juvenile life stages. Reduced reproduction of adults is another explanation.

After a preincubation period of 1 d in moist artificial soil, test animals were introduced into 1-l glass jars containing moist artificial soil (equivalent to 0.5 kg dry weight) into which different amounts of the test substance had been mixed homogeneously. In tests with juveniles, smaller glass jars with 0.1 kg (dry weight) artificial soil were used. Tests with *E. fetida* were incubated at 20 ± 2°C; tests with the three indigenous species were incubated at 15 ± 2°C or 10 ± 1°C. Each concentration level was tested with three replicates using ten animals (*E. fetida*) or three to five animals (indigenous species) per replicate. Test duration was 14 d. LC_{50} values were calculated by probit analysis.[8]

B. Reproduction Toxicity Tests

In reproduction toxicity tests, test substances were sprayed homogeneously on the soil surface (soil surface contamination, SC) of the test boxes. The recommended field application rates were used in order to investigate sublethal effects, especially on reproduction, under or near field conditions. Parathion was tested with an application rate of 210 ml/ha and a tenfold overdose. Propoxur was tested with an application rate of 900 ml/ha, a fivefold and a tenfold overdose. As a variation of this experimental design, the amount of test substance

according to the soil surface of the test box was mixed into the soil homogeneously (total soil contamination, TC, only in tests with *E. fetida*). For practical
reasons the amount of water used to apply the test substance was 20 times the
normal rate used in agricultural applications. Assuming an average bulk density
of natural soils of 1.5 g/ccm^3 and an even distribution of the pesticide in the
upper 2.5 cm of the soil, the onefold application rates correspond to a soil
concentration of 0.56 mg/kg for parathion and 2.40 mg/kg for propoxur.

In the reproduction toxicity test, 20 *E. fetida* were used per test box (surface
area: 364 cm^2) which contained 2.0 kg (dry weight) artificial soil. With *Aporrectodea caliginosa* and *Allolobophora chlorotica*, six test animals were introduced in smaller test boxes (surface area: 200 cm^2) with 0.5 kg (dry weight)
artificial soil. Finely ground cattle manure (0.5 g per worm per week) as food
source was spread on the soil surface. Test boxes with *E. fetida* were kept at
20 ± 2°C in a light to dark cycle of 16:8 h, test boxes with *Aporrectodea
caliginosa* and *Allolobophora chlorotica* were kept at 10 ± 1°C in darkness.
The test duration was 6 to 10 weeks. The numbers of juveniles were obtained
by hand sorting, cocoon numbers by washing the soil substrate through a sieve
with a mesh size of 1 mm.

Each treatment was replicated four times. Tukey's multiple *t*-test was used
to calculate significant differences ($p < 0.05$).

III. RESULTS

A. Acute Toxicity Tests

Differences in mortality were observed between *E. fetida* and *Aporrectodea
caliginosa, A. longa,* and *Allolobophora chlorotica* (Table 1). Weight development during the test period indicated sublethal effects. At low pesticide concentrations without mortality, earthworms lost up to 25% of live weight in the
course of the test, whereas control animals kept their initial weight.

B. Reproduction Toxicity Tests

In experiments with *E. fetida* (Table 2), no mortality was observed, not
even in the tenfold overdose treatment. With parathion, the control had 2.5 times
and 3.6 times more juveniles than test boxes with soil contamination, which had
been sprayed with a normal rate or a tenfold rate, respectively. These differences
were significant, whereas test boxes with total soil contamination did not differ
from the control. Regarding the number of cocoons, all treatments showed
significant lower cocoon production compared to the control.

With propoxur, significant differences in the numbers of juveniles between
control and both treatments were found. Cocoon production, in contrast, seemed
to be unaffected. In further experiments with only ten test animals and less soil
substrate, though, cocoon production also was significantly reduced.

Table 1. LC$_{50}$ (mg Formulated Pesticide/kg Dry Weight Soil Substrate) of Parathion and Propoxur For Different Earthworm Species (Acute Toxicity Test According to OECD Guideline No. 207)

Species	Parathion	Propoxur
Eisenia fetida		
Adults	554 (3)	291 (1)
Juveniles	296 (3)	153 (1)
Aporrectodea caliginosa		
Adults	232 (3)	6.8 (3)
Juveniles	213 (3)	8.5 (1)
Allolobophora chlorotica		
Adults	411 (2)	54 (1)
Juveniles	143 (1)	—
Aporrectodea longa		
Adults	119 (1)	3.8 (1)

Note: Numbers in parenthesis indicate the number of tests (means calculated by log-values).

In experiments with *Aporrectodea caliginosa* and *Allolobophora chlorotica,* some test animals died in the overdose treatments with propoxur, as could be expected from results of the acute toxicity test. Because of the low reproduction rate and the long cocoon development in these species,[9,10] no juveniles were found. Therefore, only weight development and cocoon production could be used as test parameters (Tables 3 and 4).

A reproduction toxicity test was run with juveniles of *E. fetida* to get information on the influence of pesticides on sexual development (Table 5). In the control group, animals reached sexual maturity very quickly and had produced cocoons after a test duration of 6 weeks. Only one hatched juvenile of the next generation was found in one of the replicates.

Table 2. Reproduction Toxicity Test With *E. fetida*

Treatment	Mortality of adults (%)	Body weight change (%)	Number of juveniles	Number of cocoons[a]
Parathion[b]				
Control	0	+21.5 ± 7.0	127 ± 27	57 ± 10
Parathion SC 1×	0	+22.4 ± 3.5	52 ± 30*	35 ± 9*
Parathion SC 10×	0	+18.6 ± 8.1	35 ± 7*	32 ± 7*
Parathion TC 1×	0	+22.9 ± 9.7	89 ± 42	29 ± 3*
Propoxur[c]				
Control	0	+26.5 ± 4.5	305 ± 36	37 ± 6
Propoxur SC 1×	0	+35.4 ± 13.7	150 ± 48*	52 ± 14
Propoxur TC 1×	0	+25.8 ± 12.1	188 ± 29*	38 ± 5

Note: 20 test animals per replicate, 4 replicates per treatment; SC 1× and SC 10×: soil surface contamination; TC 1×: total soil contamination.

[a] Cocoons not yet hatched. [b]Test duration: 6 weeks. [c]Test duration: 8 weeks.
* Significant difference ($p < 0.05$) against control, Tukey's multiple *t*-test.

Table 3. Reproduction Toxicity Test With *A. caliginosa*

Treatment	Mortality (%)	Body weight change (%)[a]	Number of cocoons
Control	0	−25.1 ± 2.3	27.0 ± 2.9
Parathion SC 1×	0	−17.9 ± 2.9	8.5 ± 6.0*
Parathion SC 10×	0	−32.1 ± 3.0	3.0 ± 2.3*
Propoxur SC 1×	0	−51.6 ± 8.1*	2.0 ± 2.0*
Propoxur SC 10×	100	—	0.5 ± 1.0*

Note: Six test animals per replicate, four replicates per treatment; test duration: 10 weeks.

[a] Calculated for survivors.
*   Significant difference ($p < 0.05$) against control, Tukey's multiple *t*-test.

Table 4. Reproduction Toxicity Test With *A. chlorotica*

Treatment	Mortality (%)	Body weight change (%)[a]	Number of cocoons
Control	0	−23.9 ± 3.9	9.3 ± 2.8
Parathion SC 1×	0	−19.3 ± 3.7	8.0 ± 1.4
Parathion SC 10×	0	−25.4 ± 4.6	6.3 ± 3.3
Propoxur SC 1×	0	−27.8 ± 3.3	2.3 ± 0.5*
Propoxur SC 5×	25	−37.4 ± 6.3*	2.0 ± 0*
Propoxur SC 10×	50	−45.1 ± 9.5*	0.5 ± 1.0*

Note: Six test animals per replicate, four replicates per treatment; test duration: 8 weeks.

[a] Calculated for survivors.
*   Significant difference ($p < 0.05$) against control, Tukey's multiple *t*-test.

Table 5. Effects of Parathion and Propoxur on the Sexual Development of Juvenile *E. fetida*

Treatment	Mortality (%)	Clitellate animals (%)	Body weight change (%)[a]	Number of cocoons
Control	0	92.5 ± 9.6	+3178 ± 513	13.8 ± 3.3
Parathion SC 1×	0	92.5 ± 5.0	+3490 ± 737	16.5 ± 5.0
Parathion SC 10×	0	72.5 ± 5.0*	+3283 ± 684	5.3 ± 2.8*
Propoxur SC 1×	7.5	44.8 ± 29.3*	+2388 ± 406*	0*
Propoxur SC 10×	27.5	0*	+443 ± 186*	0*

Note: Ten 2-week old juveniles per replicate, four replicates per treatment; test duration: 6 weeks.

[a] Calculated for survivors.
*   Significant difference ($p < 0.05$) against control, Tukey's multiple *t*-test.

Table 6. Effects of Parathion on Juveniles of *A. caliginosa* and *A. chlorotica*

	Mortality (%)	Body weight change (%)[a]
Aporrectodea caliginosa		
Control	0	+334 ± 19
Parathion SC 1×	0	+316 ± 54
Parathion SC 10×	6.2	+195 ± 46*
Parathion TC 1×	0	+346 ± 22
Allolobophora chlorotica		
Control	0	+278 ± 38
Parathion SC 1×	0	+217 ± 39
Parathion SC 10×	6.2	+237 ± 72
Parathion TC 1×	12.5	+319 ± 7

Note: Four 4- to 6-week old juveniles per replicate, four replicates per treatment; test duration: 6 weeks.

[a] Calculated for survivors.
* Significant difference ($p < 0.05$) against control, Tukey's multiple *t*-test.

Propoxur appeared to be very toxic to juvenile *E. fetida*, because mortality was observed in both treatments and sexual maturation was retarded (SC 1×) or absolute suppressed (SC 10×).

With juveniles of *Aporrectodea caliginosa* and *Allolobophora chlorotica*, the influence of parathion on mortality and weight development was investigated (Table 6). Mortality was observed in both test species, but weight development showed significant differences only for the tenfold overdose treatment in *Aporrectodea caliginosa*.

IV. DISCUSSION

Comparing acute toxicity of organic chemicals to different earthworm species, Neuhauser et al.[11] detected only little sensitivity differences among the test species. Pizl[12] investigated acute toxicity of some herbicides towards earthworms and did not find significant differences.

In the study presented here, the three indigenous species were found to be more susceptible to the tested insecticides than *E. fetida*. Especially, propoxur turned out to be very toxic to them in acute toxicity tests. Heimbach[5] assumed that sensitivity differences among various earthworm species are sufficiently taken into consideration with a factor of 2 to 3. Reviewing the literature, Van Gestel[13] concluded that sensitivity differences of a factor of 10 have to be considered as normal. In contrast to this, results with propoxur indicated factors ranging between 40 (*A. caliginosa*) and 80 (*A. longa*) when compared to *E. fetida*. For this reason, results with propoxur do not fit into the regression equation ($y = 0.83x + 0.27$; correlation coefficient $= 0.81$) proposed by Heimbach[5] when comparing LC_{50} of *E. fetida* and *L. terrestris*. This does not mean that *E. fetida* is not appropriate for laboratory testing as stated by Martin,[14] but emphasizes the importance of reasonable uncertainty factors when extrapolating from laboratory results to field conditions in hazard assessment. The suitability of *E. fetida* as a representative test organism has been demonstrated in several studies.[5,15-17]

Reproduction toxicity tests provide more ecologically relevant information, since effects on weight development and reproduction can be evaluated.[18] In the present study, the information obtained with reproduction toxicity tests turned out to be valuable for the understanding of data from field studies which usually show a high variability, and therefore are difficult to interpret.[19] Another advantage can be seen in the use of field application rates and an application mode near to practical conditions, thus improving ecological relevance of the test.

Significant effects on reproductional success could be demonstrated at these concentrations. Propoxur was very toxic, suppressing the sexual maturation of juvenile *E. fetida* as well as the cocoon production of all species tested. However, the cocoon production in the controls of experiments with *A. caliginosa* and *Allolobophora chlorotica* was suboptimal. Lofs-Holmin[20] reported about two cocoons per worm per week for these species. In the present study, *Aporrectodea caliginosa*, for example, produced approximately 0.5 cocoons per worm per week only. Low activity and/or quiescence, which is also indicated by body weight loss, might be an explanation for this. Similar experiments with *A. caliginosa* in natural soil showed the same results. Thus a negative influence of the artificial soil substrate on reproduction can be excluded.

Body weight development turned out to be a rather unsensible parameter, since no significant differences could be found in *E. fetida*. In mineral soil species, significant differences occurred with propoxur only, which caused severe overall effects. When testing parathion, Van Gestel et al.[21] and Römbke,[22] on the contrary, found weight changes to be even more sensitive than cocoon production. Further investigation of this exception, so far occurring only with parathion, is necessary.

It can be concluded that *E. fetida* is more appropriate as a test organism in reproduction toxicity tests than the tested indigenous mineral soil species. Because of its high reproductive potential[2,3] in contrast to other species, besides weight development, cocoon production and especially the number of surviving juveniles can be evaluated. The last two parameters were found to be most sensitive in reproduction tests.[21]

In hazard assessment, the observed sensitivity differences have to be taken into account when establishing adequate uncertainty factors.

REFERENCES

1. Anon., OECD-Guidelines for testing chemicals No. 207. Earthworm, acute toxicity tests. Adopted April 4, 1984.
2. Hartenstein, R., Neuhauser, E.F., and Kaplan, D.L., Reproductive potential of the earthworm Eisenia foetida, *Oecologia,* 43, 329, 1979.
3. Tomlin, A.D. and Miller, J.J., Development and fecundity of the manure worm, Eisenia foetida (Annelida: Lumbricidae), under laboratory conditions, in *Soil Biology as Related to Land Use Practices,* Dindal, D.L., Ed., Syracuse, 1980, 673.
4. Haque, A. and Ebing, W., Toxicity determination of pesticides to earthworms in the soil substrate, *Z. Pflanzenkr. Pflanzenschutz,* 90, 395, 1983.
5. Heimbach, F., Comparison of laboratory methods using Eisenia foetida and Lumbricus terrestris, for the assessment of the hazard of chemicals to earthworms, *Z. Pflanzenkr. Pflanzenschutz,* 92, 186, 1985.
6. Van Gestel, C.A.M. and Ma, W., An approach to quantitative structure-activity relationships (QSARs) in earthworm studies, *Chemosphere,* 21, 1023, 1990.
7. Bouché, M., Strategies lombriciennes, *Ecol. Bull.,* 25, 122, 1977.
8. Finney, D.J., *Statistical Methods in Biological Assay,* Griffin Press, London, 1971.
9. Satchell, J.E., Lumbricidae, in *Soil Biology,* Burges, A. and Raw, F., Eds., London, 1967, 259.
10. Meinhardt, U., Vergleichende Beobachtung zur Laboratoriumsbiologie einheimischer Regenwurmarten. II. Biologie der gezüchteten Arten, *Z. Angew. Zool.,* 61, 137, 1974.
11. Neuhauser, E.F., Durkin, P.R., Malecki, M.R., and Anatra, M., Comparative toxicity of ten organic chemicals to four earthworm species, *Comp. Biochem. Physiol.,* 83C, 197, 1986.
12. Pizl, V., Interactions between earthworms and herbicides. I. Toxicity of some herbicides to earthworms in laboratory tests, *Pedobiologia,* 32, 227, 1988.
13. Van Gestel, C.A.M., Earthworms in Ecotoxicology, Ph.D. thesis, Utrecht, 1991, chap. 11.
14. Martin, N.A., Toxicity of pesticides to Allolobophora caliginosa (Oligochaeta: Lumbricidae), *N.Z. J. Agric. Res.,* 29, 699, 1986.
15. Edwards, C.A., Development of a standardized laboratory method for assessing the toxicity of chemical substances to earthworms, Report Nr. EUR 8714 EN, Commission of the European Communities, Luxembourg, 1983.
16. Heimbach, F., Correlations between three different test methods for determining the toxicity of chemicals to earthworms, *Pestic. Sci.,* 15, 605, 1984.
17. Van Gestel, C.A.M., Van Dis, W.A., Van Breemen, E.M., and Sparenburg, P.M., Development of a standardized reproduction toxicity test with the earthworm Eisenia andrei using copper, pentachlorophenol and 2,4-Dichloroaniline, *Ecotoxicol. Environ. Saf.,* 18, 305, 1989.
18. Neuhauser, E.F. and Callahan, C.A., Growth and reproduction of the earthworm Eisenia fetida exposed to sublethal concentrations of organic chemicals, *Soil Biol. Biochem.,* 22, 175, 1990.
19. Kula, H. and Kokta, C., Side effects of selected pesticides on earthworms under laboratory and field conditions, *Soil Biol. Biochem.,* 24, 1711, 1992.

20. Lofs-Holmin, A., Reproduction and growth of common arable land and pasture species of earthworms (Lumbricidae) in laboratory cultures, *Swed. J. Agric. Res.,* 13, 31, 1982.

21. Van Gestel, C.A.M., Dirven-Van Bremen, E.M., Baerselman, R., Emans, H.J.B., Janssen, J.A.M., Postuma, R., and Van Vliet, P.J.M., Comparison of sublethal and lethal criteria for nine different chemicals in standardized toxicity tests using the earthworm Eisenia andrei, *Ecotoxicol. Environ. Saf.,* 23, 206, 1992.

22. Römbke, J., personal communication, 1992.

CHAPTER 17

Usefulness of Testing With *Eisenia fetida* For the Evaluation of Agrochemicals in Soils

G. Diaz-Lopez and R. Mancha

TABLE OF CONTENTS

ABSTRACT

The LC_{50} of copper salts used in agriculture as fungicides and nitrate fertilizers were determined by the 48 h contact filter paper test on the earthworm *Eisenia fetida* as described by OECD guidelines. The LC_{50} for copper sulfate was 10.3 $\mu g/cm^2$, for copper chloride 12.8 $\mu g/cm^2$ and for Bordeaux mixtures (I and II) 38.69 $\mu g/cm^2$. The formulation additives for the Bordeaux mixtures, calcium hydroxide, and potassium permanganate, showed LC_{50} values of 520 $\mu g/cm^2$ and 600 $\mu g/cm^2$, respectively. The nitrate fertilizers tested showed LC_{50} values of 195 $\mu g/cm^2$ for ammonium nitrate, 230 $\mu g/cm^2$ for manganese nitrate, 235 $\mu g/cm^2$ for potassium nitrate, and 240 $\mu g/cm^2$ for sodium nitrate. When ammonium nitrate and copper salt were tested in combination at or below their LC_{50} levels, an enhanced toxic effect of nitrate and copper salt was shown for *Eisenia fetida* in terms of lethality in the 48 h contact toxicity test (potentation).

I. INTRODUCTION

With time, chemicals in soil have been built up from various sources, a major source being the use of chemicals in agriculture. Some of these chemicals seem to be of low toxicity, but because of massive and repetitive application they accumulate and still reach high levels and become toxic for the natural environment.

Environmental studies are usually carried out for known, major toxic pollutants. Nevertheless, the massive use of some chemicals that are considered to be less toxic may be just as important because of their long-term effect in the ecosystems. Moreover, some chemicals can increase in toxicity when used in combination.

In this work we studied the toxic effect of some nitrates used as fertilizers and two copper salts used as fungicides in a biological model for soil pollution. The earthworm[1,2] *Eisenia fetida* was used as a test species. The contact toxicity test, described in Neuhauser et al.,[3] was used for the first-level screening of chemical substances to detect toxicity for the earthworm.

II. METHODS AND MATERIALS

The earthworm *E. fetida* used in the test was grown in our laboratory at 20 ± 1°C and light of 800 lux, on a consistent organic food source (sheep manure) at pH 7 and an overall moisture content of about 35%. Adult earthworms with clitellum, each weighing 300 to 600 mg, were used in the test.

The contact filter paper test following Organization for Economic Cooperation and Development (OECD) guidelines (1984)[4] for evaluating chemical toxicity was used. Glass cylinder vials (8 cm long × 3 cm in diameter) were used. Strips of filter paper (Whatman® No. 1) 9.42 × 7.7 cm were placed in the vials so as to cover the inside without overlap; contact area = 72.5 cm². The vials were covered with punctured plastic film (Parafilm M). The test chemicals were dissolved in deionized water and the filter paper in the test vial was treated with a 1-ml solution of the test substance. Chemical solutions in a concentration range from 0.01 to 1000 μg/cm² were tested with 3 × 10 replicates for each concentration. One earthworm was added to each vial. The vials were placed in a climate room at 20 ± 1°C in the dark in a horizontal position for 48 h. The earthworms were considered to be dead if they did not respond to gentle mechanical stimulation of the front part of the body.

The LC_{50} value for each chemical tested was calculated using Litchfield and Wilcoxon's method,[5] and reported as μg of test chemical per cm² of filter paper (μg/cm²). Confidence intervals ($p = 0.05$) were calculated.

The test chemical mixtures selected for this study were:

A. Metallic salts
  1. Two forms of *Bordeaux mixture*: $CuSO_4 \cdot 5H_2O$ + $KMnO_4$ (10 g + 12.5 g/1 1 H_2O) and $CuSO_4 \cdot 5H_2O$ + $Ca(OH)_2$ (10 g + 12.5 g/1 1 H_2O)
  2. *Two copper salts*: $CuSO_4 \cdot 5H_2O$ and $CuCl_2 \cdot 2H_2O$
B. Nitrates
  NH_4NO_3; KNO_3; $Mn(NO_3)_2 \cdot 4H_2O$; $NaNO_3$
C. Nitrates and copper salt mixtures
  NH_4NO_3 + $CuSO_4 \cdot 5H_2O$ and NH_4NO_3 + $CuCl_2 \cdot 2H_2O$

The chemicals were of reactive quality purchased from Merck® and Panreac Co.

III. RESULTS

Table 1 lists the individual LC_{50} values and 95% confidence intervals for the copper mixtures used as fungicides. Their active ingredient, copper sulfate, was tested separately. The toxicity of the two formulation additives and copper chloride are also shown.

Copper sulfate is the active ingredient in the Bordeaux mixtures, which are used in two formulations, one with potassium permanganate and the other with calcium hydroxide, as additive. The two mixtures were tested in the proportion commonly used in agriculture (10 g + 12.5 g/l H_2O) for active ingredient and additive, respectively. The LC_{50}, expressed in terms of copper ion for all the active products tested, is also calculated.

Copper shows a more toxic response in the sulfate form than in the Bordeaux mixtures. The two formulation additives reduce the toxic effect of copper in the mixtures.

Table 2 list the results for the nitrates, used as fertilizer, showing a very narrow range for toxicity according to the LC_{50} and the no effect level found. The LC_{50} results are listed for all the nitrates tested and expressed in nitrate ion.

Table 1. Toxicity of Copper Salts and Its Formulation Additives to *Eisenia fetida* in the 48-h Filter Paper Contact Test

Chemical form	LC_{50} $\mu g/cm^2$	Confidence limits 95%	Metal ion LC_{50}[a]
Bordeaux mixture (I) $CuSO_4 + KMnO_4$	152	100–231	38.69
Bordeaux mixture (II) $CuSO_4 + Ca(OH)_2$	152	91–254	38.69
$CuSO_4 \cdot 5H_2O$	10.3	5.9–18.2	2.62
$CuCl_2 \cdot 2H_2O$	12.8	2.1–78	4.72
$Ca(OH)_2$	520	391–692	
$KMnO_4$	600	435–828	

[a] In $\mu g/cm^2$.

Table 3 shows the toxicity results for mixed samples of ammonium nitrate with copper sulfate and ammonium nitrate with copper chloride, taking into consideration that they are often used in combination in soils. The toxic response is expressed as a percentage of observed lethality in the worms.

The amount (Q) of each chemical in the mixed sample is expressed in $\mu g/cm^2$ and is also shown as a percentage of the obtained LC_{50} (Tables 1 and 2) for the same chemical when tested individually.

Thus, the mixed sample of 195 $\mu g/cm^2$ of ammonium nitrate (%LC_{50} = 100%) plus 0.72 $\mu g/cm^2$ of copper sulfate (%LC_{50} = 7%) gives a 100% lethal response in the contact test. A stronger toxic effect can be seen for the combined use of nitrates and copper salts than for the chemicals tested individually.

IV. DISCUSSION

The aim of this study was to determine the acute toxicity of nitrates and copper salts individually and in combination. We chose these chemicals because of their low toxicity and large-scale use in agriculture.

The research on the effects of Bordeaux mixture on *E. fetida* is based on previous work in our laboratory with copper compounds. Bordeaux mixture is used in several countries as a fungicide for foliage applications. Major uses

Table 2. LC_{50} to *Eisenia fetida* in the 48-h Contact Filter Paper Test For Nitrates

Chemical form	LC_{50} $\mu g/cm^2$	Confidence limits 95%	NOLC[a]	NO_3^- ion LC_{50}[a]
NH_4NO_3	195	224–160	110	151
KNO_3	235	241–229	210	144
$Mn(NO_3)2 \cdot 4H_2O$	230	251–210	160	114
$NaNO_3$	240	265–218	160	175

[a] In $\mu g/cm^2$.

Table 3. Toxic Response in Terms of Lethality For Combinations of Ammonium Nitrate and Copper Salts in 48-h Paper Filter Contact Test

NH_4NO_3 + $CuSO_4$				% Lethal observed
Q^a	$\%LC_{50}$	Q^a	$\%LC_{50}$	
195	100	10.3	100	100
195	100	1.44	14	100
195	100	0.72	7	100
156.6	80	1.16	11	90
130.5	67	0.96	9	60
97.9	50	1.45	14	60
97.9	50	0.72	7	0
19.5	10	0.14	1.4	0

NH_4NO_3 + $CuCl_2$				% Lethal observed
Q^a	$\%LC_{50}$	Q^a	$\%LC_{50}$	
195	100	12.8	100	100
97.6	50	6.4	50	100
19.5	10	1.3	10	0
9.7	5	0.65	5	0

[a] Q = amount in $\mu g/cm^2$.

include the control of *Phytophthora infestans* on potatoes, *Venturia inaequalis* on apples, *Pseudoperonospora humuli* on hops, and mildew in vineyards. Nitrate salts are widely employed as fertilizers and have been shown to accumulate in soils with excessive use.[6]

In this study, the toxicity of combinations of nitrates and copper salts was increased, indicating that the interaction between nitrate and copper salt may enhance the toxicity of these chemicals. A toxic effect of nitrates in earthworm was shown at the presence of sublethal doses of copper salts, indicating potentiation.

The paper contact toxicity test is described as an optional initial screening test to indicate those substances likely to be toxic to earthworms in soil and therefore requiring further, more detailed testing in an artificial soil.[4] The paper contact toxicity test, however, can also be an excellent system for studying the mechanism of chemical toxicity, unraveling the combined effects of chemicals and relating this to individual toxicity, as described. In view of these results we are continuing this research using the artificial soil test and *E. fetida*, to obtain data which are more representative for natural exposure of earthworms to chemicals.

ACKNOWLEDGMENTS

We thank Dr. A. Garcia-Lorente for his critical comments on this research. We thank J. Aranda, R. Folgado, and A. Laborda for their assistance in conducting the experiments.

REFERENCES

1. Bouché, M.B., Ecotoxicologie des lombriciens, *Ecol. Appl.*, 5, 291, 1984.
2. Edwards, C.A. and Lofty, J.R., *Biology of Earthworms*, 2nd ed., John Wiley & Sons, New York, 1977.
3. Neuhauser, E.F., Loehr, R.C., Malecki, M.R., Milligan, D.L., and Durkin, P.R., The toxicity of selected organic chemicals to the earthworm *Eisenia fetida*, *J. Environ. Qual.*, 14, 383, 1985.
4. OECD, Earthworm, Acute Toxicity Test. The Contact Filter Paper Test. *Guidelines for testing of chemicals*. Section 2. Effects on biotic systems, 1984.
5. Litchfield, J.T. and Wilcoxon, F., A simplified method of evaluating dose-effect experiments, *J. Pharmacol. Exp. Ther.*, 96, 99, 1949.
6. Ramos, C. and Varela, M., Nitrate leaching in two irrigated fields in the region of Valencia (Spain), in *Proc. Int. Symp. Nitrates, Agriculture, Water*, Paris, France, 1990, 335.

CHAPTER 18

A Prolonged Laboratory Test on Sublethal Effects of Pesticides on *Eisenia fetida*

C. Kula

TABLE OF CONTENTS

0-87371-530-6/94/$0.00 + $.50

© 1994 by CRC Press, Inc.

ABSTRACT

A laboratory test was developed to investigate sublethal effects of pesticides on *Eisenia fetida* (Savigny, 1826). Adult earthworms are exposed to the pesticide for 6 weeks. After another 4 weeks juvenile earthworms are extracted from the artificial soil substrate. To achieve an exposure which is related to field conditions the pesticide is sprayed onto the soil surface. Results with several pesticides indicate a suitable test method. In addition to the number of juveniles body weight and mortality of adults should be investigated.

I. INTRODUCTION

Tests on acute toxicity are well established for the investigation of side effects of chemicals on earthworms in the laboratory. A common test using the earthworm species *Eisenia fetida* is the artificial soil test according to the Organization for Economic Cooperation and Development (OECD) guideline No. 207.[1] In this test the chemical is mixed homogeneously into the soil substrate. The toxicological end point is mortality after 14 d of exposure.

Besides mortality, sublethal effects might occur and are of importance for the maintenance of earthworm populations in the field. Especially, effects on reproduction should be investigated for means of hazard assessment of chemicals to earthworms.

In the case of pesticides, it has to be considered that the effects on earthworm populations in the field are dependent on the intrinsic toxicity of the product and on the exposure of earthworms. For possible use within a tiered test system, a laboratory test was developed and is described in the following. Special attention was paid to an exposure situation which might be similar to field conditions. For the pesticides tested this was achieved by spraying the pesticide onto the soil surface.

II. MATERIALS AND METHODS

The test design is closely related to the OECD test on acute toxicity.[1] As in the acute test, *E. fetida* is an appropriate test organism as well. This species was chosen because of its short generation time and because it is easy to breed in a synchronized culture in the laboratory. However, possible differences in sensitivity compared to species of arable land have to be taken into account when using this test species.[2-4]

Ten earthworms are placed in test boxes with a surface area of 200 cm^2 containing 600 g dry weight of artificial soil. This amount of artificial soil corresponds to a substrate height of 5 to 6 cm. Moisture content of the substrate is about 35% of dry weight and should be 40 to 60% of the water-holding

Table 1. Effects of Pesticides on Reproduction, Body Weight, and Mortality of *Eisenia fetida*

Active ingredient	Mortality of adults (%)	Body weight of adults (% of initial weight)	Number of juveniles/alive adult
Benomyl			
Control	0	104.8 ± 4.6	12.4 ± 4.3
0.25 kg/ha	0	114.7 ± 7.1	9.4 ± 2.3
1.25 kg/ha	0	74.1 ± 5.1*	0.6 ± 0.4*
Parathion			
Control	0	142.5 ± 11.7	5.3 ± 1.0
0.21 l/ha	0	135.3 ± 5.7	7.6 ± 2.7
1.05 l/ha	0	72.7 ± 1.2*	4.5 ± 2.3
Tebuconazole			
Control	0	173.7 ± 5.2	6.4 ± 2.4
1.5 l/ha	0	180.4 ± 8.4	4.3 ± 1.6
7.5 l/ha	0	180.8 ± 9.1	2.4 ± 0.7*
Copper oxychloride			
Control	0	157.0 ± 12.3	16.5 ± 1.3
20 kg/ha	0	155.0 ± 8.1	18.8 ± 2.3

* Significant difference ($p < 0.05$) against control, Tukey's multiple t-test.

capacity. Application is done as close as possible to practical conditions, for example, with a plot sprayer. A recommended dose and a fivefold dose are applied with a water amount corresponding to 600 to 800 l/ha. Food is given following the application of the pesticide once a week. Finely ground dry cattle manure is a suitable food. It is spread onto the soil surface and moistened. If no or low feeding activity is observed during the test, food should only be given when required because moldering can occur. After 6 weeks of exposure the adults are removed from the substrate. The juveniles are allowed to hatch and grow for another 4 weeks.

The toxicological end points in this test are the number of juveniles after 10 weeks of exposure, body weight changes, and mortality of the adult earthworms.

III. RESULTS

The test design was tested with several pesticides. Table 1 shows results from tests with pesticides with the active ingredients benomyl (fungicide), parathion (insecticide), tebuconazole (fungicide), and copper oxychloride (fungicide). In all trials, one common application rate was tested and in some cases also a fivefold dose.

In these trials, no mortality of the adult earthworms was observed in the treated boxes 6 weeks after treatment. In copper oxychloride there was obviously no effect on reproduction or body weight development. Concerning reproduction, there was a slight increase in the number of juveniles. In benomyl, parathion,

**Table 2. Effect of Benomyl on Reproduction of *Eisenia fetida* in Three
Different Trials (Number of Juveniles per Alive Adult)**

Trial number	Control	0.25 kg/ha	1.25 kg/ha
1	12.4 ± 4.3	9.4 ± 2.3	0.6 ± 0.4*
2	5.3 ± 1.0	5.7 ± 1.4	0.3 ± 0.3*
3	4.8 ± 1.4	4.2 ± 1.7	0.1 ± 0.1*

* Significant difference ($p < 0.05$) against control, Tukey's multiple t-test.

and tebuconazole there was an effect either on body weight (parathion) or on reproduction (tebuconazole) or on both parameters (benomyl) in the fivefold rate. The fact that the effects occurred either in body weight or in reproduction shows the necessity to look at both parameters. Tables 2 and 3 show results of three trials with benomyl. The application was done with a plot sprayer with an amount of water corresponding to 800 l/ha. The results show that significant effects on reproduction occurred in all trials at the rate of 1.25 kg/ha. With 0.25 kg/ha a slight, but not significant decrease could be observed in trial 1. Body weight change of the animals showed significant effects at 1.25 kg/ha in two out of three trials. There was a slight increase in body weight at 0.25 kg/ha in two tests.

Table 4 shows a comparison of a surface contamination with benomyl compared to a total contamination. The application rates used in this trial corresponded to 0.5 kg/ha and 2.5 kg/ha. The control boxes have been the same for both parts of the test. Whereas in the surface-contaminated boxes the application took place as usual, in the totally contaminated boxes the same amount was mixed into the soil homogeneously. The results for the juveniles show the same tendency in both contamination types, but in the totally contaminated boxes the effect on reproduction was stronger. In body weight development there was an increase at 0.5 kg/ha. An explanation for this might be an inverse relationship between cocoon production and body weight development.[5]

IV. DISCUSSION

The results indicate a suitable and sensitive testing method which can give information on sublethal effects of pesticides near to field exposure conditions. This design might be especially helpful for granular pesticides or baits when a

**Table 3. Effect of Benomyl on Body Weight of *Eisenia fetida* in Three
Different Trials (Body Weight of Adults in % of Initial Weight)**

Trial number	Control	0.25 kg/ha	1.25 kg/ha
1	104.8 ± 4.6	114.7 ± 7.1	74.1 ± 5.1*
2	142.5 ± 11.7	139.8 ± 7.1	146.5 ± 14.3
3	140.8 ± 3.4	152.8 ± 7.5	103.0 ± 8.5*

* Significant difference ($p < 0.05$) against control, Tukey's multiple t-test.

Table 4. Effects of Benomyl on *Eisenia fetida* After Surface and Total Contamination of the Soil Substrate

	Control	0.5 kg/ha	2.5 kg/ha
Juveniles[a]			
Surface contamination	11.8 ± 3.6	8.7 ± 1.7	1.8 ± 0.4*
Total contamination	11.8 ± 3.6	6.9 ± 2.5*	0*
Body weight[b]			
Surface contamination	124.4 ± 6.3	130.1 ± 5.3	65.0 ± 1.3*
Total contamination	124.4 ± 6.3	136.7 ± 3.6*	49.9 ± 2.0*

[a] Number of juveniles per alive adult.
[b] Body weight of adults in percent of initial weight.
* Significant difference ($p < 0.05$) against control, Tukey's multiple *t*-test.

test with homogeneous distribution of the pesticide does not give the information wanted.

In some pesticides, for example in benomyl, there is a good knowledge about effects occurring in the field.[6] The results shown with benomyl might fill the gap between laboratory toxicity studies and effects found in the field. The repeated trials with benomyl show that the results were reproducible. The trial with surface and total contamination for benomyl showed stronger effects with the total contamination probably because the test animals had no possibility to escape from the contaminated test substrate.

However, there are some points which have to be focused on in future tests. A ring test using this test design showed a great variability of juvenile numbers in the control from one laboratory to the other and from one test to the other.[7] Therefore, a minimum number of juveniles per test box should be given to allow results to be interpreted. Another proposal resulting from this ring test was to shorten the period of adult exposure from 6 to 4 weeks. This approach is followed in a second phase of the ring test which is running at the moment. In this test the number of juveniles is the main end point. There is no possibility to obtain information on cocoon production and hatching rate, because for this purpose the substrate has to be destroyed to wash out cocoons. In some cases, however, this information might be required, for example, when testing chemicals. There is a method by Van Gestel et al.[8] using *E. andrei* and the artificial soil substrate for a reproduction test. In this test the hatching rate of juveniles is determined in an uncontaminated substrate. However, as a contaminated substrate might be an important exposure route of pesticides for hatched juveniles, this should be taken into account when doing a higher-tier test in the laboratory.

REFERENCES

1. Anon., OECD-Guidelines for testing of chemicals No. 207. Earthworm, Acute toxicity tests. Adopted April 4, 1984.
2. Heimbach, F., Comparison of laboratory methods using Eisenia foetida and Lumbricus terrestris, for the assessment of the hazard of chemicals to earthworms, *Z. Pflanzenkr. Pflanzenschutz,* 92, 186, 1985.
3. Van Gestel, C.A.M., Earthworms in Ecotoxicology, Ph.D. thesis, Utrecht, 1991, chap. 11.
4. Kula, H., Species-specific sensitivity differences of earthworms to pesticides in laboratory tests, in *Ecotoxicology of Soil Organisms,* Eijsackers, H., Heimbach, F., and Donker, M., Eds., Lewis Publishers, Chelsea, 1992.
5. Van Gestel, C.A.M., Dirven-Van Breemen, E.M., and Baerselman, R., Influence of environmental conditions on the growth and reproduction of the earthworm Eisenia fetida in an artificial soil substrate, *Pedobiologia,* 36, 109, 1992.
6. Stringer, A. and Lyons, C.H., The effect of benomyl and thiophanate-methyl on earthworm populations in apple orchards, *Pestic. Sci.,* 5, 189, 1974.
7. Kokta, C., A laboratory test on sublethal effects of pesticides on Eisenia fetida, in *Ecotoxicology of Earthworms,* Becker, H., Edwards, P.J., Greig-Smith, P., and Heimbach, F., Eds., Intercept, Dorset, 1992, 213.
8. Van Gestel, C.A.M., Van Dis, W.A., Van Breemen, E.M., and Sparenburg, P.M., Development of a standardized reproduction toxicity test with the earthworm species Eisenia andrei using copper, pentachlorphenol, and 2,4-dichloraniline, *Ecotoxicol. Environ. Saf.,* 18, 305, 1989.

CHAPTER 19

The Development of Soil Toxicity Test Systems With Lumbricides to Assess Sublethal and Lethal Effects

W. Kratz and R. Pöhhacker

TABLE OF CONTENTS

0-87371-530-6/94/$0.00 + $.50

ABSTRACT

Earthworms are important components in terrestrial ecosystems. In soil ecotoxicology two guidelines for the determination of the acute toxicity of environmental chemicals for earthworms have been published in recent years. Today it is generally accepted that a survival test system alone does not fulfill the ecological demands in soil protection strategies. Therefore, the developed soil toxicity test systems with lumbricides are able to demonstrate sublethal effects of different pesticides on a subspecies level.

I. INTRODUCTION

Soil organisms are important components in terrestrial ecosystems. Two international guidelines for the determination of the acute toxicity of chemicals for earthworms have been published by the Organization for Economic Cooperation and Development (OECD) in 1984[1] and the European Economic Community (EEC) in 1985.[2] The main purpose of these guidelines is the determination of lethal concentration (LC_{50}). Today, however, it is generally accepted that the survival of a species is a rather insensitive parameter. From the ecological point of view, vitality characteristics such as growth, fertility, and reproduction are often more important parameters for the maintenance of populations.[3]

In the above-mentioned OECD and EEC guidelines, an artificial soil substrate for the acute toxicity tests is selected for reasons of standardization and reproducibility. The adsorption capacity of this artificial soil, based on the cation exchange capacity (CEC), is similar to that of a loamy soil. CEC is important for the sorption of cations, but not for apolar organic chemicals.

The acute toxicity test for earthworms[1,2] requires adult individuals (with clitellum) with a minimum age of 2 months and a maximum age of 1 year. The body weight should be between 300.0 to 600.0 mg. So a rough standardization of the body weight is required. However, in our opinion this range is too wide to determine the physiological stage of maturity. Most of the cultivated adults of *Eisenia foetida* which we tested had a weight between 270.0 to 470.0 mg. To fulfill the increasing demands of standardization, two weight classes were included in our test procedure.

II. MATERIAL AND METHODS

A. Growth and Cocoon Production Test With the Earthworms *Eisenia f. foetida* and *E. f. andrei*

1. Test Organisms

The OECD[1] and EEC[2] guidelines recommend both subspecies of *E. foetida* (Savigny, 1826): *E. f. foetida* and *E. f. andrei* for soil ecotoxicological studies.

So, adult earthworms of *E. f. foetida* and *E. f. andrei* were obtained from our own culture at an ambient temperature of 20.0 ± 5°C. These species are easy to breed in mass cultivation. Covering genetical differences in physiological metabolisms of long-term cultivated *E. foetida*, we contaminated both subspecies. Simultaneous measurements of sublethal effects, such as a change in biomass and cocoon production, require a constant weight class. The weight of adults determines the cocoon weight and the number of emerged juveniles.[4] Two weight classes (I = 300.0 mg and II = 400.0 mg) were tested and the relative biomass change compared to initial weight was calculated.

2. Test System and Chemicals

The test was carried out in artificial soil which was composed by dry weight of 10% peat finally ground (<0.5 mm), 20.0% kaolin clay, 69.0% sand (particle size 0.2 to 1.3 mm) according the OECD guideline,[1] and 1.0% $CaCO_3$ to adjust the soil pH value. The artificial soil was contaminated homogeneously with two different pesticides:

A fungicide (Carbendazim/Derosal, LC_{50} = 21.0 mg*kg^{-1} for *E. foetida*)

An herbicide (Phenmedipham/Betanal®, LC_{50} = 899.0 mg*kg^{-1} for *E. foetida*)

A control population was also included in the test with individuals of both subspecies and the two weight classes in noncontaminated artificial soil.

The used concentrations of 0.1, 1.0, and 10.0 mg*kg^{-1} Derosal and 1.0, 10.0, and 100 mg*kg^{-1} Betanal® approached the amounts applied normally in the field. The concentrations in the beginning of the test were 0.1, 1.0, and 10.0 mg*kg^{-1} Derosal and 1.0, 10.0, and 100.0 mg*kg^{-1} Betanal®. There was no possibility to measure the real concentration of the used pesticides during the test procedure. However, from the literature we know that the phenylcarbamates like Betanal® can be easily decomposed by soil microorganisms and that 50%

of the compound can be degraded during 28 to 56 d, depending the soil physical and chemical conditions.[5] Each test had four replicates per concentration.

Amounts of artificial soil were placed in 1-l glass jars. Ground cow manure was used as a food source. During the incubation time the average temperature was 20.0 ± 2°C and the soil moisture content was 65.0 to 70.0% (w/w). Soil pH (0.01 m $CaCl_2$) was 6.7 to 7.0 ± 0.3 at the beginning and did not change during the incubation period of 56 d.

At the start of the test procedure the individual weight of the worms was determined, and to each jar ten individuals were added.

3. Test Parameters

At the beginning of the test, the average gut content of the worms was determined with 12.0% (after 3 d of defecation) of the fresh weight of the worms. After this procedure, adults of both subspecies of *E. foetida* were preincubated for 7 d in artificial soil.

Recording the biomass as weight increase per time compared to initial weight, all worms were sorted out after 14, 28, and 56 d. They were washed with tap water, dried on filter paper, and weighed again.

For the determination of fertility, cocoons from each sample were incubated in an artificial soil and the production rate calculated as cocoon per living worm per week, and juveniles emerging from each cocoon were recorded.

Each experimentation vessel contained ten adults of each subspecies. Individual weights were not determined. After every period of measurement, so-called independent spot checks were carried out. The U-Test of Wilcoxon, Mann, and Whitney was used as a suitable nonparametric statistical method ($p = 0.01$). Samples with significant biomass differences to the initial weight are marked in Figures 1 and 2 with asterisks.

III. RESULTS

A. Growth and Cocoon Production Test With the Earthworm Species *E. f. foetida* and *E. f. andrei*

1. Biomass Change

Figures 1 and 2 demonstrate the different reactions of both subspecies and weight classes under the influence of the fungicide Derosal und the herbicide Betanal®.

Individuals of both subspecies showed a slightly higher increase in biomass with an initial weight of 300.0 mg than adults of weight class II (400.0 mg). This result is true for the control population and contaminated earthworms with the lowest concentration of Derosal and Betanal®.

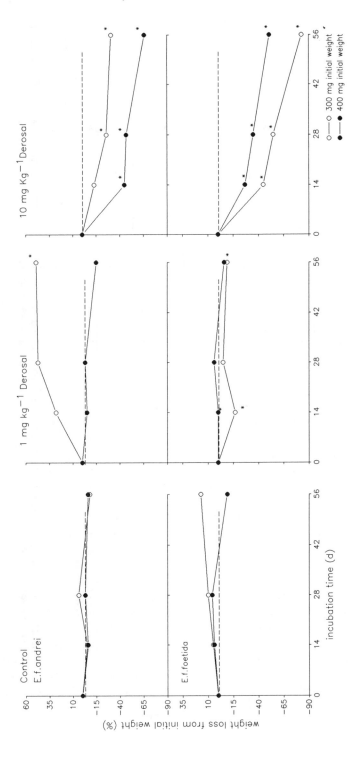

FIGURE 1. Impact of Derosal (fungicide) on biomass development of the subspecies *E. f. andrei* (upper figure) and *E. f. foetida* (lower figure) in different weight classes. (○——○ = 300.0 mg; ●——● = 400.0 mg; * = U-Test, p = 0.01)

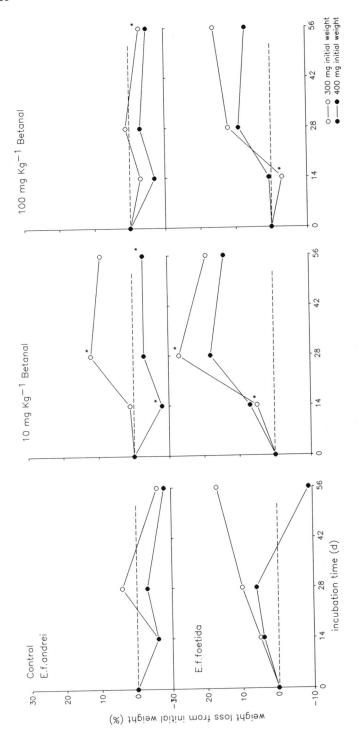

FIGURE 2. Impact of Bentanal® (fungicide) on biomass development of the subspecies *E. f. andrei* (upper figure) and *E. f. foetida* (lower figure) in different weight classes. (○—○ = 300.0 mg; ●—● = 400.0 mg; * = U-Test, *p* = 0.01)

A biomass increase up to 43% (indicated as a relative measure) 28 d after the pesticide application was observed only in the case of *E. f. andrei* (initial weight 300.0 mg) in the 1.0 mg kg^{-1} Derosal treatment.

The contamination level of 10.0 mg kg^{-1} Derosal leads to a mortality of 10% for both subspecies in weight class II after 14 d. Also, the applied manure was not consumed by the animals and a large decrease in the earthworm biomass (31 to 83%) was observed. The pigmentation of the worms was also lost.

Under the influence of 1.0 and 10.0 mg kg^{-1} of the low toxic herbicide (Figure 2), *E. f. foetida* and *E. f. andrei* showed different increases of biomasses. Individuals of the 300.0 mg initial weight class of *E. f. andrei* increased their biomass during the investigation period up to 22.0% at the contamination level of 1.0 mg kg^{-1}. In difference, the same initial weight class of *E. f. foetida* reacted only at 10.0 mg kg^{-1} with a biomass increase of 28.0%.

The worms of the two subspecies, belonging to the initial 400.0 mg weight class, showed a lower increase of biomass during the incubation of 56 d.

Generally, differences in biomass change between the two subspecies were obvious for the control and the contaminated worms for both pesticides.

2. Cocoon Production

Figures 3 and 4 show the total number of cocoons produced in all weight classes and used subspecies. Cocoons that failed the emergence were defined as "no juveniles". Individuals with an average weight of 300.0 mg showed a larger increase in biomass, but less cocoon production than individuals of weight class II (400.0 mg) under the influence of both pesticides.

Below the application rate of 10.0 mg kg^{-1} Derosal, all the produced cocoons were dead.

In general, no differences in cocoon production of the two subspecies of *E. foetida* were found with increasing concentrations of Betanal®. The numbers of juveniles emerging from each living cocoon were not influenced by the contamination under sublethal concentrations of the pesticides, nor was the ratio of cocoons per living worm per week.

IV. DISCUSSION

A. Biomass Change

It can be concluded that there are differences in biomass change between the two studied weight classes of both subspecies *E. f. andrei* and *foetida* in all contamination levels of the pesticides and in the controls, except for *E. f. andrei* in the control of the Derosal test. Another result is that at the lowest pesticide concentrations both subspecies in the investigated weight classes demonstrate a stimulated growth in comparison to the controls. An explanation of this growing reaction could be given by a study of Knowles and Benezet,[5] in which they

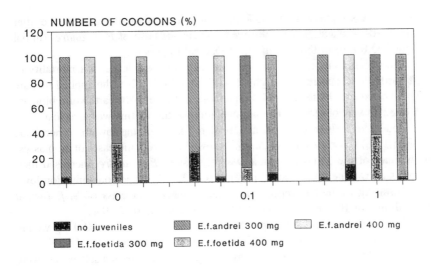

FIGURE 3. Effect of Derosal on cocoon production of the species *E. f. andrei* and *foetida*.

demonstrated clearly an increase of the soil microorganism biomass up to 64.0% below an 10^{-3} *M* pesticide solution.

Soil microorganisms play an important role as a food source for earthworms, and furthermore they are also responsible for many decomposition processes in the soil, making the substrate more digestible for the worms.[6]

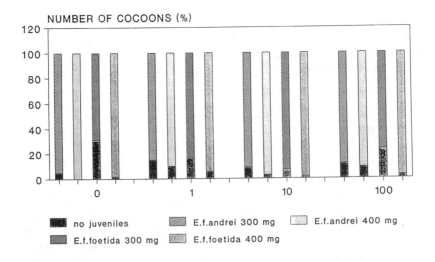

FIGURE 4. Effect of Betanal® on cocoon production of the species *E. foetida andrei* and *foetida*.

The highest concentration, especially of the toxic fungicide Derosal, caused a large weight loss. So, Keogh and Whitehead[7] found a strong decrease of orchard litter consumption by earthworms under the influence of Carbendazim.

The presented results show that the biomass change is a very sensitive parameter recording sublethal effects, including also growth stimulation at low concentration levels. According to van Gestel,[8] who prefers *E. f. andrei* for ecotoxicological tests, we suggest both subspecies as suitable test organisms. Taking into account the initial weight of individual worms, the developed test system for sublethal effects on the subspecies level is a good supplement for LC_{50} under the OECD guidelines.

B. Cocoon Production

Only the highest tested concentration of 10.0 mg kg^{-1} Derosal showed effects in cocoon production and emergence rate. Comparing on one hand the emergence rate and the cocoon production under the influence of sublethal concentrations of both pesticides, parameters of fertility were not sensitive in all cases. On the other hand, the initial weight of adults influenced cocoon production in the control and contaminated populations.

So, we can make the conclusion that for the integration of earthworm fertility in ecotoxicological test studies, we need more knowledge about the abiotic and biotic mechanisms that influence this biological result.

C. Summary

Earthworms are important components in terrestrial ecosystems. In soil ecotoxicology, two guidelines for the determination of the acute toxicity of environmental chemicals for earthworms were published during the last decade. Today it is generally accepted that a survival test system alone does not fulfill the ecological demands in soil protection strategies. So the developed soil toxicity test systems with lumbricides are able to demonstrate sublethal effects of different pesticides on the subspecies level.

REFERENCES

1. OECD, Guideline for testing of chemicals No. 207, Earthworm, acute toxicity tests, Paris, 1984.
2. EEC, EEC Directive 79/831. Annex V. Part C: Methods for the determination of ecotoxicology. Level I. C 4: Toxicity for earthworms. Artificial soil test. DG XI/128/28. Final, 1985.
3. van Gestel, C.A.M., Influence of cadmium, copper, and pentachlorophenol on growth and sexual development of *Eisenia andrei, Biol. Fertil. Soils,* 12, 117, 1991.

4. Hartenstein, R. et al., Reproductive potential of the earthworm *Eisenia foetida, Oecologia,* 43, 329, 1979.
5. Knowles, Ch.O. and Benezet, H.J., Microbial degradation of carbamate pesticides Desmedipham, Phenmedipham, Promecarb and Propamocarb, *Bull. Environ. Contam. Toxicol.,* 27, 529, 1981.
6. Lee, K.E., *Earthworms, Their Ecology and Relationship with Soils and Land Use,* Academic Press, Australia, 1985.
7. Keogh, R.G. and Whitehead, E., Observations on some effects of pasture spraying with Benomyl and Carbendazim on earthworm activity and litter removal from pasture, *N.Z. J. Exp. Agric.,* 3, 103, 1974.
8. van Gestel, C.A.M., Development of Terrestrial Ecotoxicity Test and Research on QSARs with Earthworms, Workshop Ecotoxicology Netherland, April 1989.

CHAPTER 20

Assessment of Pesticide Safety to the Carabid Beetle, *Poecilus cupreus,* Using Two Different Semifield Enclosures

U. Heimbach, P. Leonard, A. Khoshab, R. Miyakawa, and C. Abel

TABLE OF CONTENTS

0-87371-530-6/94/$0.00 + $.50

© 1994 by CRC Press, Inc.

ABSTRACT

A study was initiated, as part of an on-going program at the BBA, Braun-schweig, to develop a semi-field micro-plot method for assessing the safety of pesticides to the carabid beetle, *Poecilus cupreus*. The impact of two variables was investigated: enclosure design and application technique. Chlorpyrifos was selected as a model soil insecticide having contact, ingestion, and vapor activity. It was sprayed at 960 g·ha^{-1} as a 300 l·ha^{-1} overall application and as a 320 g·ha^{-1} banded (15 cm) application with a spray volume of 160 l·ha^{-1} to assess the impact of application technique on the system. The product was applied at drilling of a sugar beet crop during the spring of 1990. Beetles were introduced to enclosures on treated and untreated soil at intervals after application. Soil samples were also taken from the area surrounding the enclosures throughout the study and analyzed so that the fate of chlorpyrifos could be correlated with beetle survivorship. Soil samples were taken from within the enclosures at the end of the study so that the impact of enclosure design on insecticide loss rate could be assessed. Chlorpyrifos was most toxic to beetles on the day of appli-cation. This level of toxicity declined relatively rapidly with little or no mortality being observed when beetles were introduced three weeks after application. The banded application was safer to *P. cupreus* than the overall application. Soil analysis indicated that the rate of loss of chlorpyrifos from within enclosures was significantly less than that from the surrounding exposed soil. It was probable but not statistically proven that enclosure design had an impact on the observed product safety.

I. INTRODUCTION

The safety of agricultural products to beneficial insects, spiders, and mites is being viewed with increasing importance by registration authorities, farmers, conservationists, and by the agricultural chemical industry. A registration re-quirement for data on safety to beneficials was introduced by German authorities in December 1989. Guidelines and methodologies for European harmonization of these data requirements are being developed. Guidelines have been proposed by the International Organization for Biological Control (IOBC)[1,2] for evaluation of product safety to beneficials. A sequence of experimentation is proposed. This starts with use of a relatively simplistic laboratory model. If further evaluation is required, experimentation may be taken to a more realistic "semifield" environment. Ultimately, if the safety of a product is still unclear, experimentation under field conditions may be necessary.

The benefits of such a sequential testing philosophy are appealing. If a product is safe at the recommended field rate under laboratory conditions, it is probable that it will be safe in the field. In this case, a conclusion may be made about the safety of a product while avoiding the cost of a semifield or field study. However, where such a conclusion may not be made, the semifield situation offers an opportunity to evaluate product safety under simulated field conditions in a partially controlled environment. The interpretation of semifield studies is typically less complex than that of larger-scale field studies. Despite these advantages in reliability and reduced cost, relatively few semifield methods have been developed.

This study was initiated as part of an on-going program at the BBA in Braunschweig to develop a method for evaluating the hazard of agricultural products to the carabid beetle, *Poecilus cupreus*, under semifield conditions. A microplot system for enclosing *P. cupreus* adults on treated soil was under evaluation and had been reported.[3,4] The system had potential, but the impact of the enclosure microclimate on the fate of volatile products was not known.

Dursban® 4 (480 g chlorpyrifos per liter ec) is an insecticide which, while having other applications, is used for control of soil pests in Germany. Chlorpyrifos is relatively volatile with a vapor pressure of 2.5 mPa at 25°C and works by contact, ingestion, and by vapor action. Dursban® 4 was therefore selected as a model product for development of the semifield *P. cupreus* microplot method. The product was sprayed as a band and overall to assess the impact of application technique on the microplot system. Two enclosure designs were used to assess the impact of microclimate. Arrangements were made to take soil samples throughout the study so that the fate of chlorpyrifos could be correlated with beetle survivorship.

II. MATERIALS AND METHODS

A. Site Description and Experimental Enclosures

The experimental site consisted of a 10-ha cultivated field located in Wendhausen near Braunschweig in Germany. The soil, classified as a clay loam containing 2% organic matter, was relatively uniform. Sugar beet (var: Kawetina) was drilled with 45 cm between rows on April 2, 1990, using a Rau Betasem combined drill and spray boom. Mean air temperatures, soil temperatures at 5-cm depth, and rainfall readings were recorded on a daily basis at Braunschweig. Air and soil temperatures declined to 3.4 and 7.9°C, respectively, 2 d after application. These temperatures steadily increased to a peak of 19.7 and 20.4°C, respectively, 34 d after application. A total of 7.9 cm of rainfall was recorded, most of which fell between 7 and 25 d after application.

B. Application of Chlorpyrifos

Chlorpyrifos was applied to 1-ha plots using two methods: as an overall spray and as a band. A 1-ha plot was left untreated as a control. The recommended rate of 960 g ai $\cdot$ ha^{-1} was applied as a 300 l $\cdot$ ha^{-1} overall application using TeeJet 11006 GLP nozzles, a speed of 8 km $\cdot$ h^{-1}, and a pressure of 2.9 bar.

The recommended rate for a banded application of 320 g ai $\cdot$ ha^{-1} was sprayed at drilling as a 15-cm band over the row using the Rau Betasem boom. TeeJet 10080 nozzles, a speed of 4.8 km $\cdot$ h^{-1}, and a 1.6-bar pressure were used to produce a volume of 160 l $\cdot$ ha^{-1}.

Applications were conducted during and immediately after drilling, between 10.00 h and 15.30 h with a clear sky and a mean air temperature of 14.6°C. The mean soil temperature was 11.5°C and a wind speed of approximately 7.5 km $\cdot$ h^{-1} was recorded.

Two designs of experimental enclosure were used to contain beetles. A "high" construction (H) had previously been evaluated and was subsequently reported.[3,4] It was comprised of a 1-m-square galvanized steel frame of 25 cm in height. The frame was designed to be dug into the ground to a depth of 5 cm, thus leaving an exposed height of 20 cm above ground level. Beetles were prevented from escape by use of a 1-mm nylon mesh which was fixed over the top of the enclosure.

The H construction was modified to produce a "flat" version (F) which was produced to minimize possible microclimate effects within the enclosure. The F enclosure, like the H enclosure, consisted of a 1-m-square galvanized steel frame. However, the wall height was reduced to 3 cm above the ground and a wire mesh size of 10 mm was used to prevent predation by birds and rodents. The wall, like the H cage, was dug into the ground to a depth of 5 cm below the soil surface. The top of the wall was folded inwards to prevent beetles escaping.

Two ceramic tiles of 1 × 10 × 25 cm were placed flat on the soil surface in each enclosure as refuges. Four of each type of enclosure were randomly distributed and dug into each plot immediately after application.

C. Beetles

Laboratory-reared carabid beetles (*P. cupreus*) were held under simulated winter conditions (8°C and 8-h day length) for at least 2 months prior to exper-

Table 1. Summary of Results Obtained With *Poecilus cupreus* and Two Designs of Semifield Enclosures

| Treatment | Enclosure | Released | Percentage of adults | | | % Total recaptured |
			Alive	Dead	Missing	
Chlorpyrifos	H	180	15	63	18	
Overall		(Unassigned)	+2	+2		82
Chlorpyrifos	F	180	24	56	14	
Overall		(Unassigned)	+3	+3		86
Chlorpyrifos	H	100	39	37	12	
Banded		(Unassigned)	+10	+2		88
Chlorpyrifos	F	100	41	31	8	
Banded		(Unassigned)	+16	+4		92
Untreated	H	120	63	8	4	
		(Unassigned)	+22,5	+2,5		96
Untreated	F	120	53	3	16.5	
		(Unassigned)	+25	+2,5		83.5

imentation to ensure that females were in reproductive state. Beetles were held at 10°C and a day length of 14 h for a further 48 h to prepare them for transfer to field enclosures. Their elytra were bonded with wax to prevent escape by flying.[5] Colored 1-mm metal plates, attached by wax, were used so that the release date of each beetle could be identified.

Equal numbers of male and female beetles were transferred to enclosures in each plot at intervals after application. A total of ten beetles were introduced to each enclosure immediately after application. At subsequent dates either five or ten were introduced, depending on the number of dead beetles found at that assessment. The reduced number of five beetles was used, where low mortality had occurred, to prevent overcrowding and cannibalism.

D. Mortality

Beetle mortality was assessed on 21 occasions following application from April 2 to May 16, 1990 (see Figure 1). The way mortality was determined was dictated by weather and soil conditions. When wet, the search for dead beetles was confined to the soil surface. When dry, a more intensive search for dead beetles was made in the top centimeter of soil. Dead beetles were removed. Damaged beetles were left within the enclosures and not recorded as mortalities. Recapture of introduced beetles was not complete. A proportion escaped and were classified as "missing". Others were found with no color label and classified as "unassigned". The problem of escape was worst during a dry spell at the end of May, when deep cracks developed in the soil. At the end of the study the recapture rate was calculated for each treatment (see Table 1).

E. Soil Sampling and Analysis

Soil samples were taken from the ground surrounding cages in the overall treated plot immediately after application and 4, 10, 18, 28, and 38 d later. The samples were selected randomly and replicated six times. Two samples were taken from within each of the F enclosures 38 d after application at the end of the study. Each sample was extracted as a 6-cm-deep cylindrical core with a diameter of 15 cm. The samples were individually sealed in polythene bags, frozen, and transported to the DowElanco European Research and Development Centre at Letcombe Regis in England.

Individual "replicated" samples were air dried and prepared using a Tector homogenizer with dry ice. Subsamples from each replicate were bulked for analysis. Chlorpyrifos was extracted in an acetone/water solvent mixture. Additional water was added and chlorpyrifos was partitioned into hexane.

Chlorpyrifos concentration was quantified by packed column gas chromatography with flame photometric detection. The method has a validation limit of 0.01 mg ai $\cdot$ kg^{-1} dry weight of soil with a standard deviation of $\pm 9\%$.

F. Statistical Analysis

The LIFETEST procedure in SAS[6] was used to estimate survival functions for insects in H and F enclosures following inoculation 4 d after application. Two nonparametric methods (methods that do not assume that the data came from any particular distribution), the log-rank test, and the Wilcoxon test were employed. These procedures test the hypothesis that the groups being compared have the same survival function. This hypothesis is rejected if the significance level for either test is smaller than 0.05.

III. RESULTS

A. Beetle Mortality

The total recovery of test animals at the end of the study varied between 82% for the overall treatment with H enclosures and 92% for the banded application with F enclosures, respectively (see Table 1). The total number of dead and live beetles which were not assignable to a particular beetle introduction date varied between 4% for the F enclosures on the overall treatment and 27.5% for the F enclosures on the untreated plots, respectively. The percentage of missing or escaped beetles varied between 4 and 18% for the H enclosures on untreated plots and H enclosures on overall plots, respectively.

Mortalities in the untreated plots did not occur until the 17th assessment, 30 d after application. The total mortality observed was 10 and 6% for beetles in the untreated H and F cages, respectively.

Beetle mortality following each inoculation is expressed as a series of cumulative graphs (see Figure 1). These results were not corrected for untreated mortality, but missing and unassignable beetles were disregarded when calculating cumulative mortalities.

Most mortalities occurred when beetles were introduced on the day chlorpyrifos was applied (see Table 1 and Figure 1). This level of toxicity declined progressively following each beetle introduction until little or no effects were observed when beetles were introduced 18 d after application.

The overall application induced greater mortalities than did the banded application following the first three beetle introductions. This trend is still evident, but not so clear following the final inoculation.

As a trend, mortalities observed in the H enclosures were equal to or greater than those observed in the F enclosures. The greatest difference between cumulative mortality observed in the two enclosure types was seen following the Day 4 and Day 10 beetle introductions of the overall and banded applications, respectively.

Survival functions were estimated for beetles enclosed on the overall and banded applications 4 d after treatment. The analysis for beetles in H and F enclosures on the overall application is summarized (Table 2) and shown graphically (Figure 2). When all the survival times from all four subsample cages are combined for each enclosure type, the results show that the survival curve for H enclosures is significantly different from that of the curve for F enclosures. It appears that beetles have a tendency to survive for longer in the F enclosures than in the H enclosures when chlorpyrifos is applied overall. However, when the survival estimates for each enclosure were analyzed individually, it was apparent that this difference was largely produced by an extended survivorship in one enclosure.

The analysis for beetles in the two enclosure types on the banded application is summarized (Table 2) and shown graphically (Figure 3). The data recorded for this application are different from that of the overall application because only five insects were released in each enclosure in this case as opposed to ten for the overall application. Thus, not as much information is available to construct survival functions. The same analysis, however, was carried out for data sets from both applications. Because of the limited and highly ''sensored'' data base, results from individual enclosures were combined for analysis of H vs. F enclosure types. When analyzed in this way, the analysis of the two enclosure types shows no significant differences.

B. Soil Analysis

Residues of chlorpyrifos in soil sampled from the overall plots declined from 0.72 on Day 0 to 0.06 mg · kg^{-1} dry weight of soil 38 d after application (see Figure 4). No observable peaks were found in the untreated control. Soil samples were taken from within the F enclosures 38 d after application. A concentration of 0.15 mg chlorpyrifos per kilogram dry weight of soil was observed in these samples. This concentration is more than two times greater than the concentration of chlorpyrifos observed in soil sampled from the surrounding area at the same date.

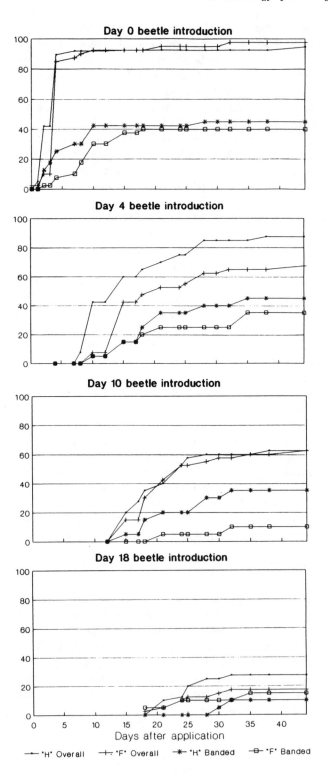

Day 0 beetle introduction

Day 4 beetle introduction

Day 10 beetle introduction

Day 18 beetle introduction

Days after application

→ "H" Overall   —+— "F" Overall   —*— "H" Banded   —▫— "F" Banded

Table 2. Analysis Summary of Toxicity of Chlorpyrifos on Carabid Beetles, Second Insect Release 4 d after Application. Test for Differences Between Enclosure Types

Testing Homogeneity of Survival Curves over Enclosure Type Overall Application

Test of Equality over Enclosure Type

Test	Chi-square	DF	Pr> Chi-square
Log-rank	10.6954	1	0.0011
Wilcoxon	10.3141	1	0.0013

Banded Application

Test of Equality over Enclosure Type

Test	Chi-square	DF	Chi-square
Log-rank	0.1485	1	0.7000
Wilcoxon	0.1532	1	0.6955

IV. DISCUSSION

The transition from laboratory bioassays to the semifield environment introduces the need to ask several fundamental questions such as, "What is a replicate?". Should it be a beetle, an enclosure, or an area which is sprayed with a particular treatment? This study was set up as a split-plot design with chemical treatment and enclosure type as the whole- and subplot factors, respectively. To be properly replicated, each chemical treatment should have been sprayed on more than one area. (The replicate enclosures within a treated area are merely subsamples.) Thus the study was nonreplicated and an estimate of experimental error could not be made, which severely hinders the ability to find statistically significant differences between chemical treatments and between enclosure types.

If this type of semifield study is to be used to make comparisons between toxicities of agricultural products to carabid beetles, this fundamental limitation must be addressed. One way around this problem would be to apply products to smaller plots with each chemical treatment being replicated.

The "hazard" of a product can be considered to be a function of both its toxicity and potential exposure to an organism. The results of this study show that the toxicity of chlorpyrifos to *P. cupreus* declines with time after application. The "hazard" that chlorpyrifos may pose to carabids which are active in the treated areas can therefore be considered to decline with time after application.

The exact estimate of how long a product represents a "hazard" can be influenced by a combination of the residuality of the product in the test envi-

---

FIGURE 1. (*Opposite*) Cumulative mortality of beetles (*P. cupreus*) introduced to H and F enclosures at intervals after application.

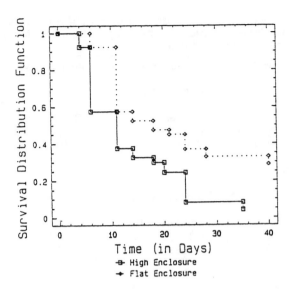

FIGURE 2. Survival function estimates for *P. cupreus* beetles introduced to H and F enclosures on soil treated with an OVERALL application of chlorpyrifos.

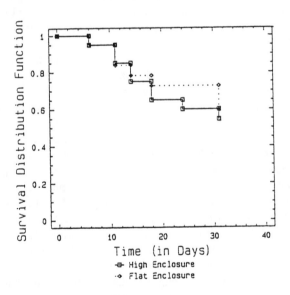

FIGURE 3. Survival function estimates for *P. cupreus* beetles introduced to H and F enclosures on soil treated with a BANDED application of chlorpyrifos.

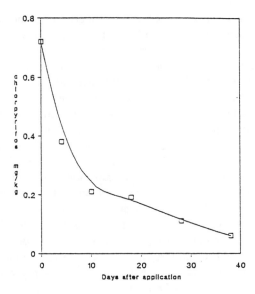

FIGURE 4. Rate of loss of chlorpyrifos applied as an overall spray from unenclosed soil.

ronment and by the nature of the test system. In this case, for example, little or no side effects were observed with beetles in the F enclosures when chlorpyrifos was applied as a band and beetles were infested 10 d after application. By contrast, appreciable levels of mortality were observed with beetles in H enclosures in the same plot and in both enclosure types on the plot treated with the overall application.

Thus it may be concluded that chlorpyrifos is most "hazardous" to the carabid beetle, *P. cupreus,* at the time of application, but this level of hazard declines relatively rapidly with little or no perceived risk from 10 to 21 d after application, depending on the experimental system used.

It is obvious from the data set that the banded application was more safe to *P. cupreus* beetles than was the overall treatment. With a 30-cm untreated strip between treated bands it is probable that the reduced hazard of the banded application was produced by a reduced exposure to the test product. The hazard of chlorpyrifos to carabid beetles may be further reduced by use of granular formulations. A recently reported large-scale field study[7] showed that Lorsban® 15 G (a 15% w/w granular formulation of chlorpyrifos) applied at planting to two 5-ha plots of maize, at the recommended rate of 11.2 g ai 100 m row^{-1}, had no significant impact on carabid activity.

It is probable that enclosure design can have a significant impact on the observed toxicity of chlorpyrifos. However, the analysis of this data produced results that are far from being conclusive. The analysis of the overall treatment tends to support the concept that beetles survived longer in F enclosures than they did in H enclosures. However, the fact that, within the same treatment, enclosures of the same design could be found to be significantly different indicates that there is the potential for large experimental error here. This is backed up by the fact that the survival of beetles in both types of enclosure in the banded treatment were not found to be significantly different from each other. Further experimentation will be required to resolve this issue.

The substantially enhanced chlorpyrifos concentration observed within the F enclosures, when compared with that of the surrounding soil, at the end of the study, is somewhat disturbing. The difference of 0.09 mg ai $\cdot$ kg^{-1} is nine times greater than the validation limit of the analytical method and is therefore considered to be highly significant. By extrapolation, the soil concentration of chlorpyrifos within the F enclosures 38 d after application was equivalent to the concentration expected 23 d after application for samples from the surrounding area. Whatever the cause, the observed result is important. The aim of the research program was to develop a method to assess pesticide side effects on *P. cupreus* under semifield conditions. This enhanced concentration of chlorpyrifos, in this case, at the end of the study within the F enclosures renders it difficult to correlate observed toxicity with the applied field dose.

It would be possible to make a more effective correlation between toxicity and field dose by setting up new enclosures on undisturbed treated soil at intervals after application. Enclosure-induced effects on the loss rate of the experimental product would therefore be limited to the post-beetle-introduction period. This would, however, require an increased number of enclosures and therefore an increased cost.

V. FURTHER RESEARCH

Use of this "worst-case" semifield enclosure system may lead to an overestimate of the "hazard" of a product to *P. cupreus*. Future studies should aim to reduce overestimates of "hazard" in order to reduce subsequent requirements for further expensive field testing.

Future studies should incorporate true treatment replication and introduction of new enclosures at each inoculation date. This will increase the cost of the study, but it would enhance statistical power. Most importantly, the significance of differences between treatments could be calculated. Additional enclosures should be incorporated to facilitate sampling and residue analysis from within the "enclosure environment" throughout the trial. This is required if side effects in the "semifield" environment are to be correlated with field exposure and the true hazard or safety of products is to be determined.

REFERENCES

1. Hassan, S.A., Standard methods to test the side-effects of pesticides on natural enemies of insects and mites developed by the IOBC/WPRS Working Group "Pesticides and Beneficial Organisms", *EPPO Bull.*, 15, 214–255, 1985.
2. Hassan, S.A., Guidelines for testing the effects of pesticides on beneficials: short description of test methods, 10BC/WPRS Bulletin XI.(4), 1988.
3. Heimbach, U. and Abel, C., Nebenwirkungen von Bodeninsektiziden in verschiedenen Applikationsformen auf einige Nutzarthropoden, *Verhandl. Ges. Ökol.*, Vol. XIX/III, 163–170, Osnabrück 1991.
4. Büchs, W., Heimbach, U., and Czarnecki, E., Effects of snail baits on non-target carabid beetles, slugs and snails in world agriculture, *BCPC Monogr.* No. 41, 245–252, 1989.
5. Abel, C., in preparation.
6. SAS/STAT User's Guide, GLM-VARCOMP. Version 6, SAS Institute, Inc., 4th ed., 1027, 1990.
7. Dintenfass, L.P., Effects of LORSBAN 15 G on arthropod activity in large plots of conventionally tilled corn. Paper presented at the annual meeting of the Entomologial Society of America, December 1990.

CHAPTER 21

An Index of the Intrinsic Susceptibility of Nontarget Invertebrates to Residual Deposits of Pesticides

J.A. Wiles and P.C. Jepson

TABLE OF CONTENTS

287

ABSTRACT

A full methodology for estimating the relative susceptibilities of non-target invertebrates to pesticide residues is given. Susceptibility is determined as an index, which is the ratio between an exposure function (based on walking track width, walking speed, and the proportion of the area covered by the insect that is contacted) and susceptibility (expressed as the species' topical tolerance distribution at endpoint). The exposure parameters for the index were derived for seven coleopteran predators which colonize temperate cereal crops. Susceptibility data for the synthetic pyrethroid insecticide, deltamethrin, was then used to calculate the index of susceptibiity for each species. The aphid predators were *Pterostichus melanarius* (Illiger), *Nebria brevicollis* (F.), *Agonum dorsale* (Pont.), *Demetrias atricapillus* (L.), *Bembidion lampros* (Herbst.) (Coleoptera, Carabidae), *Tachyporus hypnorum* (F.) (Coleoptera, Staphylinidae) and *Coccinella septempunctata* (L.) (Coleoptera, Coccinellidae). The indices indicated that the staphylinid beetle *T. hypnorum* would be most affected by deltamethrin residues because of its high susceptibility and high level of surface contact relative to body size. The least affected species would be the carabid beetle *D. atricapillus,* because it had a low susceptibility to deltamethrin and a small contact area with the surface. These predictions of extremes were confirmed in two residual bioassays. The uses and limitations of the susceptibility index are discussed.

I. INTRODUCTION

The need to understand the ecological effects of toxic substances in the environment has become increasingly clear with the rapid development of the chemical industry over the last 50 years. The agrochemical industry is one example where the number of compounds and the uses to which they are put has expanded rapidly during this period.[1] Problems such as how to select organisms for monitoring and clearance testing, how to determine what is an acceptable risk from pesticides, and how this can be measured and interpreted in the field face pesticide regulatory bodies and the agrochemical industry.[2-5]

Currently, agrochemicals are tested against beneficial invertebrates in several European countries.[6] Tests tend to progress from the laboratory to the field, providing data which are judged against safety criteria[7] (for a full review of testing methodology see Jepson).[8] The ability of most current assessment techniques to give an accurate estimate of hazard is questionable, however, because of the complex chemical, toxicological, ecological, and operational factors that mediate pesticide side effects.[3]

The susceptibility index described in this paper took a novel approach that aimed to predict the relative susceptibilities of seven species of predator, which inhabit temperate cereal crops, to pesticide residues. Intrinsic species characteristics were used to derive a function of exposure which was corrected for susceptibility. The index may provide a basis for estimating the relative susceptibilities of species, thus aiding selection of organisms for registration testing and helping the interpretation of field studies. Some of the predators studied are nocturnal and likely to be hidden in soil refuges during spraying. They therefore may avoid being directly contacted by spray. Thus the main routes of exposure for these species is likely to be via contact with residual deposits and possibly by consumption of contaminated prey. A residual susceptibility index may therefore be helpful for the selection of species for side effects testing.

A. Background to the Susceptibility Index

Salt and Ford[9] investigated factors that determine the residual toxicity of pesticides to insects using a stochastic simulation model. Their model simulates the encounter and transfer of insecticide from treated plant surfaces to lepidopteran larvae and predicts the proportion of insects responding. Jepson et al.[10] developed a reductionist approach based upon a sensitivity analysis of this model for short-term hazard prediction for terrestrial invertebrates exposed to pesticides. They postulated that the walking velocity of the insect, the proportion of pesticide transferred per encounter, and the insects' area of contact with the treated surface had an important influence on its susceptibility to pesticide residues. The index (given below), proposed by Jepson et al.,[10] consists of the ratio between an exposure function (based on walking track width, walking speed, and the proportion of the area covered by the insect that is contacted) and susceptibility (expressed as the topical tolerance distribution of the species at endpoint). Susceptibility is taken as the topical LD_{50} in terms of dose per insect, as this takes into account variations in body size between species. The susceptibility index value gives an estimate of the dose encountered per unit tolerance of the species. A high index value would indicate that lethal doses may be readily acquired by that species, which may therefore be highly susceptible to pesticide residues, whereas a relatively low susceptibility index value may indicate slower uptake and thus lower susceptibility. The index only provides a comparative measure of susceptibility under the given experimental conditions and is not intended to be used to predict effects in the field, as these will be related to complex biological and operational factors such as the behavior and distribution of the organisms, the environmental conditions (i.e., temperature and humidity), and the nature of the pesticide.[3,11] The index may, however, be used to compare groups of organisms with similar habits. It is an intrinsic characteristic that modifies basic susceptibility measurements by a function of potential exposure.

$$\text{Susceptibility Index} = (v \times w \times a)/LD_{50}$$

Susceptibility parameter: topical LD_{50} (μg ai/insect)
Exposure function parameters:

  v = mean walking speed (cm/s)
  w = mean track width (distance between tarsi)(cm)
  a = mean contact area (proportion of area covered by insect that is
       contacted)

II. MATERIALS AND METHODS

A. Test Invertebrates

The predators that were tested are highly ranked in importance as cereal aphid predators.[12,13] They included polyphagous predators such as the large ground beetles *Pterostichus melanarius* (Illiger) and *Nebria brevicollis* (F.), the medium-sized ground beetle *Agonum dorsale* (Pont.), and the small ground beetles *Demetrias atricapillus* (L.) and *Bembidion lampros* (Herbst.) (Coleoptera, Carabidae). Also included were the small rove beetle *Tachyporus hypnorum* (F.) (Coleoptera, Staphylinidae) and the aphid-specific ladybird beetle *Coccinella septempunctata* (L.) (Coleoptera, Coccinellidae).

The predators were captured between October 1989 and September 1990 in cereal fields and field margins at Leckford, near Stockbridge, Hampshire, U.K., by dry pitfall trapping, Dietrick vacuum suction sampling, hand-held air aspirator, and surface searching.[14] The Carabidae and Staphylinidae were kept in plastic aquaria, containing a layer of moist soil. They were fed on ground, moist cat biscuits (''Delicat'', Quaker® Latz GmbH). The coccinellids were placed in perspex boxes with barley plants infested with *Sitobion avenae* (F.) (Hemiptera: Aphididae). All invertebrates were kept in a controlled environment room in an insectary, maintained at 19 to 22°C, 55 to 70% relative humidity, and photoperiod 16:8 L:D, prior to treatment.

B. Determination of Index Parameters

1. Topical Bioassays to Determine Species Susceptibility to Deltamethrin

Formulated deltamethrin (Decis®, 25 g/l; Hoechst UK Ltd.) was used as the test chemical. Distilled water was used as the diluent and for control treatments. Topical applications were performed using a 250-μl Hamilton® gas-tight syringe mounted in either a Burkard hand or automatic microapplicator (Burkard Manufacturing Co. Ltd.).[15] The syringe was calibrated to deliver drops of 0.1, 0.5, or 1 μl (depending on the size of the test organism). The syringe and needle were cleaned with detergent ("Decon® 90", Decon® Manufacturing Ltd.) and thoroughly rinsed in tap and then distilled water between treatments. Prior to treatment, test organisms were anesthetized with CO_2 from a cylinder supply. The period of exposure to CO_2 was between 30 to 60 s, again depending on the size of the test species. Pesticide drops were placed at the junction of the pronotum and elytra. Initially, range-finding tests were carried out, with five logarithmically spaced doses and a water control and five to ten test organisms per dose. From the results of these treatments, a definitive dose range was determined with four to seven doses spanning the range of >10 to <90% mortality. The number of individuals of each species tested per pesticide dose was dependent on their abundance in the field. We tested 30 insects per dose for *P. melanarius, N. brevicollis, A. dorsale, D. atricapillus, T. hypnorum,* and *C. septempunctata* and 20 insects per dose for *B. lampros.*

After treatment, all test invertebrates were placed in clean, ventilated containers with food and returned to the insectary where responses were recorded at 24-h intervals for the next 4 d. Individuals were classified as unaffected (moving as normal), or affected, either knocked down or dead.

Probit analysis was performed on the 72-h mortality data to obtain dose-response statistics.[16] Abbott's[17] formula was used to correct the data for control mortality.

2. Exposure Function Parameters

a. Mean Walking Speed   A Panasonic® video camera (WVP-A1E) and video cassette recorder were used to record the movement of the seven coleopteran species on a lightly compacted, sieved, sandy loam soil surface. Video recordings were made for two batches of five individuals of each species in a plastic arena 56 × 29 × 9 cm. The sides of the arena were coated with Fluon Grade GP1 (PTFE) (Whitford Plastics, Runcorn, Cheshire, U.K.) to prevent beetles from climbing the arena sides. The video recordings were made in light for the diurnal species (*C. septempunctata, D. atricapillus,* and *B. lampros*) and in the dark with a red light source for the nocturnal species (*P. melanarius, N. brevicollis, A. dorsale,* and *T. hypnorum*) in a controlled environment room in an insectary, maintained at 19 to 22°C, 40 to 60% relative humidity, and photoperiod 16:8 L:D. Griffiths et al.[18] have shown that red light does not affect the nocturnal activity of *A. dorsale.* An estimate of the walking speed of each species was obtained by analyzing a 1-h period of the video recording for each species. A time was chosen when all beetle species were active. This was from 11:00 to 12:00 h Greenwich Mean Time (GMT) for diurnal species and 23:00 to 00:00 h GMT for nocturnal species. Ten walking paths were measured for individuals of each species over five 10-min intervals by tracing the walking path of insects on a sheet of acetate overlying the monitor screen. An ipsometer (map measurer) was calibrated, allowing for screen curvature, to measure the length of path (cm). The time taken to cover the distance was determined from a digital counter on the screen and an estimate of the mean walking speed was calculated for each species. A total of 50 readings were made per species. These measurements were later verified by Micromeasure (Wye College Software), which is a computerized system enabling analysis of video recordings.

b. Measurement of Contact Area and Track Width   White kymograph paper (R.A. Brand Ltd., Paper Manufacturers, Hedge End, Hampshire, U.K.) was attached to a kymograph apparatus[19] in a fume cupboard. The paper was coated with a layer of soot by igniting gas passed through a Drechsel bottle assembly (Quickfit® Ltd.) containing toluene (general purpose reagent, GPR) (Merck® Ltd.) connected to a Johnson burner. The paper was smoked uniformly by rotating the kymograph drum at a constant rate over the burner. After it had been smoked, the paper was carefully removed from the drum and attached to a flat surface. Individual test invertebrates were allowed to walk across the paper, sweeping away the layer of soot on the areas they contacted and leaving a track. Clearer tracks were obtained when the paper was coated with a thin uniform soot layer. The tracks were semi-permanently fixed through immersion in a 2% solution of Shellac (Merck® Ltd.) in methanol GPR (Merck® Ltd.). An IBAS image analysis computer (Kontron Ltd.) was calibrated to measure the proportion that the area of smoke removed by the insect occupied within its walking track. A total of 30 tracks were measured for each species.

C. Validation of Index Predictions

Laboratory residual bioassays were carried out to test the index predictions of the relative susceptibilities of the beetle species to deltamethrin. Only five of the seven beetle species were tested because the other species were absent from the field collection sites at the time of the test. The species tested included four species of carabids, *P. melanarius, N. brevicollis, B. lampros,* and *D. atricap-*

illus, and one staphylinid, *T. hypnorum*. Batches of 14 to 16 beetles of each species were exposed to deltamethrin deposits in all three bioassays and batches of ten insects were used in the controls. Plastic trays (56 × 29 × 9 cm) with Fluon (PTFE)-coated sides were used as arenas. The tops of each arena were covered with a glass sheet to prevent beetles that were capable of flying from escaping. The chambers were ventilated via hypodermic needles placed under the glass sheet and connected to an aquarium pump via rubber tubing. Bioassays were for 2 h on glass and 24 h on soil. The same arenas were used in both types of bioassay. In the first tests a sheet of glass was fitted to the bottom of the arena and in the second set of tests a 2-cm-deep layer of sieved sandy loam soil was lightly compacted into the base of the arena. The surfaces were then sprayed with deltamethrin at a rate of 6.25 g ai/ha and a volume of 200 l/ha with a CP15 knapsack sprayer (Cooper Pegler Ltd.) fitted with a 1-m boom and three Lurmark FO2-80 nozzles. The spray deposit was allowed to dry for approximately 30 min before the beetles were introduced into the arenas. The control arenas remained untreated. After exposure, the beetles were removed from the test arenas and placed in clean, ventilated containers with food and kept in the insectary where responses were recorded at 24-h intervals for the next 4 d. Individuals were classified as unaffected (moving as normal), or affected, either knocked down or dead. Individuals that displayed uncoordinated movement when stimulated were recorded as knocked down and individuals that remain immobile and showed no response to stimulation were recorded as dead. Treatment mortality was corrected to control mortality using Abbott's formula.[17]

III. RESULTS

The parameters used to calculate the exposure function are given in Table 1. The mean walking speed of the beetle species varied from 0.46 cm/s for *C. septempunctata* to 2.87 cm/s for *A. dorsale*. Frequency distributions of walking speeds are plotted for each species in Figure 1. The frequency distributions obtained for all species were found to be normally distributed (χ^2 test for goodness of fit, $p > 0.05$). The small staphylinid beetle *T. hypnorum* and the small carabid beetle *B. lampros* had the narrowest mean track widths (0.23 and 0.21 cm, respectively), and the large carabid *P. melanarius* had the widest mean track width (1.41 cm). The species with the largest mean proportional contact area with the surface were the large carabid *P. melanarius* and the small staphylinid *T. hypnorum*, contacting 6 and 5.9% of the area swept (Table 1). The species that had the lowest mean proportional contact areas were the small carabid *D. atricapillus* (1.3%) and the ladybird *C. septempunctata* (1.7%). Example of walking track patterns of three of the test species can be seen in Figure 2 (a to c). The track patterns varied considerably between the species tested. For example, the ground beetle *P. melanarius* tended to drag its tarsi over the surface (Figure 2a), whereas the rove beetle *T. hypnorum* tended to contact the surface

Table 1. Parameters For the Estimation of the Exposure of Seven Species of Aphid Predators to Pesticide Residues

Family and species	Mean walking speed (cm/s) (and 95% confidence intervals [CI])	Mean track width (cm) (and 95% CI)	Mean proportional contact area (and 95% CI)
Carabidae			
P. melanarius	1.85 (1.76–1.94)	1.41 (1.36–1.46)	0.060 (0.057–0.062)
N. brevicollis	2.12 (2.04–2.19)	1.37 (1.32–1.41)	0.039 (0.038–0.041)
A. dorsale	2.87 (2.80–2.95)	0.73 (0.71–0.75)	0.024 (0.021–0.026)
D. atricapillus	1.51 (1.47–1.56)	0.36 (0.34–0.38)	0.013 (0.011–0.015)
B. lampros	1.68 (1.63–1.72)	0.21 (0.19–0.23)	0.031 (0.029–0.033)
Staphylinidae			
T. hypnorum	1.85 (1.78–1.92)	0.23 (0.21–0.24)	0.059 (0.056–0.062)
Coccinellidae			
C. septempunctata	0.46 (0.40–0.51)	0.77 (0.74–0.80)	0.017 (0.014–0.020)

with its abdomen, and the ladybird *C. septempunctata* (Figure 2c) tended to leave a more delicate track and did not drag its tarsi or contact the surface with its abdomen. The differences in track patterns and contact areas between the beetle species may be related to differences in body posture and walking action and differences in leg morphology,[20,21] for example, between those species which are adapted for climbing and those which are adapted for running.

The susceptibilities, calculated exposure functions, and susceptibility indices for the seven species of aphid predators are given in Table 2. The most susceptible species to deltamethrin were the small carabid *B. lampros* and the staphylinid *T. hypnorum*, which had 72 h LD_{50} values of 0.013 μg ai per insect. The least susceptible species were the large carabid *N. brevicollis* and the small carabid *D. atricapillus*, which had 72 h LD_{50} values of 0.219 and 0.232 μg ai per insect. The values for the exposure function were greatest for the larger beetles than for the smaller beetles, with the exception of the ladybird *C. septempunctata*, which had the lowest exposure function of all.

When the exposure function was divided by the susceptibility to give the susceptibility index prediction (Table 2), the small staphylinid *T. hypnorum* had the largest value (1.93), which indicated that it may be the most susceptible species to deltamethrin residues. The second highest susceptibility index value was that of the large carabid *P. melanarius* (1.15), which had the largest exposure function value, due to its size and high proportional contact area. The small carabid *D. atricapillus* had the lowest susceptibility index value (0.03), indicating that this species may be least at risk from deltamethrin residues of the seven tested.

The mortality rankings from the two bioassays that tested the predictions of the susceptibility index are given in Table 3. The mortality rankings for the five species tested corresponded to the susceptibility index predictions for the most

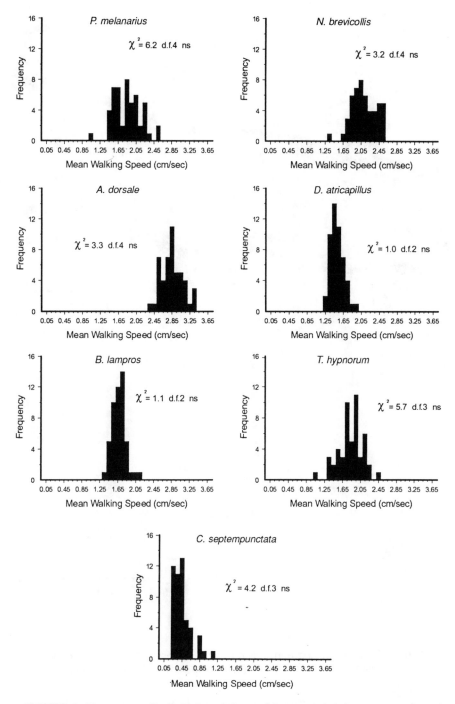

FIGURE 1. Frequency distributions of the walking speeds of seven species of aphid predators (χ^2 values indicate goodness of fit to normal distribution; ns = not significant).

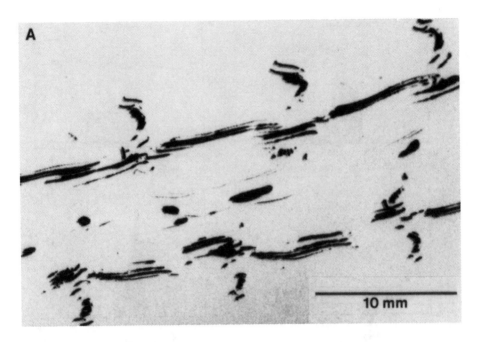

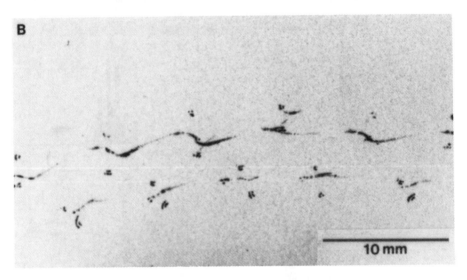

FIGURE 2. Example kymograph walking tracks for three of the predator species tested. (a) *P. melanarius*; (b) *A. dorsale;* (c) *C. septempunctata.*

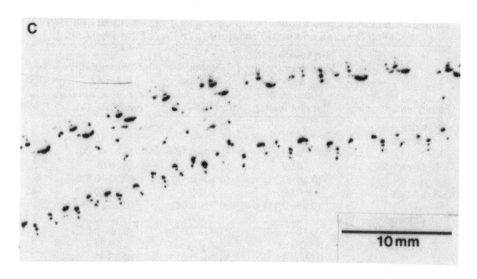

FIGURE 2. *Continued*

susceptible species, *T. hypnorum,* and the least susceptible species, *D. atricapillus.* The intermediate rankings were as predicted by the susceptibilities in the 2-h glass bioassay; however, they did not match the 24-h soil bioassay results. This may be because of differences in behavior on the different substrates, for example, some beetles may burrow into the soil and therefore be exposed to lower levels of pesticide residues.

IV. DISCUSSION

The susceptibility index (Table 2) indicated that, under the given conditions, the staphylinid beetle *T. hypnorum* would be the most susceptible (rank = 1) to deltamethrin residues, because it had a high exposure index and was relatively susceptible to the deltamethrin. The index predicted that the small carabid *D. atricapillus* would be least susceptible to deltamethrin residues (rank = 7) because it had a low proportional contact area and a high LD_{50} value. This may suggest that the susceptibility of the species to residues is strongly related to the intrinsic susceptibility of the insect to the pesticide. However, trends within the five carabid species tested indicated that susceptibility to pesticide residues may not be so easy to predict. For example, the large carabid *P. melanarius* was predicted to be more susceptible than several smaller, more susceptible species.

The index predictions for the most and least susceptible species agreed well with the mortalities observed in the two residual bioassays. However, the species falling between these two extremes conformed less consistently with the predictions. The 95% confidence intervals of the susceptibility index for some of

Table 2. Susceptibilities, Exposure Functions, and Susceptibility Indices of Seven Species of Aphid Predators to Deltamethrin Residues

Family and species	72-h topical LD_{50} for deltamethrin (μg ai/insect) (and 95% fiducial limits)	Exposure function (v × w × a) (and 95% C.L.)	Susceptibility index (v × w × a)/LD_{50} (and 95% C.L.)	Rank[a]
Carabidae				
P. melanarius	0.14	0.16	1.15	2
	(0.09–0.21)	(0.14–0.18)	(0.67–2.00)	
N. brevicollis	0.22	0.11	0.52	5
	(0.16–0.31)	(0.10–0.13)	(0.33–0.82)	
A. dorsale	0.08	0.05	0.60	4
	(0.06–0.11)	(0.04–0.06)	(0.36–0.96)	
D. atricapillus	0.23	0.007	0.03	7
	(0.18–0.30)	(0.005–0.009)	(0.02–0.05)	
B. lampros	0.013	0.011	0.84	3
	(0.009–0.018)	(0.009–0.013)	(0.53–1.44)	
Staphylinidae				
T. hypnorum	0.013	0.025	1.93	1
	(0.010–0.016)	(0.021–0.029)	(1.43–2.96)	
Coccinellidae				
C. septempunctata	0.10	0.006	0.06	6
	(0.07–0.14)	(0.004–0.008)	(0.03–0.12)	

[a] Rank = 1, predicted to be most susceptible; rank = 7, predicted to be least susceptible.

the species tested showed considerable overlap, indicating that the intermediate rankings may be sensitive to small variations in susceptibility or walking speed. When the mean walking speeds were excluded from the exposure function, however, the susceptibility values still matched the ranking from the 2-h glass bioassay. This may indicate that saturation points for the uptake of deltamethrin from the glass surface were reached for the species tested. Saturation of pyrethroid

Table 3. Predicted and Actual Rankings of the Susceptibilities of Five Species of Aphid Predators to Deltamethrin Residues

Family and species	Susceptibility index[a]	Mortality ranking from 2-h glass bioassay (% mortality in parenthesis)[a]	Mortality ranking from 24-h soil bioassay (% mortality in parenthesis)[a]
Carabidae			
P. melanarius	2	2 (60)	3 (13)
N. brevicollis	4	4 (38)	4 (0)
D. atricapillus	5	5 (20)	4 (0)
B. lampros	3	3 (57)	2 (27)
Staphylinidae			
T. hypnorum	1	1 (71)	1 (38)

[a] Rank 1 = most susceptible; rank 5 = least susceptible.

uptake with distance moved was not found to occur with lepidopteran larvae by Salt and Ford,[9] but has been found to occur in a short period of time with linyphiid spiders by Jagers op Akkerhuis and Hamers.[22]

The validation of the susceptibility index is not conclusive. However, the index appears to be capable of predicting extremes of susceptibility. This may be useful for regulatory bodies who wish to select the species most at risk for further registration testing and also for the interpretation of field trial results. The susceptibility index is applied to plant and soil surface-active invertebrates only and not subterranean species. Loose or fissured soil, or soil with a surface flora are likely to present different risks of contamination and toxic effects. Thus the nature of the index as a corrective factor for laboratory bioassay data must be emphasized. The index does not take into account the following factors which will also affect the impact of a given residual pesticide deposit in the field:

a. The species diel activity cycles (i.e., one species may walk rapidly, but may only be active for a short period each day and may therefore pick up a small amount of pesticide: a slower-moving species may spend a larger proportion of the day active and pick up a larger amount of pesticide).
b. The distribution of the species through the crop (i.e., some species may be active on plant surfaces where the pesticide may be at higher residue levels or be more available than on the soil).
c. The proportion of the population that is likely to be in the crop during and shortly after spraying.
d. Behavioral responses, such as activation or repellency, which may modify exposure dramatically.

To develop models that may accurately predict mortality resulting from exposure to residual deposits of pesticides it is necessary to quantify the amount of pesticide picked up in a given exposure period. Salt and Ford[9] and Jepson et al.[10] established that the toxicity of residual deposits of pesticides against crawling insects is influenced by the droplet size and density, the mass of active ingredient, and concentration-dependent behavioral responses. The extent of pesticide transfer will depend on the probability of the insect encountering the pesticide and the proportion of the insecticide which adheres to the insect cuticle.[23] To attempt an analysis of all these parameters for numerous species would be extremely time consuming. Therefore, simple indices, of the form described in this paper, may be used in testing frameworks to select species for *in situ* bioassays within the appropriate crop.[24] These may then feed into more complicated models of hazard, which could be verified with semifield trials if safety criteria are not met.

ACKNOWLEDGMENTS

J.A.W. was in receipt of a MAFF CASE award with The Game Conservancy Trust. We would like to thank Dr. N.W. Sotherton of The Game Conservancy

Trust, Dr. M.G. Ford and Dr. D.W. Salt from Portsmouth University, and Dr. G.A. Jagers op Akkerhuis from the Agricultural University of Wageningen for their comments and advice. J.A.W. would also like to thank Mr. W.J. Parks for his assistance with some of the experimental work.

REFERENCES

1. Hess, C., Presentation, Proc. of Ohio State University, Battelle Endowment for Technology and Human Affairs, 1–27, 1987.
2. Brown, R.A., Pesticides and non-target terrestrial invertebrates: an industrial approach, in *Pesticides and Non-Target Invertebrates*, Jepson, P.C., Ed., Intercept, 1989, 19.
3. Jepson, P.C., The temporal and spatial dynamics of pesticide side-effects on non-target invertebrates, in *Pesticides and Non-Target Invertebrates*, Jepson, P.C., Ed., Intercept, 1989, 95.
4. OECD Report of the workshop on environmental effects assessment, *Environ. Monogr.*, May 1989, Paris.
5. van Leeuwen, K., Ecotoxicological effects assessment in the Netherlands: recent developments, *Environ. Man.*, 14(6), 779, 1990.
6. Bode, E., Brasse, D., and Kokta, C., Auswirkungen von Pflanzenschutsmitteln auf Nutzorganismen und Bodenfauna: Uberlegungen zur Prufung im Zulassungsverfahren, *Gesunde Pflanz.*, 40, 239, 1988.
7. Hassan, S.A., Testing methodology and the concept of the IOBC/WPRS Working Group, in *Pesticides and Non-Target Invertebrates*, Jepson, P.C., Ed., Intercept, 1989, 1.
8. Jepson, P.C., Insects, spiders and mites, in *Handbook of Ecotoxicology*, Calow, P., Ed., Blackwell Scientific, Oxford, 1993, 299.
9. Salt, D.W. and Ford, M.G., The kinetics of insecticide action. Part III. The use of stochastic modelling to investigate the pick-up of insecticides from ULV-treated surfaces by larvae of Spodoptera littoralis Boisd, *Pestic. Sci.*, 15, 382, 1984.
10. Jepson, P.C., Chaudhry, A.G., Salt, D.W., Ford, M.G., Efe, E., and Chowdhury, A.B.M.N.U., A reductionist approach towards short-term hazard analysis for terrestrial invertebrates exposed to pesticides, *Functional Ecol.*, 4/3, 339, 1990.
11. Everts, J.W., Sensitive Indicators of Side-Effects of Pesticides on the Epigeal Fauna of Arable Land, Ph.D. thesis, University of Wageningen, 1990.
12. Sunderland, K.D. and Vickerman, G.P., Aphid feeding on some polyphagous predators in relation to aphid density in cereal fields, *J. Appl. Ecol.*, 17, 389, 1980.
13. Sopp, P. and Wratten, S.D., Rates of consumption of cereal aphids by some polyphagous predators in the laboratory, *Entomol. Exp. Appl.*, 41, 69, 1986.
14. Southwood, T.R.E., *Ecological Methods: with Particular Reference to the Study of Insect Populations*, Chapman and Hall, London, 1987.
15. Arnold, A.J., Hand-operated micro-applicator to deliver drops of five sizes, *Lab. Prac.*, 16, 56, 1967.

16. Finney, D.J., *Probit Analysis,* 2nd ed., University Press, Cambridge, 1971.

17. Abbott, W.S., A method of computing the effectiveness of an insecticide, *J. Econ. Entomol.,* 18, 265, 1925.

18. Griffiths, E., Wratten, S.D., and Vickerman, G.P., Foraging by the carabid *Agonum dorsale* in the field, *Ecol. Entomol.,* 10, 181, 1985.

19. Jackson, D.E., A new kind of long paper kymograph, *J. Lab. Clin. Med.,* 2, 426, 1917.

20. Forsythe, T.G., Running and pushing in relationship to hind leg structure in some Carabidae (Coleoptera), *Coleopt. Bull.,* 35, 353, 1981.

21. Forsythe, T.G., Locomotion in ground beetles (Coleoptera: Carabidae): an interpretation of leg structure in functional terms, *J. Zool., London,* 200, 493, 1983.

22. Jagers op Akkerhuis, G.A.J.M. and Hamers, T.H.M., Substrate-dependent bioavailability of deltamethrin for the epigeal spider Oedothorax apicatus (Blackwall) (Aranaea, Erigonidae), *Pestic. Sci.,* 36, 59, 1992.

23. Ford, M.G. and Salt, D.W., Behaviour of insecticide deposits and their transfer from plant to insect surfaces, in *Pesticides on Plant Surfaces: Critical Reports in Applied Chemistry,* Cottrell, H., Ed., John Wiley & Sons, London, 1987, 26.

24. Jepson, P.C., Duffield, S.J., Thacker, J.R., Thomas, C.F.G., and Wiles, J.A., Predicting the side effects of pesticides on beneficial invertebrates, *Br. Crop Protect. Conf.,* 3, 957, 1990.

CHAPTER 22

Effects of Different Input of Pesticides and Fertilizers on the Abundance of Arthropods in a Sugar Beet Crop: An Example for a Long-Term Risk Assessment in the Field

W. Büchs

TABLE OF CONTENTS

I. INTRODUCTION

Effects of different input of pesticides on various soil parameters have been investigated by an interdisciplinary working group at the Federal Biological Research Centre at Braunschweig.[1] Topics of this research program have been: behavior of residuals of fungicides, insecticides, herbicides,[2,3] as well as effects on soil-inhabiting microorganism,[4] soil-inhabiting fungi and algae,[5,6] earthworms,[7] endogaeic collembola,[8] soil mites,[9] and soil-inhabiting nematoda.[10] In this research program until now no research work has been carried out for all taxa which have a nascent stage developing in the soil (e.g., Diptera, Staphylinidae, Carabidae, and other Coleoptera) and which are in permanent contact with the soil during their active live stages (e.g., Araneae, Heteroptera, various taxa of Coleoptera, some [wingless] parasitic Hymenoptera, epigaeic Collembola, and mites).

Except for the carabid beetles and some selected species of Staphylinidae and other Coleoptera, we know only little particularly about pterygote taxa with nascent stages developing in the soil. Especially there is a lack of knowledge to be considered about abundance (i.e., emergence or hatching rates which can be related to a defined area).

It is assumed that soil arthropods with an epigeic life stage are more affected by pesticide applications as true soil organisms, because their exposure to the pesticides is more direct.

The aim of the investigation was to record the long-term effects of different pest management intensities on the arthropod species community at least for one crop rotation (sugar beet, winter wheat, winter barley).

Derived from reflections regarding forest ecosystems,[11] it was also the aim to create a kind of "minimum program for the analysis of ecosystems". That means, to assess a simple experimental design with the lowest possible amount of field work and to use standardizable methods which are suitable for the detection of long-term effects. Therefore, only "automatically" working sampling methods were used which do not require a frequent presence in the field.

The results for the sugar beet crop are presented here in a general point of view. Further detailed evaluations on the species level are done especially for the Staphylinidae,[12] Brachycera,[13] Nematocera (Weber and Büchs in press), Araneae,[14] Acari, Collembola (Kampmann and Büchs, in prep.), parasitic Hymenoptera (Vidal and Büchs, in prep.), Thysanoptera (Patrzich and Büchs, in prep.), and several other taxa.

II. MATERIAL AND METHODS

In each pest management intensity five emergence traps[15,16] covering an area of 1 m² have been used. In the soil covered by the emergence trap five pitfall traps have been dug in in order to catch all arthropods which are not positive phototactic. As conservation fluid a 5% solution of sodium benzoic acid was used. Additionally, six pitfall traps[17] per intensity were installed outside the emergence traps. Since April 1989 all traps were checked at intervals of 14 d. Each month the location of the emergence traps was changed ad random. Although the sampling is continued until November 1992, in this paper only the results of the period of the sugar beet crop (from April to October 1989) are presented.

The investigations have been carried out on a field of 12 ha about 15 km south of Braunschweig. The field was divided in four plots of different size (I_0 = about 1.2 ha, I_1 = about 3.2 ha, I_2 and I_3 = about 3.8 ha) which have been managed with a different input of pesticides and fertilizers since 1982.

I_0 = Crop production without input of pesticides (except seed dressing), minimum input of fertilizers.

I_1 = Extensive crop management with a suboptimum input of nitrogen and low input of pesticides. Renunciation of a high yield.

I_2 = Integrated crop management with the aim to come to a high yield by minimizing the input of pesticides and fertilizers as far as possible.

I_3 = Intensive crop management by using all registered and required agents (means) to obtain a maximum yield. Prophylactic use of pesticides.

For the amount of pesticides and fertilizers used in the different intensities see Table 1. In sugar beet, a clear difference between the pest management intensities can only be seen for the input of insecticides, whereas in the 3-year period of the full crop rotation the differences in the input of herbicides, growth regulators, fungicides, insecticides, and fertilizers are very obvious regarding the total amount as well as the number of applications.

III. RESULTS AND DISCUSSION

The emergence of the investigated field soil in the sugar beet crop was nearly 13,000 ind./m^2 in the average. This is a quite remarkable abundance in comparison to the emergence rates recorded in other, less disturbed ecosystems.[18-20]

Great differences in the emergence rate are obvious between the management intensities (Table 2): in I_0 nearly 2.5 times more individuals emerged as in I_3, so that a clear relation can be registered between the pest management intensities and the presence of arthropods. That means, the more the farming intensity increases, the more the natural development of arthropod populations is disturbed. In the sugar beet crop nearly 80% of the taxa listed in Table 2 reached in I_0 a higher emergence than in I_3. A graduated decrease of the emergence according to the increase of pest management intensity has been recorded for the Acari, Araneae, some carabid beetles (e.g., *Bembidion* spp.), the Cecidomyiidae, Chironomidae, Collembola, Heteroptera, the Staphylinidae, Anotylus spp., Lathrobium spp., Tachyporus spp., and the Sciaridae.

Table 1. Application of Pesticides and Nitrogen in Sugar Beet, 1989

Date	I_0	I_1	I_2	I_3
Herbicides				
3.31.89 4.21.89	— —	Chloridazone 65%: 2 kg/ha Ethofumesate 50% + Phenmedipham 90%: 2 l/ha Metamitrone 70%: 1 kg/ha	Chloridazone 65%: 2 kg/ha Ethofumesate 50% + Phenmedipham 90%: 2 l/ha Metamitrone 70%: 1 kg/ha	Chloridazone 65%: 2 kg/ha Ethofumesate 50% + Phenmedipham 90%: 2 l/ha Metamitrone 70%: 1 kg/ha
5.8.89	—	Chloridazone 65%: 1 kg/ha Ethofumesate 50% + Phenmedipham 90%: 2 l/ha Metamitrone 70%: 2 kg/ha	—	
5.17/18.89	—	Glyphosate: 5 l/ha	Fluazifop-butyl 250 g/l: 1 l/ha Glyphosate: 5 l/ha	Metamitrone 70%: 2 kg/ha Glyphosate: 5 l/ha
5.24/29.89	—	Ethofumesate 50% + Phenmedipham 90%: 3 l/ha Chloridazone 65%: 2 kg/ha	Ethofumesate 50% + Phenmedipham 90%: 3 l/ha Metamitrone 70%: 2 kg/ha	Ethofumesate 50% + Phenmedipham 90%: 3 l/ha Chloridazone 65%: 2 kg/ha
Insecticides				
3.30/31.89	Methiocarb (seed dressing)	Methiocarb (seed dressing) Carbofuran 5% (granules): 5 g/m	Methiocarb (seed dressing) Carbofuran 5% (granules): 5 g/m	Methiocarb (seed dressing) Carbofuran 5% (granules): 5 g/m
5.27/29.89	—		Pirimicarb 50%: 300 g/ha	Pirimicarb 50%: 300 g/ha
6.8/13.89	—	Pirimicarb 50%: 200 g/ha	Pirimicarb 50%: 300 g/ha	Oxydemetone-methyl 256 g/l: 0.8 l/ha
6.12/23/29.89	—	Pirimicarb 50%: 300 g/ha	Pirimicarb 50%: 300 g/ha	Pirimicarb 50%: 300 g/ha
7.6/10.89	—	—	Pirimicarb 50%: 300 g/ha	Pirimicarb 50%: 300 g/ha
Nitrogen				
	50 kg/ha	280 kg/ha	288 kg/ha	295 kg/ha

Obviously, higher emergence rates of the intensity I_0 (without input of pesticides) in comparison to the intensities treated with insecticides (I_1 to I_3) showed the Aleocharinae, Aphidina, *Atomaria* spp., some carabid beetles such as *Bembidion quadrimaculatum, B. tetracolum,* some coccinellid and spider species such as *Coccinella undecimpunctata* or *Meioneta rurestris,* the Coccinellidae larvae, the Symphypleona, Lepidoptera, Psocoptera, and Psyllina.

A difference in the emergence of the lower intensities (I_0, I_1) on the one hand and of the higher intensities (I_2, I_3) on the other hand was registered for the Araneae, the genus *Cryptophagus*, the Elateridae, Formicidae, Hymenoptera, Sciaridae, and some Staphylinidae. For the Cicadina, Diplopoda, and the Phoridae the emergence in I_3 was obviously suppressed.

No influence of the pest management intensity could be stated for the Cantharidae, Carabidae (especially *Trechus quadristriatus* and *Pterostichus melanarius*), the Chironomidae, Empididae, Lathridiidae, Rhagionidae, Symphyta, Thysanoptera, Coleoptera- and Planipennia-larvae. Among these are a lot of predatoric taxa which were obviously not affected by the management intensity. Even an increase is recognizable for the Bibionidae and the carabid beetles of the genuses *Clivina* and *Calathus*.

Because of the monthly moving of the emergence traps the abundances listed in Table 2 might be higher as in reality for a certain percentage. This discrepancy is lower the higher the percentage of the emergence rate of one control period (1 month) is in relation to the total amount of all individuals of each taxon. This relation is listed in Table 2 together with the control period in which the highest emergence rate was registered for each taxon. These columns show that in the average nearly 60% of all individuals emerge in 1 month and that the overall highest emergence is registered between mid-May and mid-June.

The above mentioned results lead to the questions: Which factors are influencing the presence of individuals in the different management intensities? What role is played, for example by insecticide applications between the end of May and mid-July, a time when no other pesticides have been applied?

The aphids are the target of the insecticide spray applications. Figure 1 shows that the increase of the population development was stopped in the intensities I_1 to I_3 immediately when insecticides have been applied. If similar effects can be observed also for nontarget organisms, one is able to suggest a side effect of the used insecticides.

The larvae of the Sciaridae contribute in a decisive way to the decomposition of organic litter remaining after harvesting.[21] This also applies to the larvae of other families of the Nematocera. Dipterous larvae may take over the function of earthworms or other decomposers.[22] The enormous emergence of the Sciaridae in sugar beet, with a maximum in the second half of May (Figure 2), depends obviously on the growing of winter barley in the year before and the catch-crop before the sugar beet was sown. From these crops a great amount of organic litter remains after harvesting rsp. wielding of the catch-crop into the soil. The Sciaridae showed a decrease of their emergence rate up to 75% when the management intensity increased (Figure 2). In I_0 and I_1 their population development

Table 2. Abundance of Different Arthropod Taxa Recorded With Emergence Traps (ind./m²) in Sugar Beet (April to October 1989)

Taxon	Cultivation Intensity				Average of I_0 to I_3	Dominance %	Period with Maximum Abundance	% of total no. of individuals $\bar{x}$ (I_0 to I_3)
	I_0	I_1	I_2	I_3			Date	
Sciaridae	4,100	4,000	2,000	1,100	2,825	22.1	V/VI	86.9
Acari	3,000	2,300	1,400	750	1,860	14.4	X	31.7
Collembola, Arthropleona	3,000	2,000	2,000	1,400	2,100	16.3	IV/V	50.8
Aleocharinae	1,700	1,000	750	950	1,100	8.5	V/VI	58.6
Phoridae	780	1,600	790	310	870	6.7	IV/V	59.3
Thysanoptera	1,250	500	1,100	480	830	6.4	VI/VII	37.7
Aphidina	1,600	350	200	650	700	5.4	VI/VII	48.0
Chironomidae	590	480	800	400	570	4.4	VII/IX	77.0
Hymenoptera	440	550	280	330	425	3.3	VII/VIII	54.6
Araneae	450	450	250	200	350	2.7	VII/VIII	30.3
Atomaria spp.	340	130	170	230	220	1.7	V/VI	52.3
Brachycera (rest)	50	230	300	210	200	1.6	VI/VII;IX/X	35.1
Carabidae	100	190	170	85	135	1.0	V/VI;IX/X	46.5
Empididae	110	60	75	110	90	0.7	V/VI	42.7
Collembola, Symphypleona	400	14	44	7	116	0.9	VI/VII;X	31.9
Cecidomyiidae	125	100	90	43	90	0.7	IV/V	27.8
Anotylus spp.	130	40	35	35	60	0.5	IV/V	61.7
Staphylinidae (rest)	80	75	40	40	60	0.5	IV/V	31.7
Lathridiidae	55	35	52	55	50	0.4	IX/X	32.0
Heteroptera	75	47	35	30	47	0.4	VIII/IX	40.4
Cicadina	53	40	52	13	40	0.3	VII/VIII;IX/X	40.0
Coccinellidae	46	7	8	7	17	0.13	VII/VIII	84.1
Coleoptera (rest)	22	25	9	11	17	0.13		
Psocoptera	21	14	12	13	15	0.12	IX/X	56.7
Chironomidae-larvae	35	10	5	1	13	0.10	IV/V	86.9
Diplopoda	15	21	14	2	13	0.10	IX/X	42.3
Elateridae	12	16	7	3	10	0.08	VI/VII	46.0
Nematocera (rest)	6	4	7	19	9	0.07		
Coleoptera-larvae	11	5	12	4	8	0.06	V/VI;IX/X	37.5

Taxon								
Larvae (rest)	11	5	8	8	8	0.06	VII/VIII	7.13
Lepidoptera	13	5	6	4	7	0.05	VII/VIII	21.4
Planipennia-larvae	6.4	5	3	11	6.5	0.05	VII/VIII	100.0
Carabidae-larvae	6.2	8.8	4.8	5.4	6.3	0.05	IX/X	49.2
Lepidoptera-larvae	11	2.8	5	3.2	5.5	0.04	VII/VIII	45.4
Psyllina	1	2.6	0	1	5.7	0.04	VII/X	73.7
Coccinellidae-larvae	7.4	1.8	3.4	4.6	4.3	0.03	VII/VIII	60.5
Formicidae	5.8	5.4	0.6	2	3.5	0.03	IV/V;VII/IX	37.1
Planipennia	2.2	4.2	1.8	4.4	3.2	0.02	VII/IX	46.9
Scatopsidae	2.4	1.0	3.4	5.8	3.2	0.02	V/VI	59.4
Diptera-larvae	0	0.4	5.8	3	2.3	0.02	IX/X	52.2
Symphyta	0.6	0	0.6	0	0.3	0.0	IV/V	83.3
Siphonaptera	0.2	0.2	0	0	0.1	0.0	IV/V	100.0
Trechus quadristriatus[a]	56	160	150	53	105	0.8	V/VI;IX/X	41.0
Bembidion quadrimaculatum[a]	30	2.4	1.4	2.6	9.1	0.07	VI/VII	51.8
Bembidion tetracolum[a]	13	3	7	1.2	6	0.05	VI/VII	40.3
Pterostichus melanarius[a]	0.8	10.4	1.2	13	6.4	0.05	VI/VII	65.2
Calathus spp.[a]	1.6	3.6	3.2	6.4	3.7	0.03	VII/VIII	55.2
Clivina fossor[a]	0	0.2	0.2	3.6	1	0.01	V/VI	83.3
Lathrobium spp.[a]	17	16	12	7	13	0.10	V/VI	40.7
Tachyporus spp.[a]	14	11	9	8	11	0.09	VII/IX	46.4
Coccinella 11-punctata[a]	26	0.2	0.2	0.2	6.7	0.05		
Propylaea 14-punctata[a]	4.8	4.4	3.6	4.4	4.4	0.03		
Adrastus rachifer[a]	8.8	5.6	2.4	1.8	4.7	0.04	VI/VII	50.0
Athous bicolor[a]	1.8	1.6	1.6	0.2	1.4	0.01	VI/VII	100.0
Pseudathous niger[a]	0.4	2.6	1	1	1.3	0.01	V/VI	100.0
Adrastus pallens[a]	0.4	1.2	0.2	0.2	0.7	0.01	VI/VII	100.0
Cryptophagus spp.[a]	6.4	11	3.4	2.2	5.8	0.04	IX/X	55.4
Cantharidae[a]	0.4	3.8	0.6	2.2	1.8	0.01	V/VI	94.4
Bibionidae[a]	1.8	0.4	3	13	4.6	0.04	IV/V;VII/VIII	87.0
Rhagionidae[a]	0.2	1.4	4	0.4	1.5	0.01	V/VI	56.7
Total no. of individuals	18,700	14,350	10,750	7,550	12,900			57.8

Note: Greater values are rounded out

[a] Taxa which are already included in the total amount of individuals.

ind./m²

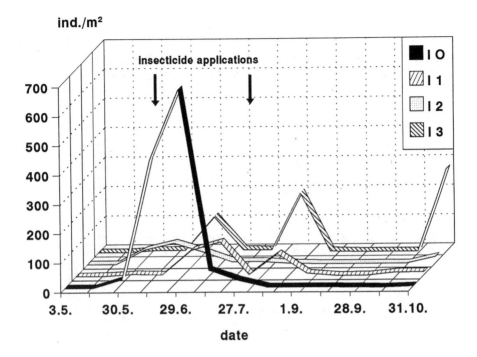

FIGURE 1. Phenology of the aphids (Homoptera: Aphidina) in different pest management intensities (I_0–I_3) in sugar beet 1989.

is established exactly on the same level, whereas in I_2 and I_3 the maximum reached a significantly lower level. It is assumed that this phenomenon is not only caused by the application of insecticides in the spring, but also by long-term effects of the different management intensities in the previous crops.

Terrestrial species of Chironomidae are mainly known from forest ecosystems.[23] The imagines occurred in the sugar beet crop in the first half of September and their emergence showed obviously no relations to the applications of pesticides and fertilizers. The abundance of the larvae in spring, however, was strictly correlated with the intensity of the pest management (Table 2). The gall midges (Cecidomyiidae) reached their highest abundance in all intensities in May. Altogether their emergence rate was comparatively low. But there seems to be a correlation to the pest management intensity (Table 2).

For the Phoridae the highest hatching rates have been recorded from May to mid-June (95% of all individuals). In I_2 and I_3 the Phoridae showed obviously lower emergence rates after the beginning of the insecticidal treatments (29th of May) in comparison to I_0 and I_1 (first insecticide application in I_1 at 13th of June).

Even the Collembola: Arthropleona showed a decrease of emergence related to the increase of the pesticide input (Figure 3). Their effect in grazing tissues

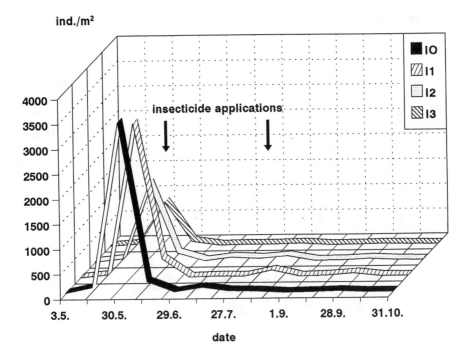

ind./m²

FIGURE 2. Phenology of the Sciaridae (Diptera) in different pest management intensities (I_0–I_3) in sugar beet 1989.

of fungi consists in catalyzing the decomposition of the organic litter.[24] The Arthropleona showed their maximum abundance in spring (April to June), whereas in the time period with applications of insecticides (May 5 to July 10) in I_1 to I_3 the emergence decreased drastically in comparison to I_0 (Figure 3).

This phenomenon can be seen more clearly for the Collembola: Symphypleona, which reached their maximum abundance in the second half of June (Figure 4). As pollen eaters, they live absolutely epigeically on the vegetation where they are obviously more exposed to pesticides than euedaphic Collembola. Their abundance may be influenced also by the density and species composition of the weeds. Figure 4 shows that the abundance of the Symphypleona was apparent only in I_0. In the intensities treated with pesticides (I_1 to I_3) the Symphypleona were not able to build up a population over the whole vegetation period (Figure 4).

The population of Psocoptera, which graze on tissues of algae and fungi, was increasing relatively late in the early autumn (September, October). By a wide margin the highest emergence rate was registered in I_0. The emergence rate of the Diplopoda as primary decomposers of the litter layer was heavily reduced in I_3.

ind./m²

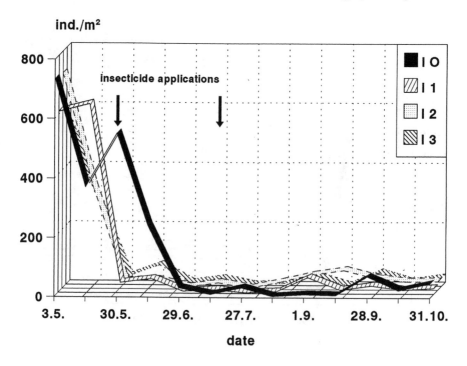

FIGURE 3. Phenology of the Arthropleona (Collembola) in different pest management intensities (I_0–I_3) in sugar beet 1989.

Also, the mites — a taxon which covers a wide ecological spectrum (predators, parasites, decomposers, etc.) — show a 75% reduction of the number of
individuals between the lowest (I_0) and the highest intensity (I_3). In I_0 several
peaks are recognizable: in the second half of May, in July to August, and in
October. Especially, the peak in July, which occurred immediately after the end
of the insecticide applications, is missing in I_1 to I_3. Only in I_1 a full regeneration
of the population of the mites could be recorded in autumn.

The Cryptophagidae of the genus *Atomaria*, which are classified as pest
insects of sugar beet,[25] came to higher abundances only after the four- to six-
leaf stage was reached. But at that time the sugar beet did overcome the most
sensitive stage of development, which is the seedling stage. This result is also
stated by the observations of König.[26] The differences between I_0 and the intensities (I_1 to I_3) treated with insecticides are very clear, although in I_3 high
abundances have been registered too. The Staphylinidae reached their highest
emergence rate in May. This applies especially to the genus *Anotylus* spp.
(mainly *Anotylus rugosus*) and *Lathrobium* spp. (mainly *Lathrobium fulvipenne*).
The reduction of the emergence of these species in I_1 to I_3 (Figure 5) is probably
not connected with the spraying of insecticides in the early summer (May 29 to
July 10) *Lathrobium* spp. and *Anotylus* spp. are both said to be zoophagous,

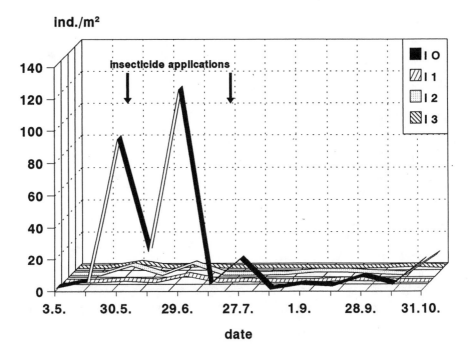

ind./m²

FIGURE 4. Phenology of the Symphypleona (Collembola) in different pest management intensities (I_0–I_3) in sugar beet 1989.

but associated with decaying plant material.[27] Although the preference of carnivorous food can be assumed, the above mentioned Staphylinidae may possibly need decaying plant material and, arised from that, a special microclimate for egg laying, larval development, etc. Their emergence rate indicates that the intensities I_1 to I_3, where herbicides had been used frequently, could not produce these special microclimatic conditions, whereas these conditions are present in I_0 due to the higher density of the weeds.[28] On the other hand, especially, through the results of Zimmermann and Büchs,[12] hints grow stronger that effects of the extremely toxic and systemic acting insecticide Curaterr Granulat (active ingredient: carbofuran), which was applied at the end of March, is the main reason for the decrease of the staphylinid emergence.

Following newer scientific findings[29] the Aleocharinae, a subfamily of the Staphylinidae, live more predatoric, although they are mycetophilic. Additionally, there are some species of the genus *Aleochara* which are parasitoids of fly pupae.[30] They have a potential as antagonists of pest insects, which cannot be estimated at the present status of knowledge.

For the Aleocharinae the emergence rate of I_1 to I_3 was established on nearly the same level (Figure 6). Only at I_0 a considerably higher amount of hatching specimen was registered. According to this the phenology of the Aleocharinae

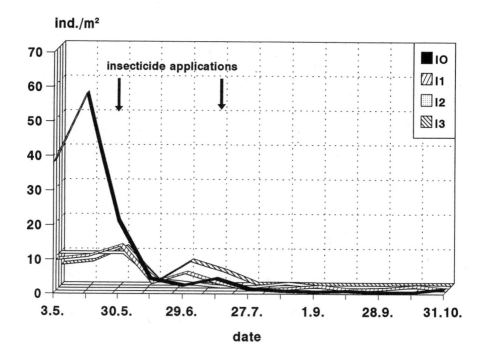

FIGURE 5. Phenology of the genus *Anotylus* (Coleoptera: Staphylinidae) in different pest management intensities (I_0–I_3) in sugar beet 1989.

developed in a synchronous way. However, in all intensities (I_1 to I_3) treated with pesticides, generally lower abundances have been registered compared to I_0. This applies especially to the maximum in the first half of June (Figure 6).

The emergence rate of the parasitic hymenoptera did show higher values in I_0 and I_1, but in general no clear differences between the intensities were recognizable. However, it has to be considered that extreme values of parasitic hymenoptera were recorded in single emergence traps of the intensities I_1 and I_3. At the 15th of August in I_1 in a single emergence trap 1153 individuals have been counted and in one of I_3 at the same time 953 individuals. The emergence of all other emergence traps at this time was about 50 ind./m² in I_1 rsp. 23.5 ind./m² in I_3. The reason for the above mentioned extreme values were distinct species of Encyrtidae which tend to be polyembryonic. That means, the extremely high abundance in single emergence traps can be qualified as an extraordinary phenomenon, which is not connected with the management intensity. So a wrong impression will be produced if these extreme values are included in the evaluation. If these values are not taken into account we have $I_0 = 433$ ind./m², $I_1 = 297$ ind./m², $I_2 = 280$ ind./m², $I_3 = 140$ ind./m², a clear correlation of the emergence of the parasitic Hymenoptera and the pest management intensity.

ind./m²

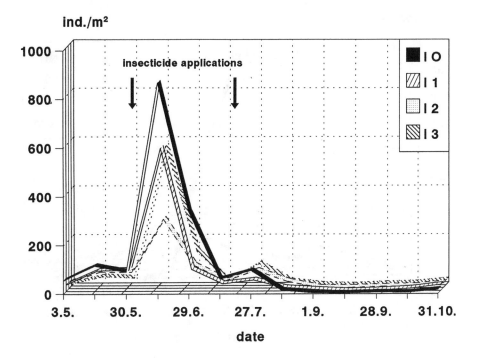

FIGURE 6. Phenology of the Aleocharinae (Coleoptera: Staphylinidae) in different pest management intensities (I_0–I_3) in sugar beet 1989.

Regarding the phenology of the parasitic Hymenoptera (Figure 7), there occurred no difference in the emergence of the different intensities during the time when insecticides have been applied. The following considerably steeper increase of the emergence in I_0, however, showed that there really exists an influence of the management intensity. This influence affects obviously not directly the imagines, but rather more the larvae rsp. their hosts. Indirectly this leads to the effect that in comparison to I_0, in I_1 to I_3 a lower hatching rate of the new generation of parasitic Hymenoptera has been registered.

Regarding the predatoric taxa, the spiders especially showed a decrease in abundance at an increasing pest management intensity. The spiders (especially the Microphantidae and Linyphiidae) reached their maximum abundance after the period of pesticide applications in July/August. At this time the most juvenile spiders have been registered also. Their phenology shows that their emergence in the lower intensities (I_0, I_1) was considerably higher as in the higher intensities (I_2, I_3) over the whole period (Figure 8). No direct effects of the pest management intensities were recognizable for some typical predatoric taxa such as the Empididae and the Carabidae (Figure 9).

For the carabid beetles, we have to consider that the species richness of intensively managed fields might be reduced[31] and restricted — especially re-

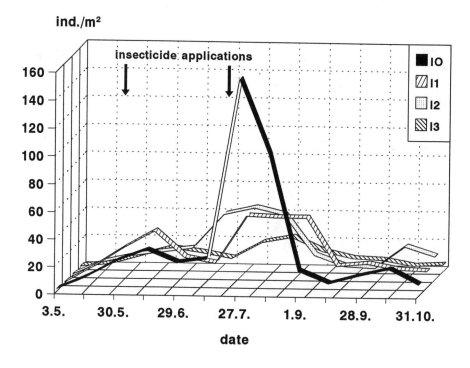

FIGURE 7. Phenology of the parasitic Hymenoptera in different pest management intensities (I_0–I_3) in sugar beet 1989.

garding the dominant species — to some very euryoecious species such as *P. melanarius* and *T. quadristriatus*, which are very tolerant against disturbances in their habitat. In the sugar beet fields discussed here the abundance of carabid beetles was predominated by *T. quadristriatus*.

Among the Brachycera there are some pest insects such as the fruit fly (*Oscinella* spp.) and the sugar beet fly (*Pegomya* spp.) which abundance increased by an increasing pest management intensity. The same phenomenon could be stated for the Bibionidae.

If the taxa are evaluated on the species level, in the same family to some extent opposite reactions are perceptible which are not recognizable on a higher taxonomic level. For example, the carabid beetle *Bembidion quadrimaculatum*, the ladybird *Coccinella undecimpunctata*, and *Adrastus rachifer* (Elateridae) obviously react very sensitively on any input of pesticides (especially insecticides) and fertilizers, whereas the ladybird *Propylaea quattuordecimpunctata* does not show any reaction and *Calathus melanocephalus* is even stimulated by them (Table 2).

From the above described effects of pesticides and fertilizers on the soil-inhabiting fauna some conclusions for evaluation assessments can be derived:

ind./m²

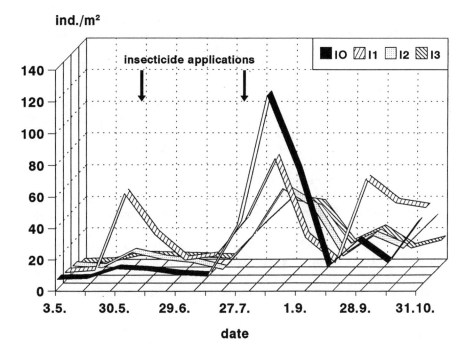

FIGURE 8. Phenology of the spiders (Araneae) in different pest management intensities (I_0–I_3) in sugar beet 1989.

1. An evaluation of the species community from the economical point of view, i.e., under the premise of farm management orientated to the yield.

   In this evaluation assessment the main topic is to determine to what extent a certain pest management intensity causes a reduction or promotion of the dominant species which are responsible for the productive capacity of the ecosystem (e.g., the dominant springtail *Isotoma viridis*), leads to a breakdown of functional groups of distinct taxa (e.g., Sciaridae as decomposer of organic litter), or stimulates the pest organisms (e.g., *Oscinella* spp., *Pegomya* spp.). Such an evaluation assessment ends in the question: how far is a reduction of the number of species and individuals tolerable in agroecosystems under the premise that the functionability of the field soil and the capacity of food and energy turnover of the soil fauna are not affected?

2. The evaluation of the species community from the pure ecological point of view without consideration of economic aspects.

   This evaluation assessment starts from the fact that the knowledge about the role and importance of each species in the agrobiocenosis is not or is only poorly understood. This assessment does not consider the asserted classification of the taxa as "useful", "harmful", or "indifferent". Therefore, the ecological evaluation of arable land is based on the species richness as the central criteria. Opposite to the number of individuals, the biomass, the respiration, or other summary parameters of the productive capacity of an ecosystem are the number and composition of species of one taxon in a distinct habitat, the only parameter which one is able to record with sufficient reliability.

The conservation of a maximum species richness can be of great importance as a provision to protect the genetic potential and therefore to represent one of the basic requirements for intact natural resources. So it seems to be absolutely necessary to change the way of farming in such a manner that the input of the means of production (e.g., pesticides and fertilizers) will be developed in a way that ecological niches for the highest possible number of species remain even on cultivated areas. Under these premises, for example, the species which are most important for the turnover capacity in the soil are automatically protected too.

Therefore, species richness is a criteria which is comparatively easy to determine and easy to apply for the evaluation of long-term effects of pesticides on the agrozoocenosis. Because of constant meshes (tillage, harvesting, etc.) arable land will never cross a distinct (pioneer) stage of succession, and therefore

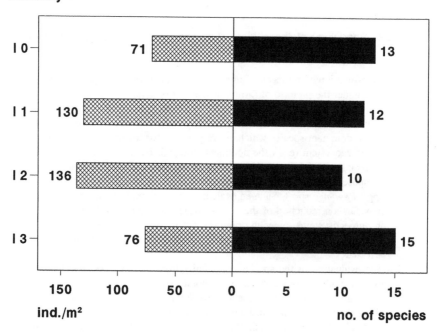

FIGURE 9. Comparison of the abundance (ind./m²) and the number of species of the Carabidae (Coleoptera) in I_0–I_3.

it contains basically a comparatively low potential number of species. However, as it could be demonstrated for the Coccinellidae,[32] it is possible to conserve a maximum of species richness even on cultivated areas by a very deliberate and careful use of pesticides and fertilizers.

ACKNOWLEDGMENTS

For various helps and technical assistance I thank very much Mrs. E. Päs, Mrs. C. Winkler, Mr. E. Czarnecki, and Mr. J. Alexander.

REFERENCES

1. Bartels, G., Auswirkungen eines langjährig unterschiedlich intensiven Einsatzes von Pflanzenschutzmitteln auf das Ökosystem Ackerboden, *Mitt. Biol. Bundesanst. Land- Forstwirtsch., Berlin-Dahlem,* H. 266, 128, 1990.
2. Ebing, W., Kreuzig, G., and Stemmer, H., Untersuchungen zum Rückstandsverhalten der im Pflanzenschutzmittel-Großversuch Ahlum angewandten Fungizide und Insektizide. Bundesminister für Forschung und Technologie: Forschungsbericht (FKZ: 0339050B), 1991, 26 pp.
3. Gottesbüren, B. and Pestemer, W., Prognose der Persistenz von Herbiziden und deren Auswirkungen auf Nachbaukulturen nach langjähriger Anwendung. Bundesminister für Forschung und Technologie: Forschungsbericht (FKZ: 0339050C), 1991, 45 pp.
4. Pohl, K. and Malkomes, H.-P., Einfluß von Bewirtschaftungsintensität und Verunkrautung auf ausgewählte mikrobielle Parameter im Boden unter Freilandbedingungen, *Z. Pflanzenkrank. Pflanzenschutz,* Sonderh. II, 373, 1990.
5. Sauthoff, W., Nirenberg, H., Metzler, B., and Gruhn, U., Untersuchungen über den Einfluß einer intensiven Pflanzenproduktion auf die Zusammensetzung der Bodenpilz-Flora. Bundesminister für Forschung und Technologie: Forschungsbericht (FKZ: 0339050F), 1991, 25 pp.
6. Sauthoff, W. and Oesterreicher, W., Untersuchungen über den Einfluß einer intensiven Pflanzenproduktion auf die Zusammensetzung der Bodenalgen-Flora. Bundesminister für Forschung und Technologie: Forschungsbericht (FKZ: 0339050), 1991, 26 pp.
7. Knüsting, E., Bartels, G., and Büchs, W., Untersuchungen zu Artenspektrum, fruchtartspezifischer Abundanz und Abundanzdynamik von Regenwürmern bei unterschiedlich hohen landwirtschaftlichen Produktionsintensitäten, *Verh. Ges. Ökol. Freising-Weihenstephan* 1990, 21, 1991.
8. Heimann-Detlefsen, D., Auswirkungen unterschiedlich intensiver Bewirtschaftungsintensitäten auf die Collembolenfauna des Ackerbodens, Thesis, TU, Braunschweig, 1991.

9. Kampmann, T., Einfluß von landwirtschaftlichen Produktionsintensitäten auf die Milbenfauna im Ackerboden, *Verh. Ges. Ökol. Freising-Weihenstephan 1990*, 13, 1991.

10. Leliveldt, B. and Sturhan, D., Untersuchungen über den Einfluß unterschiedlicher Pflanzenschutzintensitäten auf die Nematodenfauna des Bodens. Bundesminister für Forschung und Technologie: Forschungsbericht (FKZ: 0339050K), 1991, 21 pp.

11. Grimm, R., Funke, W., and Schauermann, J., Minimalprogramm zur Ökosystem-analyse: Untersuchungen an Tierpopulationen in Waldökosystemen, *Verh. Ges. Ökol. Erlangen 1974*, 77, 1974.

12. Zimmermann, J. and Büchs, W., Kurz- und langfristige Auswirkungen unter-schiedlich intensiver Pflanzenproduktion auf Kurzflügelkäfer (Coleoptera: Staphy-linidae) in der Kultur Zuckerrübe, *Mitt. Biol. Bundesanst. Land- Forstwirtsch.*, Göttingen, October 5 to 8, 1992.

13. Wehlitz, J. and Büchs, W., Langzeiteinfluß eines verschieden hohen landwirt-schaftlichen Produktionsmitteleinsatzes auf die Dipterenfauna, *Mitt. Biol. Bun-desanst. Land- Forstwirtsch.*, Göttingen, October 5 to 8, 1992.

14. Kleinhenz, A. and Büchs, W., Einfluss verschiedener landwirtschaftlicher Pro-duktionsintensitäten auf die Spinnenfauna in der Kultur Zuckerrübe, *Mitt. Biol. Bundesanst. Land- Forstwirtsch.*, Göttingen, October 5 to 8, 1992.

15. Smith, L.M., Ground emergence trap for pear thrips, *Calif. Agric. Exp. Stn. Bull.*, 562, 1933.

16. Funke, W., Food and energy turnover of leaf-eating insects and their influence on primary production, *Ecol. Stud.*, 2, 81, 1971.

17. Barber, H.S., Traps for cave-inhabiting insects, *J. Elisha Mitchell Sci. Soc.*, 46, 259, 1931.

18. Funke, W., Waldökosysteme in der Analyse von Struktur und Funktion — Un-tersuchungen an Arthropodenzönosen, *Verh. Ges. Ökol. Mainz 1981*, 1983, 13.

19. Ellenberg, H., Mayer, R., and Schauermann, J., *Ökosystemforschung — Ergeb-nisse des Solling-Projektes*, Ulmer, Stuttgart, 1986, 430 pp.

20. Dörr, L. and Eisenbeis, G., Untersuchungen zur Abundanz und Dynamik eped-aphischer Bodenarthropoden auf den Kalkflugsanden des Lennebergwaldes bei Mainz. *Mitt. Dtsch. Ges. Allg. Angew. Ent.*, 7, 54, 1990.

21. Pobozsny, M., Bradysia brunnipes (Meigen, 1804) (Diptera: Sciaridae) und ihre Bedeutung für die Streuzersetzung, *Acta Zool. Acad. Sci. Hung.*, 22, 139, 1976.

22. Altmüller, R., Untersuchungen über den Energieumsatz von Dipterenpopulationen im Buchenwald (Luzulo-Fagetum), *Pedobiologia*, 19, 245, 1979.

23. Brauns, A., Terricole Dipteren-Larven, Band 1, Göttingen, 1954, 156 pp.

24. Hanlon, R.D.G. and Anderson, J.M., The effects of collembola grazing on mi-crobial activity in decomposing leaf litter, *Oecologia*, 38, 93, 1979.

25. Küthe, K., Schäden durch Moosknopfkäfer (Atomaria linearis) an Zuckerrüben Schadensschwellen, *Gesunde Pflanzen.*, 4, 136, 1989.

26. König, K., Untersuchungen über die Auswirkungen der Anwendung von Insek-tiziden auf die epigäische Fauna von Zuckerrüben, *Bayer. Landwirtsch. Jahrb.*, 60(3), 235, 1983.

27. Koch, K., *Die Käfer Mitteleuropas — Ökologie*, Bd. 1, Goecke & Evers, Krefeld, 1989, 440 pp.

28. Aßmuth, W., Buschinger, A., Franz, J.M., Groh, K., and Tanke, W., Nebenwirkungen von Pflanzenschutzmaßnahmen auf die Agrozönose von Zuckerrübenkulturen, *DFG-Forschungsbericht Herbizide II*, 44, 1986.
29. Zimmermann, J., personal communication, 1992.
30. Peschke, K. and Fuldner, D., Übersicht und neue Untersuchungen zur Lebensweise der parasitoiden Aleocharinae (Coleoptera: Staphylinidae), *Zool. Jb. Syst.*, 104, 242, 1977.
31. Basedow, T., Pflanzenschutz und Naturhaushalt: Die Zulassung von Pflanzenschutzmitteln bedarf einer umfassenden und langfristigen Sicht, *Gesunde Pflanzen.*, 43(1), 7, 1991.
32. Büchs, W., Einfluß verschiedener landwirtschaftlicher Produktionsintensitäten auf die Abundanz von Arthropoden in Zuckerrübenfeldern, *Verh. Ges. Ökol. Freising-Weihenstephan 1990*, 1, 1991.

CHAPTER 23

Effects of Walking Activity and Physical Factors on the Short Term Toxicity of Deltamethrin Spraying in Adult Epigeal Money Spiders (Linyphiidae)

G.A.J.M. Jagers op Akkerhuis

TABLE OF CONTENTS

323

ABSTRACT

The effect of deltamethrin application on terrestrial spiders (*Linyphiidae*) was studied in the field situation in relation to walking behavior and/or physical factors. Deltamethrin was sprayed in winter wheat at an application rate of 0.5 g ai ha^{-1} in 8 experiments with replicated treatments. Spider walking activity was monitored either on a daily basis or every 1.5 h for 10 h after spraying. The deposition of deltamethrin was determined at the soil stratum using a fluorescent tracer. Physical variables were measured daily in the crop, or were obtained from a nearby weather station. Deltamethrin effects on spider activity were analyzed using short time series analysis. The species composition of the spider community was determined in May and July.

A strong positive correlation was found between spider activity on the day of spraying and the reducing effect of the pesticide on trapping success. This suggests mobility has an important role in the poisoning process of terrestrial arthropods. Similar trapping success and effect in three experiments conducted with different physical variables, implied a relative importance of walking behavior in comparison to physical factors. Circumstantial evidence was found for high bioavailability of deltamethrin on moss covered soil.

I. INTRODUCTION

A lot of information has been published on the toxic effects of pyrethroid and other insecticides under field conditions (see for instance Croft and Brown,[1] Basedow et al.,[2] Inglesfield[3]). However, there are very few studies which have identified and quantified ambient conditions which influence the behavior and response of populations in the field situation. Therefore, most of the data published cannot be used for the prediction of an effect in a given field situation.

Pesticide uptake by terrestrial arthropods is mainly residual, i.e., from substrate. Salt and Ford[4] used stochastic modeling to simulate the uptake of insecticide from plant surfaces by larvae of the lepidopteran *Spodoptera littoralis*. Sensitivity analysis indicated that the effect depended mainly on walking velocity, the contact area between insect and leaf surface, and the proportion of insecticide transferred per encounter. The relative importance, for terrestrial spiders, of residual uptake in comparison to direct or oral uptake was shown by Mullié and Everts.[5] Investigations of the residual uptake of deltamethrin by female *Oedothorax apicatus* showed saturation of pesticide uptake with distance walked (Jagers op Akkerhuis and Hamers[6]). The above shows that effects of ambient conditions on walking activity may determine the uptake of sprayed pesticide.

The present study aims at validation, in the field situation, of the assumed importance of walking activity in the poisoning process of terrestrial arthropods. A difficulty in this is that it can only be assessed properly in experiments which take into account the effect of ambient conditions on pesticide availability and arthropod vulnerability. In a project using linyphiid spiders, of which this publication is a part, uptake and effects of deltamethrin have been studied extensively, both in the laboratory and in the field, thus offering a good opportunity to study the effect of walking behavior while correcting for contributions of ambient physical factors.

Everts[7] showed that linyphiid spiders are a highly sensitive and abundant part of the epigeal arthropod fauna. These spiders are very sensitive to deltamethrin and were therefore chosen as the pesticide effect model in the present study.

Linyphiid spiders live hidden in crevices, under leaves (*Oedothorax* spp.), or sit in webs which are built in cracks and small shallow pits in the soil (*Erigone* spp.), or between plants near the soil surface (*Leptiphantes, Bathyphantes* spp.) (Thornhill).[8] Hiding in cracks, they are barely exposed to direct spray. Especially in a clay soil, in which deep cracks may be present, this offers shelter opportunities for a large number of spiders.

Spiders walk when hunting for a prey, searching for a female to mate or a place to lay eggs, or in reaction to a diurnal rhythm (Vickerman and Sunderland,[9] de Keer and Maelfait[10]) or dispersal behavior. Walking velocity mainly depends on substrate and environmental conditions (personal observations). In part, walking behavior will be performed at the soil surface, where spiders are exposed residually to sprayed substrate such as soil, moss, plant cuticula, plant debris, or spider webs. Residual uptake has been shown to be an important route of uptake for arthropods (Salt and Ford,[4] Mullié and Everts[5]). The nature of the substrate may cause changes in the bioavailability of the compound (Jagers op Akkerhuis and Hamers[6]).

After deltamethrin poisoning, spiders exhibit changes in walking behavior. Activity shows a quick rise followed by a long-lasting reduction. Walking speed is reduced immediately to decline continuously and it is possible that immobilization will occur (Jagers op Akkerhuis[11]). During prostration or quiescence the arthropod are vulnerable to desiccation and predation (Everts et al.[12]).

Based on the scenario described above, the present study aimed at evaluation of spider walking activity as a cause of pesticide side effects.

II. MATERIAL AND METHODS

A series of eight experiments were conducted during the summer of 1990 in two winter wheat fields on the terrain of the Institute for Plant Protection Research (Wageningen). A map of the fields and plots is given in Figure 1. The duration of each experiment, application date, and the number and position of plots are given in Table 1. Fields 1 and 2 differed in that field 1 had a higher and denser crop and a higher soil humidity.

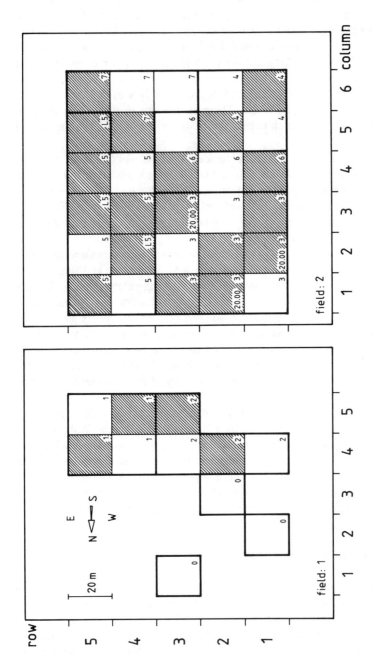

FIGURE 1. Map of experimental fields 1 and 2 showing position and treatment of plots used in the experiments. The experimental number is indicated in the lower right corner. "L" and "20.00" indicate plots sprayed with low-volume droplets or plots sprayed at 20.00 h (not in the present paper).

**Table 1. Fields, Plots, Duration and Application Dates For Experiments
1 to 8**

# Experiment	Duration	Application date	Field	Position of plots	
				Control	Treated
0	5/28–6/1	—	1	13,21,32	
1	6/16–6/25	6/20	1	44,55	45,54
2	6/23–7/31	6/26	1	41,43	42,53
3	7/1–7/9	7/4	2	11,23,32	13,22,31
4	7/7–7/18	7/13	2	51,62	52,61
5	7/12–7/21	7/16	2	14,25,44	15,34,45
6	7/21–7/31	7/26	2	42,53	41,43
7	7/26–8/3	7/30	2	63,64	54,65
8	7/26–8/1	7/27	2	21,22	11,12,31,32

Note: Column and row number giving the plot position as indicated in Figure 1.

Each experiment typically consisted of four or six plots of 18*20 m (360 m²). Field design and trapping layout assured that the density of spiders was not severely affected by trapping, and that spider migration from or into adjacent plots within a 1-week assessment period was unlikely (Jagers op Akkerhuis[13]). Half of the plots served as unsprayed controls, the other plots were sprayed (08.00 h) with 0.5 g active ingredient (ai) deltamethrin (Decis flow®, Hoechst Holland BV, Amsterdam) in 400 l water per hectare using a tractor-mounted hydraulic sprayer (Slit nozzles 11-110-06 (Douven Machinefabriek Horst), 2.3 atm, 60 cm above the crop). The deltamethrin concentration of 0.5 g ái ha^{-1} must be considered to yield sublethal effects varying from moderate to severe, depending on physical conditions (Jagers op Akkerhuis and van der Voet[14]).

In every plot six round pitfall traps were placed in a regular pattern at a distance of 7 m from the edge of the plot as described in Jagers op Akkerhuis and van der Voet.[14] A description of the pitfall traps (10 cm diameter, partly filled with an aqueous formaline solution) is given by Everts.[7] The traps were emptied daily between 08.00 and 09.30 h, and during the first 12 h after spraying every 1.5 h. Spiders were counted in the field and stored separately for each trap in 70% alcohol.

A fluorescent tracer (Tinopal®, Ciba Geigy, Arnhem) was used to measure ground level deposition of the pesticide. Each time a known quantity of Tinopal® was dissolved in the sprayed deltamethrin solution, and the deposition measured quantitatively on eight strips of window glass (50*3 cm) per field, placed flat on the soil, at an angle of 45° to the direction of the crop rows, as described in Jagers op Akkerhuis and van der Voet.[14] To minimize photodegradation of the Tinopal®, glass strips were removed from the field immediately after spraying and placed in a lightproof box. The strips were rinsed in the laboratory with a known quantity of demineralized water and the Tinopal® concentration was determined fluorimetrically (emission 450 nm, absorbtion 380 nm).

Soil humidity, air humidity, and air temperature were measured every morning (08.00 h). On the day of deltamethrin application soil humidity was also determined 8 h after spraying, while air humidity and air temperature were measured in the crop every 1.5 h for 10 h after spraying. Measurements were taken in each plot of every experiment, except for a few scattered days when less measurements were done or during experiment 3 and 5 when only plots 11, 23, 31, and 15, 34, 55 were taken into account. Soil humidity (% dryweight, oven drying overnight at 105°C) was determined for the first 2 cm of soil (a heavy illitic river clay; "medium high Nude silty clay loam", 59% lutum content, 5.3% organic matter). Minimum and maximum temperatures and relative air humidity were measured in the crop canopy at 5 cm above the soil using min-max thermometers (±0.2°C) and air hygrometers, freely exposed to the air and shielded from direct radiation.

The meteorological station of the Agricultural University Wageningen, situated 250 m southwest from field 1 and 300 m west from field 2, provided hourly measurements of the following meteorological variables. These records were used to calculate sum or mean values corresponding with the 24-h trapping periods: temperature, relative humidity and wind speed at 1.50 m, sunshine hours, net radiation, time, and quantity of rain.

The catches between 28 May and 1 June on fields 1 and 2, and between 1 and 9 July on field 2 were used for identification. Species were identified using the identification tables of Lockett and Millidge.[15]

Instantaneous effect (reduction in pitfall trap catches) and recovery (relative to the observed initial effect)[14] were calculated from the daily pitfall trap catches using before and after application on control and impact field (BACI) time series analysis (Stewart-Oaten et al.,[16] van der Voet[17]). The equations used are given in Table 2. The significance of using a more complex model was tested using the variance ratio,

$$vr = \{(SS_{res,0} - SS_{res,1})/(df_{res,0} - df_{res,1})\}/MS_{res,1} \tag{1}$$

where indices 0 and 1 refer to the simple and more complex model. Under the null hypothesis that the simple model is the true model, it has an F distribution with $df_{res,0}$ - $df_{res,1}$ and $df_{res,1}$ degrees of freedom. The initial effect, calculated as percentage reduction relative to the control, and half-life time of recovery can be calculated from g and h (Table 2) as $100*[1-exp(g)]$ and $-\ln(2)/h$.

As the hidden life style of the spiders in cracks in the clay soil hindered population density assessments in the present experiments, a tentative relative measure for spider density was used. The method was based on the assumption that comparable crop conditions during the experiments, i.e., a green crop with a moist soil, implied similar relationships between spider activity and environmental variables on different fields. These relationships were investigated for experiments 1 to 5 using a multiple regression technique (Max-R stepwise regression technique).[18] A dummy variable F_1 was introduced to account for differences

Table 2. Equations Used in the Time Series Analysis of the Pitfall Trap Data

A $Y_{it} = \mu + \alpha_i + \beta_t + \epsilon_{it}$
(No treatment effect)

B $Y_{it} = \mu + \alpha_i + \beta_t + g^*exp\{h(t - t_0)\}^*\delta_{AI} + \epsilon_{it}$
(Treatment effect, common initial effect, and recovery)

C $Y_{it} = \mu + \alpha_i + \beta_t + g^*exp\{h_i(t - t_0)\}^*\delta_{AI} + \epsilon_{it}$
(Treatment effect, common initial effect, separate recovery)

D $Y_{it} = \mu + \alpha_i + \beta_t + g_i^*exp\{h(t - t_0)\}^*\delta_{AI} + \epsilon_{it}$
(Treatment effect, separate initial effect, common recovery)

D $Y_{it} = \mu + \alpha_i + \beta_t + g_i^*exp\{h_i(t - t_0)\}^*\delta_{AI} + \epsilon_{it}$
(Treatment effect, separate initial effect, and recovery)

In which:
Y_{it}	=	$ln(N_{it})$, i.e., the log-transformed number for plot i and day t
μ	=	niveau
α_i	=	the effect term for the different plots i
β_t	=	the effect term for the different days t
g	=	the (field-dependent [g_i]) effect term for the initial effect
h	=	the (field-dependent [h_i]) effect term for recovery
δ_{AI}	=	the impact vector indicating deltamethrin application
ϵ_{it}	=	a random contribution from a normal distribution with mean 0 and variance s^2

in population density between the fields, resulting in the following general formula:

$$Y = \mu + a_j*x_j + b_j*x_j^2 + c_j*(x_{-1})_j + d_j*(x_{-1})_j^2 + e_k*(F_k) + \epsilon \quad (2)$$

in which:

Y	=	The natural logarithm of the mean number of spiders per treatment (six traps per plot)
μ	=	The intercept
a_j to e_k	=	The regression coefficients
x_j	=	The j environmental variables
x_j^2	=	The quadrats of the j environmental variables
$(x_{-1})_j$	=	The j environmental variables of the preceding day
$(x_{-1})_j^2$	=	The quadrats of the j environmental variables of the previous day
F_k	=	Dummy variable for field 1 ($k = 1$) which is one for this specific field and zero in all other cases
ϵ	=	A random contribution from a normal distribution with expectation 0 and variance sigma

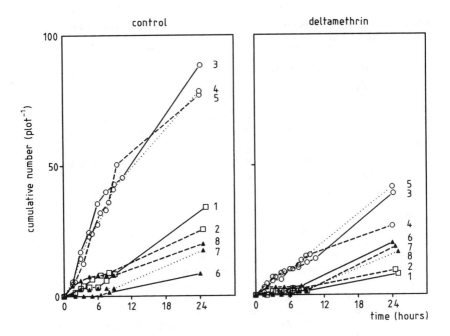

FIGURE 2. The effect of deltamethrin spraying on walking activity. Cumulative pitfall trap catches (mean of plot totals, six traps in each plot) during 24 h after deltamethrin spraying on control and impact plots. The experimental number is indicated in the figures. Experiments on field 1 are indicated with an open square, the first three experiments on field 2 with an open circle, and the last three experiments on field 2 with a closed triangle.

The environmental variables used were the daily means of the air temperature (°C) and relative humidity (%) at 150 m, and the daily sums of sunshine (h), net radiation divided in radiation gain during the day and radiation loss at night (J cm^{-2}), duration (min) and quantity (mm) of rain. Because soil humidity measurements were not available before 21 June and soil humidity stayed relatively stable during the first five experiments, soil humidity data were left out of the regression.

Close and ecologically relevant correlations have been demonstrated earlier using this method (personal observation). It is, however, realized that this method needs refinement.

III. RESULTS

Figure 2 shows that the experiments could be arranged into two groups, based on the presence or absence of effect.

Table 3. BACI Analysis of the Pitfall Trap Catches

Exp #	MSR_0	SSR_0	DFR_0	MSR_1	SSR_1	DFR_1	F	P	Initial effect	Recovery velocity
1	0.106	2.87	27	0.023	0.56	25	51.2	***	$-1.3\ (\pm0.1)$	$-0.2\ (\pm0.1)$
2	0.076	1.82	24	0.040	0.89	22	11.6	***	$-1.0\ (\pm0.2)$	$-0.7\ (\pm0.7)$
3	0.052	2.07	40	0.037	1.39	38	37.7	***	$-0.7\ (\pm0.2)$	$-0.6\ (\pm0.3)$
4	0.049	1.61	33	0.20	0.62	31	25.0	***	$-0.9\ (\pm0.1)$	$-0.4\ (\pm0.1)$
5	0.059	4.24	72	0.033	2.31	70	29.2	***	$-1.0\ (\pm0.1)$	$-0.6\ (\pm0.2)$
6	0.109	3.26	30	0.106	2.97	28	1.4	ns	—	—
7	0.066	1.60	24	0.056	1.24	22	3.2	ns	—	—
8	0.026	0.47	18	0.021	0.33	16	3.3	ns	—	—

Note: Indicated are mean square (MS), sum of squares (SS), and degrees of freedom (DF) of the residuals for the models with no pesticide effect (Equation A, Table 2) or with a common initial effect and recovery (Equation B, Table 2) respectively R_0 and R_1, the F statistic (see text) and significance of model B over model A (ns = not significant at 0.05% level, *** = significant at 0.001% level), and the calculated initial effect and recovery velocity (with standard error of mean).

A reduction in trapping success following deltamethrin spraying was found for experiments 1 to 5. The significance of the effects was demonstrated using the model with a common initial effect and recovery rate (Table 2 model B, Table 3). The other models did not improve the fit and are therefore discarded from further presentation. The calculated initial effects (Table 3) ranged from a 51% reduction in experiment 3 to a 74% reduction in experiment 1. The recovery half-lives ranged from 1.0 d in experiments 2 and 5, to 2.9 d in experiment 1 (Table 4). No clear relationship existed between the initial effect and the recovery velocity.

In the group consisting of experiments 6, 7, and 8, deltamethrin effects were not significant.

In Figure 3, cumulative spider catches of the control and impact plots are shown as a measure of arthropod activity during 24 h after application of deltamethrin. A strong reduction in trapping success shortly after deltamethrin spraying was observed in experiments 1 to 5. For experiments 6, 7, and 8, catches were extremely low during the day with slightly higher numbers at night and no apparent effects of deltamethrin on the catches during the first 24 h after spraying.

The deposition of deltamethrin at the soil stratum is shown in Table 4. Relatively little deltamethrin reached the soil on field 1.

Based on environmental variables, the experiments could be classified into three groups: the experiments on field 1 had a high and dense crop (80 cm) and a moss-covered soil (80% cover, visually estimated); experiments 3 to 5 shared a lower (60 cm), less dense crop and a moist soil without a moss cover; and experiments 6 to 8 were characterized by a shriveled, dry crop and a bare, dessicated soil which was exposed to direct weather influences.

Table 4 shows measurements of physical factors assessed during the 24 h following deltamethrin application. The following data are of special interest in relation to the observed catches. Thunderstorms during the nights following deltamethrin application caused heavy rain during experiments 1 (at 19.00 h) and 2 (from 23.00 until 01.00 h). Air humidity in field 1 was relatively high compared to field 2. Soil humidity decreased slowly during experiments 4 to 8.

Table 4. Biotic and Abiotic Variables Measured During the First 24 h Following Deltamethrin Application

| | | | | Cumulative catches (per 6 traps) | | | | Deposition |
| | Appl | Ini eff | Recov halfl | Control | | deltam | | deltam |
Exp	date	(%)	(d)	0–8	9–24	0–8	9–24	(g/ha ai)
1	6/20	74	2.9	8.5	25.6	1.5	6.5	0.015 (±0.001)
2	6/26	64	1.0	9	16.5	2.5	7	0.017 (±0.000)
3	7/4	51	1.2	45.5	43	14	24.5	0.035 (±0.003)
4	7/13	59	1.7	50.5	26.5	15.5	11	0.049 (±0.004)
5	7/16	64	1.2	41	46.5	15	26.5	0.065 (±0.001)
6	7/26	ns	ns	1.5	8	3.5	16.5	0.038 (±0.003)
7	7/30	ns	ns	8	12	1.5	17	0.073 (±0.002)
8	7/27	ns	ns	3	20.5	2	16	0.055 (±0.005)

Note: Standard deviations from the deposition measurements were calculated using the plot means (— = missing observation, ns = no significant effect found).

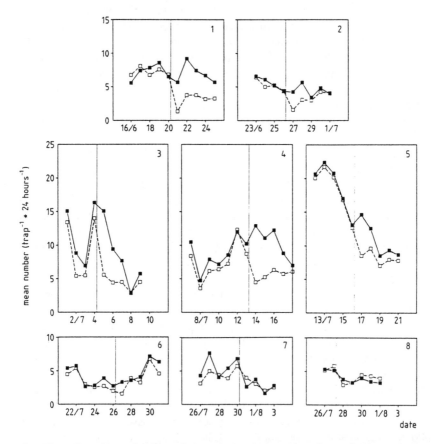

FIGURE 3. The effect of deltamethrin spraying (0.5 g ai ha⁻¹) on a field population of linyphiid spiders in relation to application date-dependent walking activity and physical factors. The experiment number is indicated in the top right corner. A vertical line shows the moment of spraying. Values represent means of two or three fields (see Table 1). Closed square = control, open square = impact.

**Table 4. Biotic and Abiotic Variables Measured During the First 24 h
Following Deltamethrin Application (*continued*)**

Air humidity (%)			Soil. Hum. (% dwt)	Temp +5 cm (Celsius)		Sunshine (h)	24 h sum net to radiation (J/cm²)		Rain	
08.00	13.00	18.00		08.00	14.00		pos.	neg.	Duration (min)	Quant (mm)
90	80	100	37	17	19	1.7	565	−165	54	16.8
100	70	100	43	20	28	3.9	1158	−47	103	30.1
100	50	70	31	14	20	4.1	916	−33	0	0
90	40	90	18	20	30	14.9	1522	−217	0	0
50	—	70	9	18	30	5.0	998	−76	0	0
80	30	60	7	17	32	10.8	1286	−179	0	0
60	30	80	6	18	33	13.0	1313	−97	0	0
80	30	70	5	20	38	8.4	1153	−70	0	0

Air temperature in the crop on the day of deltamethrin spraying fluctuated from 20°C in the morning to 30°C at noon, except on 20 June and 4 July when temperatures did not exceed 20°C.

As shown in Figure 4, the species composition of spiders caught on fields 1 and 2 from 28 May to 1 July showed great similarity. Most trapped spiders were male *E. atra* and *B. gracilis,* followed in abundance by *E. dentipalpis, Lepthyphantes tenuis,* and *Meioneta rurestris.* Few female spiders were caught, most of which were *O. fuscus.* Identification of spiders caught between 1 and 7 July on field 2 showed a lower percentage of all above mentioned male spiders in favor of *O. apicatus, O. fuscus,* and *Diplostila concolor.* Catches of females, consisting mainly of *O. apicatus* and *O. fuscus,* showed an increase.

Figure 5 shows the daily measurements of physical factors which were used to deduce changes in spider density. The multiple regression of the catches of the first five experiments resulted in a selection of the following factors:

$$Y = 3.34 + 0.0006*Pnetr - 0.002*Nnetr - 0.49*F1 \qquad (3)$$

in which:

Y = The natural logarithm of the sum of the daily catches per field
$Pnetr$ = The net radiation gain during the day (J/cm²)
$Nnetr$ = The net radiation loss at night (a negative value, J/cm²)
$F1$ = The field factor for field 1

Statistical information about these results is shown in Table 5. The catches showed a close positive correlation with net radiation during the day and eradiation at night. The negative value (-0.49) of the field factor of field 1 indicates a 26% ($=e/e^{-0.49}$) lower population density in this field.

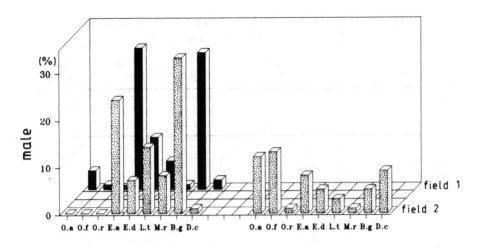

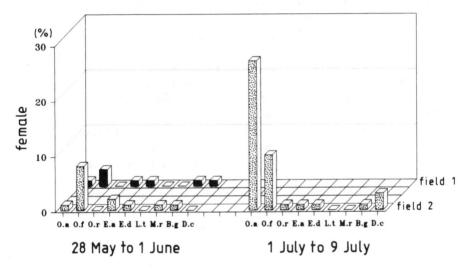

FIGURE 4. Species composition of linyphiid spiders in fields 1 and 2. Bars represent the percentage of the specified groups of the total number (N) of spiders caught per field and per period. O.a. = *Oedothorax apicatus*, O.f = *Oedothorax fuscus*, O.r = *Oedothorax retusus*, E.a = *Erigone atra*, E.d = *Erigone dentipalpis*, L.t = *Leptyphantes tenuis*, M.r = *Meioneta rurestris*, B.g = *Bathyphantes gracilis*, D.c = *Diplostyla concolor*. 28 May to 1 July; Field 1: N = 488 (408 male, 71 female), Field 2: N = 248 (225 male, 23 female); 1 July to 9 July; Field 2: N = 819 (464 male, 355 female).

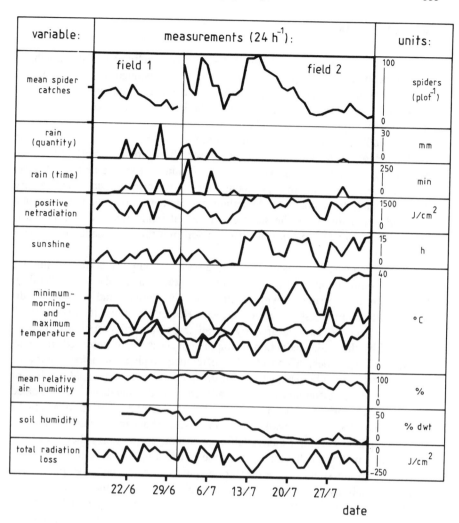

FIGURE 5. Physical factors in relation to mean daily pitfall trap catches.

Table 5. Results of the Multiple Regression Analysis of the Relationship Between Trapping Success and Physical Factors

Variable	R. coeff	STD	F	Prob
Interc	3.34			
F1	−0.49	0.09	27	0.0001
Pnetr	0.0006	0.0001	16	0.002
Nnetr	−0.002	0.001	4	0.05

Note: For explanation of abbreviations, see formula 3 (N = 49, R^2 = 0.65).

IV. DISCUSSION

The strong positive correlation between spider activity and deltamethrin-induced reduction in trapping success clearly demonstrated the importance of walking activity in determining pesticide effects based on residual uptake. This supports the assumption of Jepson[19] that "the level of activity of a species will affect residual uptake considerably".

It is unlikely that a low bioavailability of deltamethrin on the dry soil in the last three experiments was related to the observed differences in effect between experiments 3, 4, and 5 and experiments 6, 7, and 8. It should be noted that an average effect was found in experiment 5 which was also conducted with a very dry soil. In addition, Jagers op Akkerhuis and Hamers,[6] using deltamethrin and clay soil originating from the same experimental fields, showed a similar low bioavailability for *O. apicatus* at different soil moisture treatments below 63% dry weight.

The dependence of bioavailability of a pesticide on physical factors implies that each combination of physical factors is related to a specific bioavailability. Comparison of experiments with identical walking activity and physical conditions should therefore show similar effects. However, almost identical control activity of spiders, as measured by pitfall trapping, and pesticide effect, as measured by the initial reduction in trapping success, were observed in experiments 3, 4, and 5, despite considerable differences in physical factor conditions. There are two possible explanations for this phenomenon. One, spider behavior compensated for differences in bioavailability of deltamethrin. As indicated above, deltamethrin causes an increase in spider activity. The residual uptake depending on the distance walked (Jagers op Akkerhuis and Hamers),[6] this will increase uptake. Different bioavailabilities, if not too low, may therefore stimulate spider walking behavior until a dose is taken up which induces immobilization. Two, the simultaneous effect of all physical factors on spider physiology and bioavailability of deltamethrin accidentally resulted in the same reduction of trapping success. Verification of these options using a quantitative model which describes the bioavailability and effect of the compound in relation to physical factors will be published later.

The present results provide circumstantial evidence for high bioavailability of deltamethrin from moss, as was demonstrated by Jagers op Akkerhuis and Hamers[6] in controlled laboratory studies using *O. apicatus* and deltamethrin. This follows from the observation that despite low pesticide deposition at the soil stratum in experiment 1 and 2, which were the only experiments conducted on soil with a heavy moss cover, these experiments showed considerable effect, suggesting a relatively high availability of the sprayed deltamethrin.

The difference in spider density on field 1 and field 2 during experiments 3, 4, and 5 was investigated using the regression with environmental factors as described above. For the experiments on field 2 a stable population had to be assumed, as different crop circumstances made the use of the regression technique for relative density estimation impossible. The immediate rise in spider catches following the return of favorable conditions after rain on 29 August, in combination with the selective nocturnal activity of the spiders during these experiments (Figure 2), supports the assumption that low spider catches during experiments 6, 7, and 8 were caused by an adverse effect on walking activity of the dry soil and high temperatures, and not by changes in population density.

The results of the present study show the importance of arthropod walking behavior with respect to residue-based short-term pesticide toxicity. As, on species level, there is hardly any quantitative information on arthropod behavior in the field, this requires further investigations. Circumstantial evidence was found for a high bioavailability of deltamethrin from moss. There is little quantitative knowledge of substrate-dependent residual bioavailability of pesticides. With respect to modeling studies, this shows the need for investigations of the uptake of compounds from substrate in relation to substrate type, arthropod anatomy, and behavioral aspects of arthropod exposure to different substrates in the field. Additionally, the interpretation of pesticide effects in the field would profit greatly from more quantitative information on temporal changes in bioavailability of a compound in relation to ambient physical factors.

ACKNOWLEDGMENTS

These investigations were supported financially by the Program Committee for Toxicological Research. The author thanks E. van Barneveld for carrying out most of the laborious field work, I. van het Leven, D. Dekkers, T. Hamers, and R. Westerhof for help in the field, Ir. J.A. Huizinga and all personnel of the experimental farm (IPO) for excellent technical support, Prof. Dr. J.H. Koeman for discussion and/or criticism on the manuscript, and M. Aldham-Breary, M.Sc., for improving the English.

REFERENCES

1. Croft and Brown, 1975.
2. Basedow, 1981.
3. Inglesfield, C., Pyrethroids and terrestrial non-target organisms, *Pestic. Sci.*, 27, 387, 1989.
4. Salt, D.W. and Ford, M.G., The kinetics of insecticide action. Part III. The use of stochastic modelling to investigate the pick-up of insecticides from ULV-treated surfaces by larvae of Spodoptera littoralis boisd., *Pestic. Sci.*, 15, 382, 1984.
5. Mullié, W. and Everts, J.W., Uptake and elimination of c14-deltamethrin by Oedothorax apicatus (Arachnidae: Erigonidae) with respect to bioavailability, *Pestic. Biochem. Physiol.*, 39, 27, 1991.

6. Jagers op Akkerhuis, G.A.J.M. and Hamers, T., Substrate dependent bioavailability of deltamethrin for the epigeal spider Oedothorax apicatus (Blackwall) (Aranaea, Erigonidae), *Pestic. Sci.,* in press.

7. Everts, J.W., Sensitive Indicators of Side-Effects of Pesticides on the Epigeal Fauna of Arable Land, Ph.D. thesis, 1990, 114 pp.

8. Thornhil, W.A., The distribution and probable importance of linyphiid spiders living on the soil surface of sugar-beet fields, *Bull. Br. Arachnol. Soc.,* 6, 127, 1983.

9. Vickerman, G.P. and Sunderland, K.D., Arthropods in cereal crops: nocturnal activity, vertical distribution and aphid predation, *J. Appl. Ecol.,* 12, 755, 1975.

10. De Keer, R. and Maelfait, J.-P., Life history of Oedothorax fuscus (Blackwall, 1834) (Araneae, Linyphiidae) in a heavily grazed pasture, *Rev. Ecol. Biol. Sol,* 24, 171, 1987.

11. Jagers op Akkerhuis, G.A.J.M., Neveneffecten van bestrijdingsmiddelen: Linyphiide spinnen als modelorganismen. (Side effects of pesticides: Linyphiid spiders as model organisms, in Dutch), *Toxpost,* 3, 4, 1990.

12. Everts, J.W., Willemsen, I., Stulp, M., Simons, L., Aukema, B., and Kammenga, J., The toxic effect of deltamethrin on linyphiid and erigonid spiders in connection with ambient temperature, humidity and predation, *Arch. Environ. Contam. Toxicol.,* 20, 20, 1991.

13. Jagers op Akkerhuis, G.A.J.M., Walking behaviour and population density of adult linyphiid spiders in relation to minimizing the plot size in short term pesticide studies with pyrethroid insecticides, *Environ. Pollut.,* 80, 163, 1993.

14. Jagers op Akkerhuis, G.A.J.M. and van der Voet., H., A dose-effect relationship for the effect of deltamethrin on a linyphiid spider population in winter wheat, *Arch. Environ. Contam. Toxicol.,* 22, 114, 1992.

15. Locket, G.H. and Millidge, A.F., British spiders, Vol. 1, Ray Society, British Museum of National History, London, 1951, 1953, 1957.

16. Steward-Oaten, A., Murdoch, W.W., and Parker, K.R., Environmental impact assessment: "pseudoreplication" in time, *Ecology,* 67, 929, 1986.

17. Voet, H., van der, Het bepalen van behandelingseffecten op grond van korte tijdreeksen, TNO report 87 ITI B 30, 1987.

18. SAS, SAS Users Guide: Statistics, 1991. SAS Institute Inc., Box 8000, Cary, NC.

19. Jepson, P.C., The temporal and spatial dynamics of pesticide side-effects on nontarget invertebrates, in *Pesticides and Non-Target Invertebrates,* Jepson, P.C., Ed., Intercept, Wimborne Dorset, England, 1989.

CHAPTER 24

Bait Lamina as a Tool For Testing Feeding Activity of Animals in Contaminated Soils

O. Larink

TABLE OF CONTENTS

0-87371-530-6/94/$0.00 + $.50

ABSTRACT

The bait-lamina-test after v. Törne can be used as a tool to compare feeding activity of soil fauna on different plots or sites which vary in one or more factors. In the development of new pesticides it may be a helpful method for first screening of field effects. For the test several sets of small plastic strips with a number of small apertures in distances of 5 mm are exposed for some hours or days in the field. The apertures are filled with a bait mass and after the exposure the number of empty apertures is counted. The results can be shown in a matrix and treated with statistical methods.

I. INTRODUCTION

The evaluation of toxic substances often is a complicated process with sophisticated analytical methods. On the other hand, a quick procedure is frequently wanted to get first information about possible effects of toxicants in the environment. This is also necessary in soil science. The edaphon is affected by toxic substances in various ways and to a different extent. Well known is the litter-bag method,[2] which is very time consuming and afflicted with some difficulties. In 1990 von Törne published two papers in *Pedobiologia*,[5,6] where he introduced a new method for estimating influences on feeding activity of soil animals, using laminae with bait-filled apertures. This bait-lamina test seems to be a useful tool in ecotoxicology of soils. Up until now it was only used in a very few ecological investigations.[4]

II. MATERIAL AND METHODS

The main equipment is a lot of small plastic strips, about 18 cm long and 6 mm broad. The plastic material needs a thickness of 1 to 1.5 mm. We tried laminae, 1.2 mm thick, which we got from v. Törne and formed our own laminae of 1.5 mm. The laminae are perforated in 5-mm distances by 16 apertures. The single hole is an opening which is shaped like an hour glass with an external diameter of 3 mm and a diameter of 1.5 mm at the smallest point. The production of this kind of hole perhaps is the most complicated and time-consuming preparation of the whole test, because a good number of laminae is necessary. The openings are filled with the bait mass. A lot of substances can be taken for this purpose. After one of the first formulae v. Törne gave,[5] we tried powder of the nettle *Urtica dioica* and cellulose, mixed in a ratio of 3:7 with some water to form a kind of paste which is put into the holes. This can be done with a knife like putting butter on a piece of bread, but v. Törne also described a kind of machine with two rollers to do it.[5] After this the filled laminae are dried for 24 h at 40°C. To find out special or ideal bait ingredients could be one of the aims of further development of the bait test.

Test Date: 11.16.90–11.30.90/No: I2

Lam. no.	16	15	14	13	12	11	10	9	8	7	6	5	4	3	2	1	S_x
							Sequence of baits (8 to 0.5 cm below surface)										
1	+	–	–	+	+	–	–	+	+	+	+	+	+	+	+	+	12
2	–	–	–	–	–	–	+	+	+	–	–	–	+	+	–	+	6
3	–	–	–	–	–	–	–	–	–	–	–	–	–	–	–	–	–
4	–	–	–	–	–	–	+	–	+	+	+	–	–	–	–	–	4
5	–	–	–	–	–	–	–	–	–	–	–	–	–	+	+	+	3
6	+	–	+	–	–	+	+	+	+	–	–	–	+	+	–	+	9
7	+	–	+	+	+	–	+	+	–	+	+	–	+	+	+	–	11
8	–	+	–	–	–	+	+	+	+	+	+	+	+	+	+	–	11
9	–	–	–	+	–	+	+	–	–	+	–	+	–	+	+	+	8
10	–	–	–	–	–	–	+	–	–	–	+	–	–	–	–	+	3
11	–	–	–	–	–	–	–	–	–	–	–	–	–	–	+	–	1
12	–	–	–	–	–	–	–	–	–	–	–	+	–	+	–	+	3
13	–	–	–	–	–	–	–	–	+	–	+	–	–	–	–	+	3
14	–	–	–	–	–	–	–	+	+	+	–	–	+	–	+	–	5
15	–	–	–	–	–	+	–	–	–	–	–	–	–	–	–	–	1
16	–	–	–	–	–	–	–	+	–	–	–	+	+	+	+	–	5
Sy	3	1	2	3	2	4	7	7	6	7	5	6	7	9	8	8	85

FIGURE 1. Example for a matrix of one file of bait lamina from a pesticide test.

A set of 16 laminae forms a test unit or file. They are brought out into the soil in one group, arranged, for example, in four rows, each with four laminae. The distances between each laminae should be 10 to 15 cm. We mainly arranged 16 laminae in one row.

The laminae are put into the soil so that the upper hole just disappears under the soil surface. Often it is necessary to use a knife or a similar instrument to form a slit into the soil. Only under wet conditions it is possible to put the laminae just into the soil. The application of the test needs, of course, some files for each tested area. We use a set of four or five files for each plot. The inserted laminae stay outside for some days, for example, 10 d or a fortnight. This depends on the kind of soil, the activity of the edaphon, and mainly on the moisture of the soil. V Törne remarks that not more than 40% of the bait-filled apertures should be perforated at the end of the experiment.[5] At this time the bait laminae are pulled out, washed carefully in the laboratory, and then the number of either empty holes or openings one can look through are counted. This can be done easily under a stereo microscope in transparent light. Using the described laminae and files, the results form a kind of matrix (Figure 1) which is given by the binary numbers of empty or still filled openings.

There is the IBM®-compatible software "Profat" available by which the matrices can be filled easily and also the statistical work is done using the U-test after Mann-Whitney (Fa. Terra Protecta, Himbeersteig 18, D-14129 Berlin 38).

Table 1. Design of the Field Experiment with Sewage Sludge

a0   Control plot without sewage sludge and heavy metals.
a1u 4 t Dry mass of sewage sludge without heavy metals.
a1b 4 t Dry mass of sewage sludge with additional heavy metals.
a2u 12 t Dry mass of sewage sludge without heavy metals.
a2b 12 t Dry mass of sewage sludge with high amount of heavy metals.

III. RESULTS AND DISCUSSION

The method was used to test the influence of heavy metals and pesticides on arable fields.[1,3] Figure 2 shows results from some plots of a field experiment with sewage sludge, partly containing the heavy metals lead, chrome, copper, zinc, nickel, and cadmium. The different plots are characterized in Table 1. The height of the columns corresponds with the number of empty holes in the laminae. The low values on a0, a1b, and a2b and the high ones on a1u and a2u after the first test in July 1990 were in very good line with our findings about the abundances of Collembola and Acari in the area.[1] The concentrations of heavy metals in a2b correspond to those of the German limiting values for heavy metals in sewage sludge, those on a1b and a2u were about half of this amount. At the second date the differences may show this. Repetitions during the following winter were without results because of a very low feeding activity. After the second test the activity was higher, but the differentiation was not as good as in the first time. Both examples indicate the possibilities of this method.

Another example will show the possibility of testing the influence of pesticides. These experiments were done on sites of the Federal Biological Research Centre for Agriculture and Forestry (BBA) in Braunschweig, Germany. One plot was treated with Karate, a pyrethroid with the active ingredient lambda-cyhalothrin in a concentration of 200 ml/ha = 10 g ai/ha. The second insecticide was Gaucho, applied as seed dressing with the active ingredient imidacloprid with a rate of 200 ml/100 kg seeds. Eight files were used in each plot.

Figure 3 shows the results. In the average, 8.6 of the 16 baits were perforated or completely eaten in each lamina on the control plot, 6.4 on the Karate plot, and 6.5 on the Gaucho plot. Both differences are highly significant when tested by the U-test, $p < 0.01$. Numbers of Collembola, caught on the same plots using a photoeclector, did not correspond with these results. V. Törne developed different possibilities to form factors for a better comparability, for example, if the tests lasted for a different number of days. One of these factors is the mean feeding activity rate, based on the results of 100 exposed laminae.[5,6] The matrices can be utilized also with another aim. If the x-values are summed up the result is a profile of the feeding activity, as shown in Figure 4. The examples from two files with inverse positions of empty perforations show this clearly. The left one is from the control area, the right one from the area with Karate. It can be interpreted in the way that the activity of soil animals is depressed near the surface in the contaminated area.

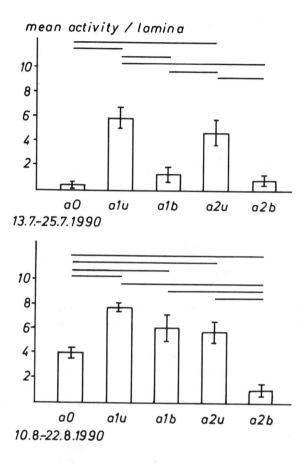

FIGURE 2. Bait-lamina test on plots with sewage sludge and heavy metals. The ordinate gives the number of open apertures = fed baits as the mean values of five files of laminae. The horizontal lines mark significant differences. For additional explanations see Table 1.

A further interpretation of the results of the bait-lamina test is not easy, for it is not certain which animals are responsible for the observed effects. Controlling the laminae Acarina, Collembola, larvae of Diptera, Enchytraeidae, and Lumbricidae were seen at the baits. On the other hand, there also will be an impact on the bait substance by the microflora, depending on the length of time of the experiment, as well as on moisture and temperature of the soil. V. Törne advises using only periods less than 15 d. In substrates rich in organic matter often some hours are sufficient.[5]

The main problem with this method is the dependency from the soil moisture. During late spring and early summer 1991 some of our series were completely without results because of dry soil. A soil moisture higher than 12 to 15% is necessary to start the test in the field.

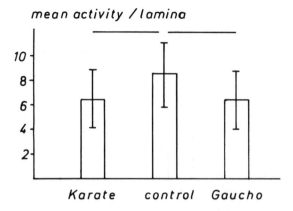

mean activity / lamina

Karate   control   Gaucho

FIGURE 3. Test of the two insecticides Karate and Gaucho. Each column represents the result of eight files of bait lamina. The differences in the control plot are highly significant, as marked by the horizontal lines.

The method can be used only to compare two or more related areas which are different in one or more factors in a given period, and the results cannot be taken as absolute values.

It is not possible to give a final judgment on the method now, as we only have a few results, but we shall continue this work. The result of the bait-lamina test is a kind of summation parameter. It only gives a first impression of the biological and especially zoological situation in the soil under investigation.

cm depth	Control file L 2	Pesticide file R 2
0.5	OOOOOOOO	OOOO
1.0	OOOOOOOO	O
1.5	OOOOOOOOO	O
2.0	OOOOOOO	OOO
2.5	OOOOOO	OOO
3.0	OOOOO	O
3.5	OOOOOOO	OOOOO
4.0	OOOOOO	OOOOO
4.5	OOOOOOO	OOOOOOOO
5.0	OOOOOOO	OOOOOOOO
5.5	OOOO	OOOOO
6.0	OO	OOOOOO
6.5	OOO	OOOO
7.0	OO	OOOO
7.5	O	OOOOOOOO
8.0	OOO	OOOOOOO

FIGURE 4. Example for a vertical activity profile of two files in a field test of the pyrethroid Karate. The number of spots indicate the number of open apertures, i.e., that the bait was fed by the soil animals.

REFERENCES

1. Lübben, B., Auswirkungen von Klärschlammdüngung und Schwermetallbelastung auf die Collembolenfauna eines Ackerbodens, Ph.D. thesis, Braunschweig, 1991.
2. Crossley, D.A.J. and Hoglund, M.P., A litter-bag method for the study of microarthropods inhabiting leaf litter, *Ecology,* 43, 571, 1962.
3. Larink, O. and Lübben, B., Bestimmung der biologischen Aktivität von Böden mit dem Köderstreifen-Test nach v. Törne: ein Erfahrungsbericht, *Mitt. Dtsch. Bodenkd. Ges.,* 66, 551, 1991.
4. Luthardt, V., Vergleich der biotischen Aktivität in extensiv und intensiv genutzten Niedermoorböden, *Pedobiologia,* 35, 199, 1991.
5. Törne, E. von, Assessing feeding activity of soil-living animals. I. Bait-lamina-test, *Pedobiologia,* 34, 89, 1990.
6. Törne, E. von, Schätzungen der Fraßaktivitäten bodenlebender Tiere. II. Mini-Köder-Test, *Pedobiologia,* 34, 269, 1990.

CHAPTER 25

Uptake, Assimilation, and Ligand Binding of Cadmium and Zinc in *Helix pomatia* After Combined Exposure to Both Metals

B. Berger, R. Dallinger, A. Gruber, and J. Moser

TABLE OF CONTENTS

0-87371-530-6/94/$0.00 + $.50
© 1994 by CRC Press, Inc.

ABSTRACT

Specimens of the terrestrial gastropod *Helix pomatia* were allowed to feed agar enriched with cadmium or zinc or with a combination of the two metals. The efficiency of metal assimilation was more than 90% for cadmium and about 80% for zinc. In cadmium-treated animals the metal concentration increased significantly in all tissues, the main site of storage being the midgut gland and the intestine. In zinc-treated snails the metal accumulated mainly in the midgut gland, reaching concentrations above 15 μmol/g (dw). No cadmium/zinc interactions were observed with regard to uptake, accumulation, and distribution of the metals. Subcellular fractionation showed that cadmium and zinc differed from each other in terms of compartmentalization and ligand binding. Most of the cadmium was bound to metallothionein which appeared to consist of two components. Zinc was found predominantly in the insoluble fraction or was associated with a homogeneous component of 1–4 kDa. Only a small fraction of zinc was bound to metallothionein. This difference in biochemical handling of cadmium and zinc appears to be the major reason for the lack of cadmium/zinc interactions in this species.

I. INTRODUCTION

Terrestrial gastropods have a high capacity for accumulating certain metals.[1-5] The hepatopancreas of these animals is the main site of storage, turnover, and detoxification of cadmium and zinc.[3,6,7] Both metals are bound to metallothionein or other low-molecular weight components.[8-13] Because of the similarities between cadmium and zinc as far as their chemistry, binding properties, and major target organ in the snail are concerned, it was the aim of this study to quantify the amount of each metal associated with subcellular fractions and to examine the effect of combined feeding of the two metals on uptake, accumulation, and distribution in *Helix pomatia*.

II. MATERIAL AND METHODS

H. pomatia were obtained from a commercial dealer (Fa. Stein, Lauingen, Germany). The animals were reared in groups of about 50 individuals in plastic boxes at a temperature of 18°C. Three times a week the snails were moistened with tap water and fed on lettuce (*Lactuca sativa*).

A set of three feeding experiments was carried out simultaneously. In feeding experiment 1, cadmium-enriched agar plates were used (0.85 μmol/g dw); in feeding experiment 2, zinc was added to the agar (7.4 μmol/g dw). The food used in feeding experiment 3 was spiked with a combination of cadmium (0.88 μmol/g dw) and zinc (7.3 μmol/g dw). In each experiment a total of 12 snails were used. After a feeding period of 30 d the snails were dissected and the metal concentration in the whole body (WB) (without shell), combined foot and mantle tissue (F/M), hepatopancreas (H), intestine (I), and remaining organs (R) was measured. Animals which had been fed uncontaminated lettuce were used as controls (n = 9). The preparation of the metal-loaded agar plates and calculation of metal budgets have been described before.[14]

Snails used for chromatographic separations were fed as in feeding experiment 3. They were dissected and the hepatopancreas homogenized and centrifuged. The supernatants were subjected to a series of chromatographic separation steps: gel chromatography (Sephacryl S-200), ion-exchange chromatography (DEAE-cellulose), and gel chromatography (Sephacryl S-100). The columns for gel permeation chromatography had been calibrated with standard proteins of known molecular weight. For details see Dallinger et al.[15]

Metal concentrations were measured by flame atomic absorption spectrophotometry (Perkin-Elmer®, model 2380) and expressed on dry weight basis.

III. RESULTS

A. Feeding Experiments

1. Feeding Experiment 1

Total cadmium increased from 14 ± 4 μg (controls) to 82 ± 20 μg per animal, the efficiency of cadmium assimilation always exceeding 90%. The increase of cadmium concentration was significant in all organs of the treated snails (Table 1a). Highest concentrations were detected in the hepatopancreas and the intestine, with values ranging between 1.8 and 3.3 μmol/g. Percentage distribution of cadmium in the soft tissues of the snails was independent of the total amount of cadmium and zinc absorbed (Table 1a). The main sites of cadmium storage were the hepatopancreas and the intestine, about 90% of the assimilated cadmium accumulating in these organs.

Table 1. Concentrations (Mean ± SD) and Percentage Distributions (%) of Cadmium (A) and Zinc (B) in Different Organs of *Helix pomatia*

A	WB	F/M		H		I		R	
Cadmium	μmol/g	μmol/g	%	μmol/g	%	μmol/g	%	μmol/g	%
Controls	0.05±0.02	0.01±0.006	10	0.34±0.15	75	0.23±0.13	11	0.01±0.006	4
FE1	0.40±0.11*	0.07±0.02*	6	2.65±1.01*	72	3.31±0.57*	20*	0.10±0.07*	4
FE3	0.38±0.08*	0.07±0.01*	6	2.13±0.13*	73	1.77±0.53*	19	0.04±0.01*	3

B	WB	F/M		H		I		R	
Zinc	μmol/g	μmol/g	%	μmol/g	%	μmol/g	%	μmol/g	%
Controls	1.5±0.4	1.1±0.4	29	6.5±1.4	58	1.4±0.2	3	0.9±0.2	11
FE2	4.4±0.9*	0.9±0.1	14*	15.9±4.2*	79*	1.6±0.3	2*	0.8±0.3	5*
FE3	3.8±0.4*	0.9±0.1	9*†	22.5±1.0*	83*	2.3±0.7*	3	0.8±0.2	5*

Note: A: Cadmium concentrations of controls and of snails from feeding experiments 1 and 3 (FE1, FE3). B: Zinc levels of controls and of animals from feeding experiments 2 and 3 (FE2, FE3). An asterisk (*) indicates a significant difference in metal concentrations or percentage distributions between controls and metal-fed snails. A dagger (†) indicates a significant difference between the metal-feeding experiments (Student's t-test, p < 0.01). Abbreviations: WB = whole body, F/M = combined foot and mantle tissues, H = hepatopancreas, I = intestine, R = rest.

Feeding Experiment 2

The uptake of zinc amounted to about 450 µg per snail, the efficiency of zinc assimilation decreasing from 94% at the beginning of the experiment to 80% after 30 d. Most of the zinc accumulated in the hepatopancreas (80%), where concentrations of 16 to 23 µmol/g were measured. In contrast to cadmium, zinc concentrations did not increase in combined F/M tissues and in the remaining organs (R) (Table 1b). In the intestine the concentration of zinc increased much less than that of cadmium (feeding experiment 1). In control animals 29% of total zinc was stored in the F/M, whereas in feeding experiment 2 this portion was only 14%. In the hepatopancreas, however, the amount of accumulated zinc increased from 58% (controls) to 79% (feeding experiment 2) (Table 1b).

Feeding Experiment 3

Compared with the controls, concentration of cadmium increased significantly in all organs of the snails, but there were no differences compared with feeding experiment 1, except a significantly lower concentration in the intestine (Table 1a). No statistically significant difference was found between the concentrations of zinc in feeding experiments 2 and 3 (Table 1b). The relative distribution of both metals after feeding experiment 3 presented the same pattern as that of cadmium after feeding experiment 1, and that of zinc after feeding experiment 2, except that in the combined F/M tissues the portion of zinc was significantly lower after feeding experiment 3 compared with feeding experiment 2 (Table 1b).

B. Subcellular Fractionation

After centrifugation, 86% of cadmium, but only 21% of zinc were found in the soluble fraction of hepatopancreas (Figure 1).

After separation of the supernatant by gel permeation chromatography, cadmium eluted in a single peak at a molecular weight of 11.5 kDa, the peak showing an elevated absorption at 254 nm. By ion-exchange chromatography and subsequent gel chromatography this cadmium peak was shown to consist of two main components with amino acid compositions typical of metallothionein-like proteins.[15] Quantification of subcellular cadmium distribution revealed 82% of this metal to be associated with the metallothionein fraction in the hepatopancreas (Figure 1).

Fractionation of supernatants of hepatopancreas homogenates by gel permeation chromatography showed that about 50% of soluble zinc was bound to a component with a molecular weight of <4 kDa. The nature of this component is not yet known, but chemical analysis is currently in progress. A second, but minor, portion of zinc was found in a peak of 11.5 kDa. After further separation steps this component was identified as the cadmium-specific metallothionein mentioned above, which also bound some zinc, the amount of which approxi-

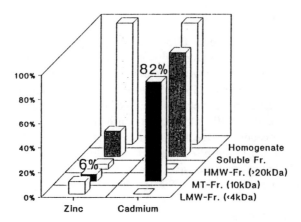

FIGURE 1. Percentage distribution of total cadmium and zinc associated with subcellular fractions of hepatopancreas homogenates from cadmium- and zinc-fed snails. Fractions were separated by centrifugation and gel permeation chromatography (see text). Abbreviations: HMW-Fr., high-molecular weight fractions; MT-Fr., metallothionein fractions; LMW-Fr., low-molecular weight fractions.

mated 6% of the total zinc in the hepatopancreas (Figure 1). The molar ratio of cadmium and zinc bound to metallothionein was 8:1.

IV. CONCLUSIONS

Compared with naturally occurring cadmium and zinc concentrations in snails, the values reached in these experiments are high.[16] Thus cellular and biochemical detoxification mechanisms are likely to have been activated in the tissues of the metal-fed snails.[8] Under the experimental conditions employed in this study, cadmium and zinc did not influence each other with regard to accumulation efficiency, storage capacity, and distribution in the body. Interferences between these two metals have been reported for many vertebrate and invertebrate species.[17-21] The apparent lack of cadmium/zinc interactions in *H. pomatia* can be explained by the different ways the two metals are handled in the cells. This is supported by the results of separation studies of hepatopancreas homogenates which revealed different subcellular patterns of ligand binding for both cadmium and zinc. The accumulation and long-term storage of cadmium in *H. pomatia* is accounted for by association of the metal with cytosolic metallothionein in the hepatopancreas. In contrast to cadmium a significant proportion of hepatopancreas zinc is bound to a low-molecular weight ligand, while only a small amount of the metal is chelated by metallothioneins. In zinc-contaminated snails, a predominant proportion of the metal is sequestered by calcium-rich pyrophosphate granules within the calcium cells.[8,22,23]

ACKNOWLEDGMENTS

The authors thank Professor W. Wieser, University of Innsbruck, for reviewing the manuscript and for helpful discussion. The study was supported by the Austrian "Fonds zur Förderung der wissenschaftlichen Forschung in Österreich", project no. P7815.

REFERENCES

1. Coughtrey, P.J. and Martin, M.H., The uptake of lead, cadmium, and copper by the pulmonate mollusc, *Helix aspersa* Müller, and its relevance to the monitoring of heavy metal contamination of the environment, *Oecologia,* 27, 65, 1977.
2. Russel, L.K., Haven, de, J.I., and Bots, P., Toxic effects of cadmium in the garden snail *(Helix aspersa), Bull. Environ. Contam. Toxicol.,* 26, 634, 1981.
3. Dallinger, R. and Wieser, W., Patterns of accumulation, distribution and liberation of zinc, copper, cadmium and lead in different organs of the land snail *Helix pomatia* L., *Comp. Biochem. Physiol.,* 79C, 117, 1984.
4. Beeby, A. and Richmond, L., Adaptation by an urban population of the snail *Helix aspersa* to a diet contaminated with lead, *Environ. Poll.,* 46, 73, 1987.
5. Greville, R.W. and Morgan, A.J., The influence of size on the accumulated amounts of metals (Cu, Cd, Zn and Ca) in six species of slug sampled from a contaminated woodland site, *J. Moll. Stud.,* 56, 355, 1990.
6. Coughtrey, P.J. and Martin, M.H., The distribution of Pb, Zn, Cd and Cu within the pulmonate mollusc *Helix aspersa* Müller, *Oecologia,* 23, 315, 1976.
7. Simkiss, K. and Watkins, B., Differences in zinc uptake between snails (*Helix aspersa* (Müller)) from metal- and bacteria-polluted sites, *Functional Ecol.,* 5, 787, 1991.
8. Dallinger, R., Strategies of metal detoxification in terrestrial invertebrates, in *Ecotoxicology of Metals in Invertebrates,* Dallinger, R. and Rainbow, P.S., Eds., Lewis Publishers, Chelsea, 1993, 245.
9. Cooke, M., Jackson, A., Nickless, G., and Roberts, D.J., Distribution and speciation of cadmium in the terrestrial snail, *Helix aspersa, Bull. Environ. Contam. Toxicol.,* 23, 445, 1979.
10. Ireland, M.P., Uptake and distribution of cadmium in the terrestrial slug *Arion ater* (L.), *Comp. Biochem. Physiol.,* 68A, 37, 1981.
11. Dallinger, R. and Wieser, W., Molecular fractionation of zinc, copper, cadmium and lead in the midgut gland of *Helix pomatia* L., *Comp. Biochem. Physiol.,* 79C, 125, 1984.
12. Dallinger, R., Berger, B., and Bauer-Hilty, A., Purification of cadmium binding proteins from related species of terrestrial Helicidae (Gastropoda, Mollusca), *Mol. Cell. Biochem.,* 85, 135, 1989.
13. Dallinger, R., Janssen, H.H., Bauer-Hilty, A., and Berger, B., Characterization of an inducible cadmium-binding protein from hepatopancreas of metal-exposed slugs (Arionidae, Mollusca), *Comp. Biochem. Physiol.,* 92C, 355, 1989.

14. Berger, B., Dallinger, R., Felder, E., and Moser, J., Accumulation of cadmium and zinc in the terrestrial gastropod *Helix pomatia,* in *Ecotoxicology of Metals in Invertebrates,* Dallinger, R. and Rainbow, P.S., Eds., Lewis Publishers, Chelsea, 1993, 291.

15. Dallinger, R., Berger, B., and Gruber, A., Quantitative aspects of zinc and cadmium in *Helix pomatia:* differences between an essential and a non-essential trace element, in *Ecotoxicology of Metals in Invertebrates,* Dallinger, R. and Rainbow, P.S., Eds., Lewis Publishers, Chelsea, 1993, 315.

16. Berger, B. and Dallinger, R., Terrestrial snails as quantitative indicators of environmental metal pollution, *Environ. Mon. Ass.,* 25, 65, 1993.

17. Elinder, C.G. and Piscator, M., Cadmium and zinc relationships, *Environ. Health Perspect.,* 25, 129, 1978.

18. Petering, H.G., Some observations on the interaction of zinc, copper, and iron metabolism in lead and cadmium toxicity, *Environ. Health. Perspect.,* 25, 141, 1978.

19. Honda, R. and Nogawa, K., Cadmium, zinc and copper relationships in kidney and liver of humans exposed to environmental cadmium, *Arch. Toxicol.,* 59, 437, 1987.

20. Wicklund, A., Norrgren, L., and Runn, P., The influence of cadmium and zinc on cadmium turnover in the Zebrafish, *Brachydanio rerio, Arch. Environ. Contam. Toxicol.,* 19, 348, 1990.

21. Ewers, E., Turfeld, M., Freier, I., Jermann, E., and Brockhaus, A., Interrelationships between cadmium, zinc and copper in human kidney cortex, *Toxicol. Environ. Chem.,* 27, 31, 1990.

22. Howard, B., Mitchell, P.C.H., Ritchie, A., Simkiss, K., and Taylor, M.G., The composition of intracellular granules from the metal-accumulating cells of the common garden snail *(Helix aspersa), Biochem. J.,* 194, 507, 1981.

23. Taylor, M.G., Graeves, G.N., and Simkiss, K., Biotransformation of intracellular minerals by zinc ions in vivo and in vitro, *Eur. J. Biochem.,* 192, 783, 1990.

SECTION IV

Bioaccumulation and Food Chain Transfer

---

26. Methodological Principles of Using Small Mammals For Ecological Hazard Assessment of Chemical Soil Pollution, With Examples on Cadmium and Lead

27. Heavy Metal Tissue Levels, Impact on Breeding and Nestling Development in Natural Populations of Pied flycatcher (Aves) in the Pollution Gradient From a Smelter

28. A Method to Assess Biorisks in Terrestrial Ecosystems

CHAPTER 26

Methodological Principles of Using Small Mammals For Ecological Hazard Assessment of Chemical Soil Pollution, With Examples on Cadmium and Lead

W.-C. Ma

TABLE OF CONTENTS

0-87371-530-6/94/$0.00 + $.50

ABSTRACT

The use of a small-mammal oriented soil ecosystem model is discussed for the ecological hazard assessment of soil pollutant chemicals with a capacity to accumulate in food chains. The critical internal pollutant burden in target organs (kidneys) is suggested as a useful sublethal toxicological endpoint. Examples for cadmium and lead show that species belonging to the Insectivora, which are linked to the decomposer food chain, are a more critical risk group with regard to their probability of toxic exposure than the largely plant-feeding Rodentia. These differences are due to the bioaccumulation in food chains and are unrelated to species differences in body size. Within species, the age of the animal is an important factor. Predictions of the internal amount of pollutant depend also on the abiotic environment, including the soil characteristics that affect the pollutant's bioavailability. Examples on cadmium and lead are given to show that soil-related variations in pollutant biovailability may have important consequences for the ecological risk of pollution hazards and hence for the setting of adequate soil quality standards.

I. INTRODUCTION

Soil is a major sink of chemical contaminants emitted from industrial and urban sources. This problem necessitates the conduction of effect-oriented ecotoxicological studies in order to estimate the impact of soil pollution on natural ecosystems and to devise effective protective measures warranting a sustainable ecosystem quality. The regulatory strategy to achieve this goal requires the establishment of certain operational criteria. These criteria include the maximum allowable amounts of pollutants in ecosystems, the setting of well-defined standards for environmental and ecosystem quality, and the monitoring of pollution trends in ecosystems.

This paper deals with some of the theoretical and practical aspects attached to the ecological hazard assessment of soil pollution. Attention is focused on the potential of small mammals as models in investigating pollutant behavior and effects in terrestrial food chains. In particular, some of the more important methodological aspects are discussed in relation to several case studies on cadmium and lead.

II. SMALL MAMMALS

Various ecological, physiological, as well as practical arguments support the use of small mammals in pollution biomonitoring and hazard assessment. One of the major ecological arguments is that these animals fulfill important functional and trophic roles in terrestrial ecosystems. They participate actively in soil bioturbation and take part in different subsystems due to their wide species variation in trophic types, which include herbivorous grazers as well as carnivorous predators of soil invertebrates. Some trophic types are thus linked to the herbivore subsystem, whereas others are linked to the decomposition subsystem.

A physiological argument in support of small mammals as bioindicators of the pollutant exposure of terrestrial wildlife is related to their small body size. Due to a more intense metabolic rate their intensity of exposure may be expected to be higher than in large mammals.

From a practical point of view, small mammals meet the basic requirements for use in biomonitoring studies.[30] They have a widespread occurrence with a limited home range or foraging area and are relatively easily collected. They also are quite accessible to both population investigation and experimental research. Effects of soil pollutants can thus be studied on individuals in the laboratory or field enclosures as well as on free-living populations present in the field.

III. CONCEPTS

A. Food Chain Model System

Figure 1 gives a simplified representation of the various possible transfer routes of pollutants within a small-mammal-based soil ecosystem. The model illustrates the possible differences in chain length and trophic type. The system can be regarded as a conceptual model for modeling the food chain behavior of pollutants in terrestrial ecosystems. This model system can be used to derive food chain-related soil quality standards with regard to maximum allowable levels of soil pollution.

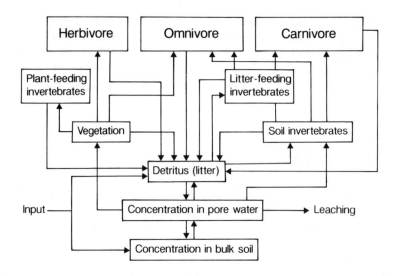

FIGURE 1. Conceptual model of the various transfer pathways of chemical pollutants within a system involving three different trophic types of small mammals.

B. Target Organ and Internal Dose

A possible approach to characterize the environmental risk of soil pollutants is based on some basic toxicological concepts. One of these concepts is the target organ, which is defined here as the tissue or organ system having the greatest sensitivity towards a pollutant chemical. The critical internal dose level is defined as the lowest internal concentration of a chemical in (part of) an organism that is associated with the occurrence of (sub)clinical toxic effects. The proposed use of the critical target organ load as an end point in hazard assessment is based upon its direct link with toxic effects. Measuring the internal target organ load also allows a more direct approach in risk assessment than the measurement of whole-body burden.

Applying the principles of dose-effect relationships, it is possible to assess the potential hazard of chemicals by estimating the likelihood to reach the critical target organ load in individuals of a given population of plants or animals. For example, pollutants may be phytotoxic above a certain critical concentration in sensitive plant tissues or they may affect normal physiological functions above a certain critical concentration in animal target organs. The application of this approach is independent of the ambient level of soil pollution. This has certain practical advantages as will be discussed below in relation to the problem of bioavailability.

A modeling study based on the critical level approach has been described for cadmium in the kidneys of small mammals.[21] The reason for focusing on the mammalian kidneys is that these are target organs for many organic and inorganic xenobiotic chemicals. Chemicals absorbed in blood are delivered in relatively high amounts to the kidneys because of the high renal blood flow. Many industrial chemicals are potent nephrotoxins by impairing important kidney functions.[11] These functions include the excretion of wastes, the formation of metabolic hormones, and the regulation of total body homeostasis by controlling, e.g., the electrolyte and acid-base balance. Proteinuria and other clinical symptoms of renal dysfunction have often been observed in mammals chronically exposed to cadmium.[25]

The lowest average renal concentration of cadmium associated with the occurrence of clinical toxicity symptoms in small mammals ranges from 120 to 150 μg/g dry weight.[6,24,29] The critical renal level for lead ranges from 25 to 40 μg/g dry weight.[10,28] These values have been necessarily derived from laboratory studies because of the paucity of data on threshold effects on individuals in field populations. For heavy metals, however, the results of field studies seem to be well in agreement with laboratory-derived data. For instance, the incidence of altered proximal kidney tubular cells in individuals of field populations of rats is associated with an average kidney lead burden of 30 μg/g.[23] In addition, the occurrence of serious pathological disorders resembling renal edema in field populations of small mammals has been associated with an average kidney lead burden of 47 μg/g in one study[27] and with 25 μg/g in another study.[17]

C. Bioaccumulation

The risk of soil pollutants to accumulate in small mammals to a toxic dose level may vary considerably between species of Rodentia and Insectivora. The difference is illustrated by the data summarized in Tables 1a and b, which give the results of a literature survey on the maximum internal levels of cadmium and lead recorded in the two taxonomic groups. The literature was selected on the condition that both groups had been sampled within the same polluted area.

It appears from Table 1a that the likelihood of cadmium to attain the critical target organ load is relatively large for the Insectivora, but only small for the Rodentia when occurring in polluted areas. Only a few reports on lead mention internal tissue concentrations in Rodentia above the critical level. The extent at which this level is exceeded in such cases is again much smaller in the Rodentia than in the Insectivora. For example, the upper range of lead measured in the kidneys of *Sorex araneus* has been reported to be as high as 1000 to 1300 μg/g,[4,17] whereas the highest values reported in Rodentia remain limited to 60 and 65 μg/g. The latter values have been measured in *Apodemus sylvaticus* (Muridae) and *Microtus agrestis* (Microtidae), respectively.[13]

It is interesting to note that the wood mouse *A. sylvaticus* can be considered as an omnivore as it takes seasonally varying amounts of seeds and invertebrates. This may place it, in trophic terms, between herbivorous voles and carnivorous shrews. However, internal levels of cadmium or lead in wood mice are equal to or even lower than those in voles (Tables 1a and b). Reasons for this may be

Table 1a. Comparison of Average Cadmium Concentrations in the Kidneys of Rodentia and Insectivora Occurring Within the Same Contaminated Area

Location	Species	Order	Cd (μg/g)	Ref.
Cu-Cd refinery	*Apodemus sylvaticus*	Rodentia	7	12
	Microtus agrestis	Rodentia	23	12
	Sorex araneus	Insectivora	193	12
Mine waste site	*M. agrestis*	Rodentia	5	2
	S. araneus	Insectivora	158	2
Smelter site	*M. agrestis*	Rodentia	2	18
	S. araneus	Insectivora	200	18

Table 1b. Comparison of Average Lead Concentrations in the Kidneys of Rodentia and Insectivora Occurring Within the Same Contaminated Area

Location	Species	Order	Pb (μg/g)	Ref.
Roadside	*Apodemus sylvaticus*	Rodentia	7	31
	Clethrionomys glareolus	Rodentia	13	31
	Microtus agrestis	Rodentia	10	31
	Sorex araneus	Insectivora	27	31
Roadside	*A. sylvaticus*	Rodentia	10	5
	C. glareolus	Rodentia	12	5
	S. araneus	Insectivora	46	5
Shooting range	*A. sylvaticus*	Rodentia	6	17
	C. glareolus	Rodentia	16	17
	S. araneus	Insectivora	269	17
Smelter site	*M. agrestis*	Rodentia	4	18
	S. araneus	Insectivora	85	18
Mine tailings	*M. agrestis*	Rodentia	21	3
	S. araneus	Insectivora	90	3
Roadside	*M. ochrogaster*	Rodentia	8	9
	Mus musculus	Rodentia	8	9
	Peromyscus maniculatus	Rodentia	8	9
	Blarina brevicauda	Insectivora	12	9

that *A. sylvaticus* feeds for a large part of its diet on plant seeds and fruits. Concentrations of cadmium or lead in plant reproductive organs (seeds, fruits) are generally lower than in roots, stems, or leafy tissues.[1] Furthermore, *A. sylvaticus* consumes soil insects rather than earthworms and do so only in relatively small quantities. As insects also contain relatively low concentrations of metals compared to other invertebrates such as earthworms,[19] they may contribute relatively little to the average daily metal intake of this small-mammal species.

An example to further illustrate the very wide species variation in the internal exposure of small mammals to soil pollutants is shown in Figure 2. The figure shows the average internal level of cadmium in adult animals belonging to four different species. These included two species of Rodentia, i.e., *M. arvalis* (Microtidae) and *Micromys minutus* (Muridae), and two species of Insectivora, i.e.,

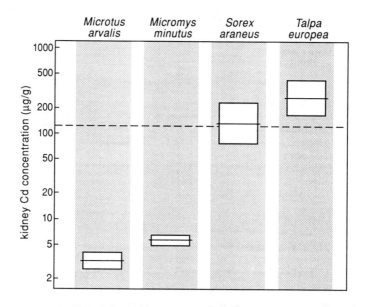

FIGURE 2. Geometric means and 95% confidence limits (N = 5 to 8) of cadmium concentrations in target organs of four species of small mammals within a cadmium-polluted area on podzolic sandy soil. Soil cadmium concentrations of the site were 23.2 mg/kg (A_o soil layer) and 2.8 mg/kg (A_1 layer). The broken line indicates the critical target organ level for cadmium.[18]

S. araneus (Soricidae) and *Talpa europea* (Talpidae). All four species had been collected within a single contaminated site on natural sandy soil.

It can be seen from Figure 2 that the target organ load of cadmium may be as much as two orders of magnitude greater in the Insectivora than in the Rodentia. It appears also that only the Insectivora show a real risk of being critically exposed to cadmium.

The relationship between exposure and the amount of metals within small mammals is related to the concentration factor (CF) in successive steps of the food chain. CF may be further distinguished into the bioaccumulation factor (BAF) and the bioconcentration factor (BCF). BAF may be defined as the ratio between the (equilibrium) concentration of a chemical in (part of) an organism and its food and environment. In terrestrial animals the principal mechanism of exposure consists of food ingestion, but some lower soil invertebrates may also take up pollutants directly from the soil to equilibrium levels. In that case the term bioconcentration factor (BCF) is more commonly applied.

The reason for the large species variation in exposure of small mammals is illustrated in Figure 3. This figure shows the range in the value of CF for each separate step of two different food chains. It can be seen that the invertebrate-based food chain directed towards *S. araneus* has categorically higher CF values for each step than the plant-based food chain directed towards *Microtus agrestis*.

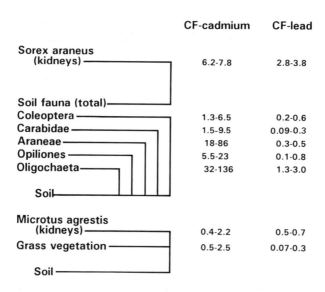

FIGURE 3. Range in concentration factors (CF) for cadmium and lead within a herbivore and a decomposer-based terrestrial food chain. Concentrations in the soil (0 to 10 cm) of the study site were 0.2 to 2.9 mg/kg for cadmium and 20 to 130 mg/kg for lead. CF for *S. araneus* were calculated from summed weighted median concentrations of individual food items.[18]

Model calculations have indeed shown that the average daily metal intake of small mammals is much greater for the Insectivora than for the Rodentia.[19]

Another interesting feature shown in Figure 3 is the difference in bioaccumulation potential that may exist between metals. Cadmium has higher CF values than lead at all steps of both the decomposer-based and plant-based food chain. Whether cadmium also poses a greater hazard to ecosystems than lead depends on the intrinsic toxicity of the pollutant. In principle, it is the combination of accumulation kinetics and critical internal concentration which ultimately determines the hazard of a pollutant to organisms.

IV. ENVIRONMENT-RELATED FACTORS

A. Bioavailability

Bioavailability may be defined as the potential of living organisms to take up chemicals from the food or from the abiotic environment to the extent that these chemicals may become involved in the metabolism of the organism. External factors influencing the bioavailability of chemicals in the abiotic soil environment include the speciation form of the chemical, its complexation kinetics with organic ligands, and the adsorption/desorption kinetics. The physi-

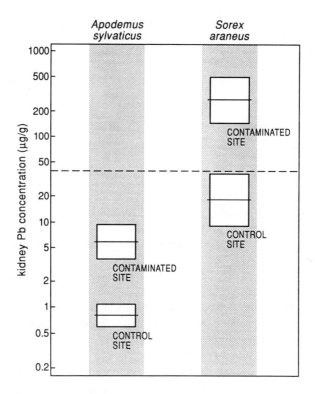

FIGURE 4. Geometric means with 95% confidence limits (N = 10 to 23) of concentrations of lead in target organs of two species of small mammals within a lead pellet-polluted site and a control situated in a chalkless coastal duneland area. The broken line indicates the critical target organ level for lead. From Ma, W.C., *Arch. Environ. Contam. Toxicol.,* 18, 617, 1989. With permission.

cochemical characteristics of the organic and inorganic soil colloids and the amount of organic matter and pH may all play an important role in determining the bioavailability of pollutants in soil.

The importance of the variation in bioavailability of pollutants due to chemical speciation is illustrated by an investigation of a lead pellet-polluted natural area. The hypothesis was tested that lead pellet pollution is ecologically not of appreciable consequence as lead from metallic pellets is unavailable to plant uptake and thus is prevented from entry into the food chain. To test this hypothesis, lead concentrations were measured in adult individuals of two different species, i.e., *A. sylvaticus* and *S. araneus*.

The results are summarized in Figure 4 and indicate that both species investigated exhibited strongly elevated internal levels of lead compared to the reference control obtained from unpolluted sites. It is concluded therefore that lead pellets in a poorly-buffered acidified soil apparently can change from an

**Table 2. Average Concentration of Lead in *Talpa europea*
(N = 4 to 5) Sampled From Two Adjacent Sites
on Sandy Soil Contaminated With Lead**

	Soil			*Talpa europea* (kidneys) Pb concentration (µg/g)	
Site	Pb conc (mg/kg)	Organic matter (%)	pH (KCl)	Mean	Range
Grassland	135	10.2	6.5	18	8–35
Heathland	149	2.0	4.1	338	238–438

inert metallic form to a highly bioavailable form. The environmental chemistry underlying such change is probably related to the chemical formation of (hydro)cerussite as a first transformation step followed by a proton-mediated formation of lead ions.[14] Lead occurring in a free ionic or some soluble complexed form in the soil solution may be available to plants through root uptake and thus enter the terrestrial food chain.

Physical and chemical properties of soil, including mineralogy, particle size distribution, organic matter content, and pH are important factors determining metal sorption and speciation. The importance of such site properties to the bioavailability of chemicals is illustrated in Table 2.

This table shows the results of a study comparing the bioavailability of lead in two different locations on natural sandy soil, i.e., a grassland and a neighboring heathland. The two sites differed in soil characteristics only with respect to the organic matter content and pH. The pollution originated from the same smelter source and was present at approximately the same level in the soil of the two sites.

The hypothesis was tested that the ecological hazard of lead pollution did not differ essentially between the two sites in view of the similarity in the extent of the soil contamination present. This hypothesis, however, had to be rejected as investigation of resident adult moles, *T. europea*, showed a much more elevated renal lead level in animals from the heathland than in those from the grassland. In fact, the lead burden of the grassland moles was rather low (Table 2). It can be concluded therefore that the lead pollution was highly bioavailable in the soil of the heathland site, but only poorly bioavailable in the grassland soil.

Earthworms are the principal food source of *T. europea* and other species of Insectivora. Predation on earthworms is a major pathway for the entry of soil contaminants into terrestrial food chains. The considerable potential of earthworms to accumulate pollutant chemicals by uptake from soil could significantly increase the exposure of their predators to a toxic level.

Earthworms accumulate cadmium and lead in levels representative of the degree of contamination. For earthworms in sandy soil the bioconcentration factor of these metals can be described in general terms as:[21]

$$BCF = \exp\{b_0 + (b_1 - 1)*\ln(C_{soil}) + b_2*pH + b_3*SOM \quad (1)$$

where BCF is the bioconcentration factor, C_{soil} is the concentration of lead in soil, pH is the soil acidity, and SOM is the percentage soil organic matter which in sandy soils is the principal matrix for metal sorption.

The degree of metal bioaccumulation in earthworms thus varies not only with the soil pollution level, but also with the organic matter content and pH. The lower the pH and quantity of organic matter the greater the potential of earthworms will be to accumulate metals like cadmium or lead. The relatively low pH and organic matter content of the heathland (Table 2) thus may have been responsible for the very high bioavailability of lead observed at this site.

The risk of hazardous effects may be strongly affected by the external conditions that determine the bioavailability of the pollutant and by the critical level of the pollutant concerned. Bioavailability and critical level affect the maximum allowable level of soil pollution in an interactive manner. Modeling studies with cadmium have thus shown that in a relative sense the maximum allowable level of soil pollution depends on the critical level independent of the bioavailability of the pollutant. In an absolute sense, however, the critical level alters the maximum allowable soil pollution level more sensitively if the pollution is little bioavailable than if the pollutant has a high bioavailability.[21]

B. Delayed Response

The above examples indicate that metal pollution of soil may represent a greater ecological hazard in natural areas with a poor buffering capacity relative to soil acidification than in areas with a well-buffered soil. The results emphasize the need to carry out risk assessments on a broader range of environmental data base than only the total concentration of a pollutant in the abiotic environment. Without a proper assessment of the bioavailability status of a pollution, the outcome of any risk assessment will be at least questionable.

It has been pointed out that a gradual alteration of certain environmental conditions such as soil acidification may cause a sharp increase in the risk of toxic exposure of wildlife in a polluted environment.[21] As can be seen from the BCF relationship given by Equation 1, such an increase may occur even if the absolute level of the contamination itself remains constant. This may cause a delayed response to pollution due to altered abiotic environmental conditions ("chemical time bomb"). To prevent the incidence of such delayed effects in susceptible areas requires the implementation of certain regulatory strategies, such as the setting of soil quality standards that take full account of the possible variation in pollutant bioavailability.

At the same time, remedial actions may follow from measures designed to reduce the bioavailability of soil pollution. For instance, for cadmium and lead it appears that the accumulation decreases with increasing soil pH, not only in soil invertebrates but also in plants.[7,22] It can therefore be recommended that soil pH in contaminated areas be maintained at about 6.5 or higher to reduce the bioavailability of these metals and thus protect the terrestrial food chain.

V. ANIMAL-RELATED FACTORS

Several animal-related factors may influence the level of exposure of small mammals to pollutants. Physiological factors include assimilation and elimination kinetics at interfaces of the gastrointestinal tract and internal organs. Other possible factors which will be discussed here include body size and age.

A. Body Size

For heavy metals, there is no evidence to suggest that the species-related variation in internal tissue burden of small mammals may be affected by body size. For example, the body weight of *T. europea* adults ranges from about 65 to 130 g fresh weight, which is about ten times higher than the body weight of adults of *S. araneus*. In spite of this large difference in size, the target organ burden of cadmium is similar in both species (Figure 2). Another example is that adult specimens of *Micromys minutus* and *S. araneus* have a common body weight of about 10 g, whereas the target organ load of cadmium is vastly different between the two species (Figure 2). These observations suggest that body size is an unlikely factor to contribute to the large species variation in the internal pollutant burden of small mammals.

B. Age Relationships

Few studies have investigated the influence of age upon the internal tissue burden of xenobiotics in small mammals. Bioaccumulation is an age-related phenomenon when changes in the accumulated amount of a pollutant do not keep up with the growth in biomass. It is also possible that uptake and elimination kinetics are not constant, but altered during the growth of an animal.

Evidence for an age-dependent accumulation of pollutants in mammals has been described for cadmium in badgers, *Meles meles* (Mustelidae, Carnivora).[20] There is similar evidence for small mammals. Figure 5 thus shows the increase in target organ load of cadmium in *S. araneus* with increasing age. It can be seen that initially there is a rapid increase which is followed by a steady-state level during the subadult and adult stages. Figure 5 thus represents an example of a situation in which the pollutant reaches an equilibrium concentration in a mammal during life-time exposure.

The practical implication of age-dependent bioaccumulation is that one should account for age as a covariable in biomonitoring. In very young animals it is likely that concentrations are measured in the nonequilibrium phase of the bioaccumulation process. It is clear that a careful estimation of age may considerably reduce the variability of pollutant loads measured in animal samples.

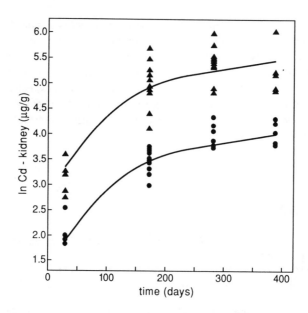

FIGURE 5. Age relationships for the target organ load of cadmium in *Sorex araneus* within two sites with a different level of soil cadmium pollution. From Ma, W.C. and Van der Voet, H., *Sci. Total Environ.*, in press.

VI. CONCLUDING REMARKS

The pollution assessment approach proposed in this paper employs the critical internal target organ concentration in high-risk species of small mammals as a hazard indicator. A similar approach can be followed for plants, by taking as a critical level the lowest pollutant concentration associated with an adverse effect on, e.g., plant biomass production.[22] An advantage of the critical level approach is the possibility to assess soil pollution hazards in a manner which includes the variation due to differences in bioavailability.

There are few generally applicable relationships between the expected pollutant concentration in plants or animals and the concentration in soil. The lowest soil level of cadmium associated with adverse effects on plant growth is 5 mg/kg.[15,22] For small mammals, the protective level has been differentiated according to bioavailability of the pollutant. Depending on the degree of bioavailability, the lowest soil level of cadmium associated with adverse effects in carnivorous small mammals may vary between 1 and 2.5 mg/kg.[21] Soil quality standards based on the small-mammal approach may thus offer a protective basis for plants as well. It should be realized, however, that the critical pathway for small mammals consists of food chains involving decomposer-linked predators rather than herbivores or omnivores.

For developing wildlife-oriented environmental quality standards the greatest challenge is in generating relevant data on dose-effect relationships. To this end, more research is needed on target organ identification, species sensitivity, and the influence of animal size, physiology, and age. In addition, clinical toxicity indices are needed that are easy to measure and have some relevance to population effects. Also, a thorough understanding of environmental chemistry in relation to pollutant availability will be of crucial importance for establishing sound environmental quality standards.

REFERENCES

1. Adriano, D.C., *Trace Elements in the Terrestrial Environment,* Springer-Verlag, New York, 1986.
2. Andrews, S.M., Johnson, M.S., and Cooke, J.A., Cadmium in small mammals from grassland established on metalliferous mine waste, *Environ. Pollut.,* 33, 153, 1984.
3. Andrews, S.M., Johnson, M.S., and Cooke, J.A., Distribution of trace element pollutants in a contaminated grassland ecosystem established on metalliferous fluorspar tailings. 1. Lead, *Environ. Pollut.,* 58, 73, 1989.
4. Beyer, W.N., Pattee, O.H., Sileo, L., Hoffman, D.J., and Mulhern, B.M., Metal contamination in wildlife living near two zinc smelters, *Environ. Pollut. (A),* 38, 63, 1985.
5. Chmiel, K.M. and Harrison, R.M., Lead contents of small mammals at a road side site in relation to the pathways of exposure, *Sci. Total Environ.,* 17, 145, 1981.
6. Chmielnicka, J., Halatek, T., and Jedlinska, U., Correlation of cadmium-induced nephropathy and the metabolism of endogenous copper and zinc in rats, *Ecotoxicol. Environ. Saf.,* 18, 268, 1989.
7. Ericksson, J.E., A field study on factors influencing Cd levels in soils and in grain of oats and winter wheat, *Water, Air, Soil Pollut.,* 53, 69, 1990.
8. Faber, J. and Ma, W.C., Observations and seasonal dynamics in diet composition of filed vole, *Microstes agrestis,* with some methodological remarks, *Acta Theriol.,* 31, 479, 1986.
9. Getz, L.L., Verner, L., and Prather, M., Lead concentrations in small mammals living near highways, *Environ. Pollut.,* 13, 151, 1977.
10. Goyer, R.A., Leonard, D.L., Moore, J.F., Rhyne, B., and Krigman, M., Lead dosage and the role of the intranuclear inclusion body, *Arch. Environ. Health,* 20, 705, 1970.
11. Hook, J.B., Toxic responses of the kidney, in *Casarett and Doull's Toxicology,* Doull, J., Klaassen, C.D., and Amdur, M.O., Eds., Macmillan, New York, 1980, 232.
12. Hunter, B.A. and Johnson, M.S., Food chain relationships of copper and cadmium in contaminated grassland ecosystems, *Oikos,* 38, 108, 1982.

13. Johnson, M.S., Roberts, R.D., Hutton, M., and Inskip, M.J., Distribution of lead, zinc, and cadmium in small mammals from ·polluted environments, *Oikos,* 30, 153, 1978.

14. Jörgensen, S.S. and Willems, M., The fate of lead in soils: the transformation of lead pellets in shooting-range soils, *Ambio,* 16, 11, 1987.

15. Kloke, A., Sauerbeck, D.R., and Vetten, H., The contamination of plants and soils with heavy metals and the transport of metals in terrestrial food chains, in *Changing Metal Cycles and Human Health,* Nriagu, J.O., Ed., Springer-Verlag, Berlin.

16. Ma, W.C., Heavy metal accumulation in the mole, *Talpa europea,* and earthworms as an indicator of metal bioavailability in terrestrial environments, *Bull. Environ. Contam. Toxicol.,* 39, 933, 1987.

17. Ma, W.C., Effect of soil pollution with metallic lead pellets on lead bioaccumulation and organ/body weight alterations in small mammals, *Arch. Environ. Contam. Toxicol.,* 18, 617, 1989.

18. Ma, W.C., unpublished data, 1992.

19. Ma, W.C., Denneman, W., and Faber, J., Hazardous exposure of ground-living small mammals to cadmium and lead in contaminated terrestrial ecosystems, *Arch. Environ. Contam. Toxicol.,* 20, 266, 1991.

20. Ma, W.C. and Broekhuizen, S., Possible influence of the polluted forelands of the river Meuse on the heavy-metal burden of badgers *Meles meles* in The Netherlands, *Lutra,* 32, 139, 1989.

21. Ma, W.C. and Van der Voet, H., A risk-assessment model for toxic exposure of small mammalian carnivores to cadmium, *Sci. Total Environ.,* in press.

22. Macnicol, R.D. and Beckett, P.H.T., Critical tissue concentrations of potentially toxic elements, *Plant Soil,* 85, 107, 1985.

23. Mouw, D., Kalitis, K., Anver, M., Schwartz, J., Constan, A., Hartung, R., Cohen, B., and Ringler, D., Lead: possible toxicity in urban vs rural rats, *Arch. Environ. Health,* 30, 276, 1975.

24. Nicholson, J.K., Kendall, M.D., and Osborn, D., Cadmium and nephrotoxicity, *Nature (London),* 304, 633, 1983.

25. Nordberg, G. and Nordberg, M., Biological monitoring of cadmium, in *Biological Monitoring of Toxic Metals,* Clarkson, T., Friberg, L., Nordberg, G., and Sager, P.R., Eds., Plenum Press, New York, 1988, 151.

26. Osweiler, G.D., Van Gelder, G.A., and Buck, W.B., Epidemiology of lead poisoning in animals, in *Toxicity of Heavy Metals in The Environment,* Oehme, F.W., Ed., Marcel Dekker, New York, 1978, 143.

27. Roberts, R.D., Johnson, M.S., and Hutton, M., Lead contamination of small mammals from abandoned metalliferous mines, *Environ. Pollut.,* 15, 61, 1978.

28. Skerfving, S., Biological monitoring of exposure to inorganic lead, in *Biological Monitoring of Toxic Metals,* Clarkson, T., Friberg, L., Nordberg, G., and Sager, P.R., Eds., Plenum Press, New York, 1988, 169.

29. Sutou, S., Yamamoto, H., Sendota, K., Tomomatsu, K., Shimizu, Y., and Sugiyama, M., Toxicity, fertility, teratogenicity, and dominant lethal tests in rat administered cadmium subchronically. I. Toxicity studies, *Ecotoxicol. Environ. Saf.,* 4, 39, 1980.

30. Talmadge, S.S. and Walton, B.T., Small mammals as monitors of environmental contaminants, *Rev. Environ. Contam. Toxicol.,* 119, 47, 1991.

31. Williamson, P. and Evans, P.R., Lead: levels in roadside invertebrates and small mammals, *Bull. Environ. Contam. Toxicol.,* 8, 280, 1972.

CHAPTER 27

Heavy Metal Tissue Levels, Impact on Breeding and Nestling Development in Natural Populations of Pied Flycatcher (Aves) in the Pollution Gradient from a Smelter

N.E.I. Nyholm

TABLE OF CONTENTS

ABSTRACT

Chemical and biological effects of heavy metals were studied on free-living breeding populations of Pied Flycatchers (*Ficedula hypoleuca*) in the pollution gradient from the sulfide ore smelter Rönnskärsverken, Sweden. The concentrations of the non-essential metals, arsenic, cadmium, mercury, and lead in the tissues of breeding females (except for cadmium) and nestlings clearly reflected the degree of pollution of the breeding site. The tissue levels of the essential elements, copper and zinc, were not influenced by the degree of environmental contamination. The inter-site variation of the concentrations of the heavy metals in insects, constituting potential food for the bird nestlings, agreed with that of the nestlings. Biological effects, which were proved as, or were potentially detrimental to the breeding success with increasing severity nearer to the smelter included: eggshell defects, embryo and nestling mortality, increased liver development, decreased muscular and brain development, reduced hemoglobin levels, and impaired calcification of the skeleton in the nestlings.

I. INTRODUCTION

A study was performed, from 1984 to 1990, which was aimed to characterize the impact of environmental heavy metal contamination on the breeding performance of the Pied Flycatcher and to evaluate the biological capacity of that species with respect to its usefulness as a practical bioindicator organism on heavy metal contamination in the terrestrial environment.[1,2] Birds in general, and the Pied Flycatcher especially, own ecological and ethological features which make them potentially well suited as bioindicators for practical environmental monitoring.

- The Pied Flycatcher occurs with high abundancy in most forest types in northern Europe.
- The densities of the breeding populations are normally restricted by the number of nesting holes, but can be made high (3 to 4 pairs per hectare) with nest boxes which offer possibilities for the otherwise nonbreeding part of the populations to breed. By introducing nest boxes in the study areas the researcher can decide the localizations of the nests and the number of breeding pairs.
- The Pied Flycatcher is uniquely insensitive to disturbances at the nest, which allows detailed studies of the breeding birds, their eggs, and nestlings.

The Pied Flycatcher is a migrating bird species, and so there could be a risk that the adults may be significantly contaminated by metals of unknown origin when they arrive in the breeding area (= study area). The metal contents of the nestlings, on the other hand, should be specifically derived from the breeding

FIGURE 1. Localization of the study areas.

area, via the food (insects). The parent mobility during food collection is restricted to the nearest 50 to 100 m around the nest.

Another advantage with using the nestlings as study objects, is that their whole life history can be known in detail, and the samplings can be standardized, e.g., with respect to the birds' age, etc.

Birds in natural populations have attracted attention as bioindicators of heavy metal pollution in several studies. Usually these studies have aimed to characterize pollution situations chemically, by analyses of bird tissues, e.g., liver, kidney,[3-5] or feathers.[6,7] Biological characterization of pollution situations by means of birds in free-living populations, which constitutes a substantial part of the present study, has so far attracted much less attention (but see, e.g., References 8, 9, 10).

II. MATERIAL AND METHODS

The studies were performed on nest box-breeding Pied Flycatcher (*Ficedula hypoleuca, L.*) populations at five localities (Figure 1), four of which — R, B, H, and A — were situated within the pollution gradient of the sulfide ore smelter Rönnskärsverken, 2.5, 5, 20, and 30 km, respectively, from the industry in the prevailing wind directions. The fifth area, T, was the reference locality, at a

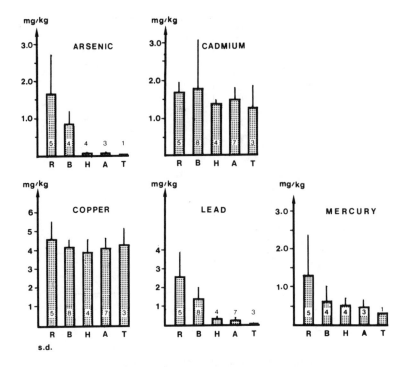

FIGURE 2. Mean concentrations of arsenic, cadmium, copper, lead, and mercury in the liver tissue of breeding females at the different localities, 1986 (w.w.). Figures in the bars denote number of females.

distance of 95 km. Clutch sizes, eggshell quality, embryo development, hatching frequency, and nestling development and survival were registered. Females were sampled at the very beginning of their incubation period (after having been in the breeding area for about a fortnight), as were fully grown nestlings, at the age of 12 to 14 d. Dissected organs (e.g., liver, brain, spleen, pectoral muscle, pectoral bone) were weighed, and their metal contents were analyzed (Ca, Cd, Cu, Pb, and Zn with AAS; As, Hg with NAA). Blood samples from the nestlings were analyzed for hemoglobin (Hemocue® Hb-photometer) and lead concentrations (AAS).

III. RESULTS AND DISCUSSION

In the breeding females, the tissue concentrations of the nonessential metals, arsenic, mercury, and lead, clearly reflected the pollution gradient from the smelter, suggesting a potentially rapid accumulation of these metals, as illustrated by the liver contents in Figure 2. The lead accumulation was extraordinarily high in the skeleton of the females — in the sternum to about 202 and 12 ppm

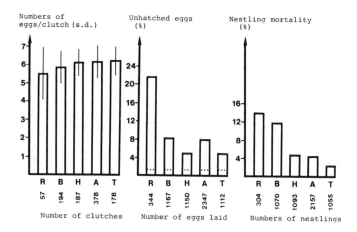

FIGURE 3. Clutch sizes, embryo and nestling mortality in the Pied Flycatcher populations at the different localities, 1984 to 1990. . . . = Frequency of unfertilized eggs. Student's *t*-test: differences of clutch sizes between R and T or A $p <0.001$; B and T or A $p <0.01$.

(dry weight) at the localities R and T, respectively. Most probably the skeletal lead, to the very greatest part, was accumulated in the medullary bone tissue, a tissue which is especially active in breeding female birds at the beginning of the breeding season. It then constitutes an important transitory source of calcium for the eggshell formation.[11] The concentrations of cadmium in the liver tissue, of the breeding females in contrast to the other nonessential metals, showed insignificant variation between the localities. This suggests that the load of that biologically highly persistent metal mainly originated from sources utilized by the birds during the preceding overwintering/migration periods. The concentrations of the essential metals, copper (Figure 2) and zinc, in the female tissues were unaffected by the different degrees of contamination of the localities. This was probably effected by homeostatic mechanisms in the organisms, the birds, and their food items, striving to keep the levels of the metals at the physiological levels.

There was strong evidence that the heavy metal contamination of the females affected the clutch sizes and the embryo survival (Figure 3), and the eggshell quality (Figure 4). The fertility of the birds was unaffected since unfertilized eggs equally often were the reason for failed hatching (in about 1.5% of the eggs) at the different localities (Figure 3).

In the tissues of the nestlings the concentrations of any of the nonessential metals (also cadmium) clearly reflected the pollution gradient (Figure 5). High lead levels occurred in the blood of nestlings in the most contaminated localities (Table 1). As in the breeding females, the highest lead accumulations in the nestlings were found in the skeleton (Table 1). Unlike the situation in the adult

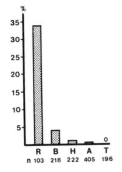

FIGURE 4. Frequency of clutches with eggshell defects, n = number of clutches.

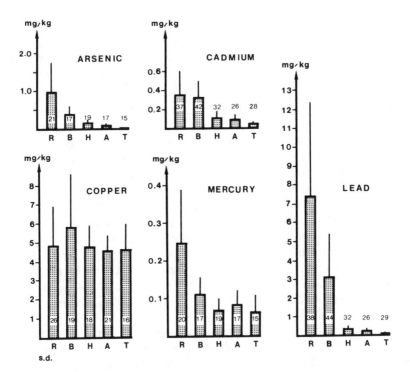

FIGURE 5. Mean concentrations of arsenic, cadmium, copper, lead, and mercury in the lever tissue of 12- to 14-d-old nestlings at the different localities (w.w.). Figures in the bars denote number of clutches represented.

females, however, in the nestlings the lead was accumulated in the structural bone tissue under formation. The tissue concentrations of essential metals in the nestlings (represented by copper in Figure 5) were similar at the different sites along the pollution gradient.

Table 1. Concentration of Lead in Blood and Sternum of 12- to 14-d-old Nestlings at the Different Localities

Locality (distance from smelter, km)	Blood lead (μg/100 ml)[a]	n	Skeleton lead (mg/kg; d.w.)[a]	n
R (2.5)	40 ± 18.4	(23)	120 ± 67.2	(6)
B (5)	22 ± 7.2	(28)	28 ± 13.0	(5)
H (20)	5.3 ± 3.51	(39)	3.6 ± 0.53	(4)
A (30)	2.7 ± 0.95	(27)	4.8 ± 1.61	(5)
T (95)	1.7 ± 0.85	(20)	2.2 ± 0.76	(4)

[a] $\bar{x} \pm SD$.

The metal contents of the food items of the nestlings (small soft insects, larvae, and spiders) sampled at the different localities, did, as concerning non-essential metals, vary in accordance with the pollution gradient (Table 2). Concerning the essential metals, copper (and zinc, though not illustrated in the table), the concentrations in the insects were not influenced by the varied degree of pollution, except at the most polluted site. This was tentatively due to the copper and zinc uptake excretion in the insects being physiologically controlled by homeostatic mechanisms, the capacity of which was exceeded only at the highest level of copper and zinc pollution. Similar mechanisms in the bird nestlings were obviously capable enough to counteract accumulation of these metals in the tissues.

Several effects, with varying degree of acuteness, and which obviously were related to metal contamination of the nestlings, were observed. The frequency of mortality of the young birds during the nesting period reflected the pollution gradient (Figure 3). Effects which were observed in the nestlings sampled close to the age of fledging were, e.g., increased liver weights (Figure 6), decreased muscular (Figure 7) and brain development (Figure 8), reduced hemoglobin levels (Figure 9). Moreover, impaired calcification of the skeleton occurred at the

Table 2. Heavy Metal Concentrations in Insects Swept at the Different Localities During the Period of Feeding the Nestlings of the Pied Flycatcher

Locality (distance from smelter, km)	Cadmium mg/kg, w.w	Copper pooled	Lead samples
R (2.5)	3.8	53	60
B (5)	2.6	10	5
H (20)	0.94	7.4	1.5
A (30)	0.71	7.9	0.74
T (95)	0.55	8.0	0.70

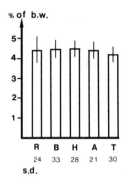

FIGURE 6. Mean liver weights in 12- to 14-d-old nestlings (percentage of body weight). Numbers of clutches represented are given below the figure. Mann-Whitney U-test, two tailed: B-T *p* <0.01; H-T *p* <0.01.

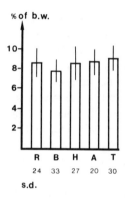

FIGURE 7. Mean weights of paired pectoral muscles in 12- to 14-d-old nestlings (percentage of body weight). Numbers of clutches represented are given below the figure. Mann-Whitney U-test, two tailed: B-T *p* <0.01.

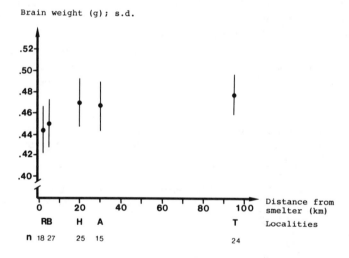

FIGURE 8. Mean brain weights in 12- to 14-d-old nestlings at the different localities, 1988 to 1990. n denotes the number of clutches represented. Mann-Whitney U-test, two tailed: R-H, A *p* <0.01; R, B-T *p* <0.001.

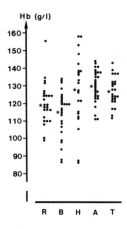

FIGURE 9. Hemoglobin concentrations in 12- to 14-d-old nestlings. * = mean concentrations. Student's *t*-test: R + B-H + A + T p <0.001.

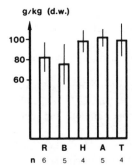

FIGURE 10. Calcium concentrations in the sternum of 12- to 14-d-old nestlings. n = number of clutches represented. Mann-Whitney U-test, two tailed: R + B-H + A + T p <0.002.

localities nearest the smelter (Figure 10), which, in an obvious way, reduced the rigidity of the bones.

Judged by the fates of the biological effects shown by the nestlings it seems improbable that but very few of those which were produced at the localities R and B could survive long after they had fledged.

The experiences from this study on the Pied Flycatcher justifies the impression that similar studies could be adequate in programs aimed at assessing ecological hazards of metal pollution in the terrestrial environment. Other observations made in the study indicate that the concentrations of arsenic, cadmium, lead, and mercury in tissues of the nestlings and thus also the biological effects observed, mainly reflected the extent of the present day depositions, and only to a minor extent the metal amounts deposited and stored in the soil during earlier time periods.[2]

REFERENCES

1. Nyholm, N.E.I., Bio-indication of Industrial Emissions of Heavy Metals by means of Insectivorous Birds, Proc. VIth Int. Conf. Heavy Metals in the Environment, New Orleans, Vol. II, 45, 1987.
2. Nyholm, N.E.I., The Pied Flycatcher (Aves) as a Bio-indicator of Aerial Heavy Metals in the Terrestrial Environment, Proc. VIIth Int. Conf. Heavy Metals in the Environment, Geneva, Vol. II, 468, 1989.
3. Dmowski, K. and Karolewski, M., Cumulation of zinc, cadmium and lead in invertebrates and in some vertebrates according to the degree of an area contamination, *Ekol. Pol., 27*, 333, 1979.
4. Hutton, M. and Goodman, G.T., Metal contamination of feral pigeons (*Columba livia*) from the London area. Part 1. Tissue accumulation of lead, cadmium and zinc, *Environ. Pollut., 22A*, 207, 1980.
5. Goede, A.A. and de Voogt, P., Lead and cadmium in waders from the Dutch Wadden Sea, *Environ. Pollut., 37A*, 311, 1985.
6. Ellenberg, H., Jr., Dietrich, J., Gast, F., Hahn, E., and May, R., Vögel als Biomonitoren für die Schadstoffbelastung von Landschaftsausschnitten. Ein Überblick, *Verh. Ges. Ökol., XIV*, 403, 1986.
7. Goede, A.A. and de Bruin, M., The use of bird feather parts as a monitor for metal pollution, *Environ. Pollut., 8B*, 281, 1984.
8. Hutton, M., Metal contamination of feral pigeons (*Columba livia*) from the London area. Part 2. Biological effects of lead exposure, *Environ. Pollut., 22A*, 281, 1980.
9. Grue, C.E., O'Shea, T.J., and Hoffman, D.J., Lead concentrations and reproduction in highway-nesting barn swallows, *Condor, 86*, 383, 1984.
10. Grue, C.E., Hoffman, D.J., Nelson-Beyer, W., and Fransson, L.P., Lead concentrations and reproductive success in European Starling (*Sturnus vulgaris*) nesting within highway roadside verges, *Environ. Pollut., 42A*, 157, 1986.
11. Simkiss, K., *Calcium in Reproductive Physiology*, Chapman Hall, London, 1967.

CHAPTER 28

A Method to Assess Chemical Biorisks in Terrestrial Ecosystems

A.M.M. Abdul Rida and M.B. Bouche

TABLE OF CONTENTS

0-87371-530-6/94/$0.00 + $.50

I. INTRODUCTION

The aim of an ecotoxicological appraisal is to assess chemical effects on living components of ecosystems.[1] In short, ecotoxicology is toxicology in ecological conditions, i.e., falsifiable in fields. Analytical studies, particularly in microcosms, are too restricted to assess biorisks directly. We must get our information from ecosystems. However, it is difficult to select in ecosystems key variables among the great number of ecosystem characteristics.

There are two kinds of chemical biorisks:

1. Risks suffered by organisms from a chemical
2. Risks coming from an organism having a toxic burden and contaminating its predators

The first one may be observed from the presence (= survival) or the absence (= intoxication) of an organism assumed as a bioindicator in an ecosystem. The second one results from the transfer of chemical from an apparently healthy (but burdened) "prey" to a poisoned "predator" in a food chain.

The assessment of these biorisks in terrestrial ecosystems are a very difficult task:

- because most of the pollutants reach soils (i.e., a great chemical and origin diversity)
- because soil is a very complex three-phase system (air/water/solid)
- because soils are very variable in space and time from a minute crumble to a landscape
- because most organisms (plants, animals, and microorganisms) are dependent on abiotic solid fractions (various minerals and dead organic matter)
- because the concept of food chain involves a clear relationship between a "prey" and a "predator" while *in fact* ecosystem compartment relations are both a succession of organic/inorganic transfers[2] and a food web
- because the soil concentration is a ratio depending on the pollutant amount *and* on the density of the matrix which could vary 15-fold on a volume basis[3]
- because the true availability of the contaminant depends of numerous soil properties and could not be measured by a chemical partial extraction

It seems easier to assess chemical biorisk directly on biotic compartments than on abiotic fractions which are questionable. We need an organism:

1. which plays a central role in ecosystems, closely connected with plant and microorganism biomass,
2. present in space and time the greatest variety of terrestrial ecosystems independent of local and seasonal conditions,
3. with a rather constant dry weight density,
4. which is not too susceptible to contaminants (to get it alive in food chain studies),
5. easy to sample and analyze.

Due to the great diversity of plant relations with their milieus (atmosphere, soil, rain) and plant composition, and the great difficulty to sample directly the microorganism biomass, earthworms as a whole seem to be the sole almost ubiquitous biological group able to fulfill the conditions described above.

Bioindications are of two basic types. The *existential bioindication* describes in ecosystems, by the presence or absence of the observed organism, the upper limit below which this organism could survive and the *physiological bioindications* observed on biological properties of surviving organisms: as, for example, respiration, reproduction, bioconcentrations, enzymatic inductions, etc. Notice that the comparison of ecosystems by physiological characters are limited by the organism survival. Among physiological bioindications, bioconcentrations have the advantage of relating the chemical directly with the organisms.

We choose to study the existential bioindications (survival) and heavy metal bioconcentrations in earthworms to estimate their potential risk in food chains. Earthworms are eaten in West Palearctic[4] by about 200 vertebrate species and an unestimated number of invertebrate predators.[5] They are the main terrestrial animal biomass available to predators (about 1 t fresh weight per hectare [average] in France).

II. MATERIALS AND METHODS

186 sites have been sampled using the "punctual" approach,[6] i.e., in each "point" a sample with closely related biotic and abiotic soil characteristics. These points were selected in six geographical areas with various levels of urban, industrial, or agricultural pollution and very different soil types and human influences (Figure 1). Research was limited to easy-to-analyze and nondegradable substances, i.e., heavy metals as models of chemical pollutants.

Existential bioindication was studied by the presence or absence of earthworms. To study earthworms as a prey compartment, earthworms and their gut contents were analyzed in 126 points. To understand the metabolic accumulation of heavy metals (= *physiological bioindication*), earthworms from 60 points were dissected, eliminating their soil gut content, and analyzed thereafter. In total 790 earthworms, belonging to different taxa and ecological categories, were analyzed. To relate soil concentrations and properties to earthworm concentrations, one total and two partial analyses were performed. Some soil properties (pH, texture, organic matter, C/N, cation exchange capacity, exchangeable cations) were also measured. All analytical data of these 186 points are managed and available in the Relational Data Base ECORDRE.[7] These data were analyzed by a great variety of different statistics.[8]

A. Analyses of Earthworms

The animals were weighed and oven dried in glass flasks at 105°C for 24 h. After reweighting nitric acid was added (5 ml HNO_3 for 100 mg dry earthworm

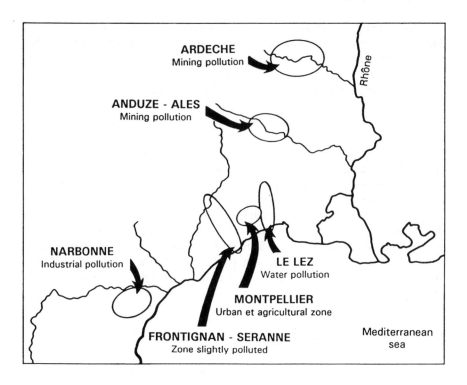

FIGURE 1. Localization of studied areas (South France).

weight), heated at 70°C for 24 h and then diluted with deionized water to 10% acid.

B. Analyses of Soils

The total extraction of trace elements was carried out by mineralizing 1 g of air-dried soil (sieved fraction <0.5 mm) with 5 ml HNO_3 for 5 h at 160°C.

The mixture was cooled before the addition of 5 ml HNO_3 and 5 ml HCl. The mixture was heated again for 5 h at 160°C, thereafter cooled and diluted with deionized water to 100 ml.

The partial extractions of trace elements were made from soils by two extractants: acetic acid and diethylene triamine pentaacetic acid (DTPA). (1) Acetic acid: 3 g air-dried soil (sieved fraction <2 mm) and 60 ml of an aqueous acetic acid solution (2.5%) were shaken for 1 h. Then, the suspension was filtered before titration. (2) DTPA: a water solution was prepared with 0.005 M DTPA, 0.01 M $CaCl_2$ and 0.1 M triethanolamine (TEA)[9] and adjusted to pH 7.3 with HCl, then 30 g air-dried soil (sieved fractions <2 mm) and 60 ml of this solution were shaken for 2 h. Then the suspension was filtered before titration.

Table 1. Comparison Between Mean and Maximum Values of Total Heavy Metal Content of Soils Where *Scherotheca* and Other Genera Are Present

	Cd mg/kg	Cu mg/kg	Ni mg/kg	Pb mg/kg	Zn mg/kg
Scherotheca					
Mean	1.9	44	29	101	168
Maximum	3.3	134	122	410	1248
Other genera					
Mean	2.8	62	23	1376	439
Maximum	19.4	237	88	10400	5808

The concentrations of elements were determined with a Varian® AA-SPECTRA atomic absorption spectrophotometer (using an air-acetylene flame) and appropriate standards.

III. RESULTS

A. Existential Bioindication and a Bias in the Sampling of Bioconcentrations

It is classical to consider the presence or absence of organisms to interpret some environmental factors, as for example the use of calcifuge plants as an indicator of soil pH. Looking at our general results we were surprised to show such an indication for heavy metal pollutions. In the general sample of 186 points of the Mediterranean South France with diverse levels of soil contamination we observed that the genus *Scherotheca* was particularly sensitive to high soil levels of Cd, Cu, Pb, and Zn in contrast to other earthworms (Table 1).

A more careful study of their distribution demonstrates that competing hypotheses (historical bias on population settlements; need of high level of Ca) does not explain such an eradication of this genus in all polluted sites.[10] From a methodological point of view, we can use such a demonstration to use this genus as a bioindicator. We need also to take into account the bias introduced by the absence of *Scherotheca* in the most polluted soils. Their susceptibility seems generally higher than others earthworms if we compare their bioconcentration and bioaccumulation (bioconcentration: ratio of metal content weight to total dry tissue weight; bioaccumulation: ratio of bioconcentration to total soil concentration) (Table 2). They concentrate particularly more Cd and Zn than other earthworms.

B. Concentrations of Metals in Soil and Earthworms

We focused our study on 60 points having a great variety of heavy metal burdens for analysis of soils (total, acetic acid and DTPA extraction) and the

Table 2. Comparison Between Mean Bioconcentration (mg/kg) and Bioaccumulation for *Scherotheca* and Other Genera

	Cd	Cu	Ni	Pb	Zn
Scherotheca					
Bioconcentration	38.9	37.8	6.6	65.2	967.0
Bioaccumulation	16.1	1.1	0.3	0.3	7.9
Other genera					
Bioconcentration	34.3	66.1	9.7	342.0	717.0
Bioaccumulation	16.0	1.1	0.6	0.4	3.1

concentration of metals in 279 dissected earthworms which were analyzed in two groups:

1. 30 points concerning the Ardèche, where soil characteristics and trace element sources are almost similar,
2. 30 points concerning five different areas labeled as "others" where soil characteristics and the trace element sources are variable (Table 3).

The comparison between levels of soil trace elements and earthworm burden (Table 4) shows discrepancies between quantities of trace elements extracted by total analysis methods and partial analysis methods. This comparison shows also a considerable difference between trace element patterns of soils and earthworms.

The concentrations in earthworms are higher for Zn and in particular for Cd, but are lower in Ni and Pb than in soil. Cu concentration is the same in both soils and earthworms. Concentrations of Pb and Zn in earthworms increase

Table 3. Mean Values of Physicochemical Characteristics for Soils From Different Areas

	Ardèche	Others
Water pH	6.8	7.4
KCl pH	6.3	7.0
Clay %	19.3	19.8
Fine silt %	15.6	21.1
Gross silt %	7.7	12.1
Fine sand %	19.5	23.7
Gross sand %	36.7	23.3
Organic matter %	2.2	3.3
Carbon %	1.3	1.9
Nitrogen %	0.1	0.2
Ratio C/N	9.9	11.5
Cation exchange Capacity meq/100 g	10.0	11.3
Calcium meq/100 g	14.2	29.1
Magnesium meq/100 g	1.1	1.3
Potassium meq/100 g	0.7	0.5
Sodium meq/100 g	0.03	0.07

Table 4. Mean Concentrations of Trace Elements in Earthworms and Total and Partial Soil Contents of Studied Points

	Cd mg/kg	Cu mg/kg	Ni mg/kg	Pb md/kg	Zn mg/kg
60 points					
Earthworms	35.3	60.1	9.1	283	770
Total	2.5	57.9	24.2	945	416
CH₃COOH	1.4	4.3	3.8	56	52
DTPA	1.3	17.1	1.1	108	42
Ardèche					
Earthworms	38.4	50.6	9.4	132	832
Total	2.4	42.3	21.8	349	521
CH₃COOH	1.2	2.3	1.8	26	70
DTPA	1.7	10.7	0.8	64	53
Others					
Earthworms	32.5	68.5	8.8	418	715
Total	2.6	73.5	26.5	1541	312
CH₃COOH	1.6	6.2	5.7	87	35
DTPA	0.8	23.5	1.5	151	32

with soil concentration while the Cd, Cu, and Ni earthworm concentrations do so only slightly, probably because of the rather constant soil contents. These differences are due to:

1. Different extraction powers of chemicals reagents vs. the different species of chemical elements. For example, the mixture HNO_3 and HCl extracts almost all trace element contents of soils,[11-13] while acid acetic extracts them from carbonates or oxides, and DTPA extracts elements from oxides and breaks bounds with the organic matter.[14]
2. Differences between chemical mineralization and physiological earthworm metabolic control. Physiological mechanisms are very complex and depend both on the control of organisms and of the soil bioavailability of trace elements. It should be noted that the species of earthworms have different sensitivities towards trace elements.
3. Soil physicochemical characteristics, in particular pH and redox potential, affect the bioavailability of elements.

In Table 5 metal concentrations in earthworms and the three types of soil analysis are compared.

The relationships vary according to elements, soil properties, and chemical analysis methods. In general, the correlations between earthworm concentrations and total and DTPA estimates are more significant than with the acid acetic extraction. The Cd, Cu, and Pb earthworm concentrations seem tightly bound

Table 5. Differences Significant at $p < 0.05$ (*), $p < 0.01$ (**), and $p < 0.001$ (***) Between Trace Element Concentrations in Earthworms and Soils of Ardèche (Ar), the Other Areas (Ot), and Both Ar and Ot (All)

Zone	$HNO_3 + HCl$			CH_3COOH			DTPA		
	Ar	Ot	All	Ar	Ot	All	Ar	Ot	All
Cd	***	**	**	***	**	**	***	***	***
Cu	**	***	***	***	***	***	***	***	***
Ni	*	—	—	—	—	—	—	—	—
Pb	***	***	***	***	***	***	***	***	***
Zn	***	—	**	*	—	*	**	—	*

to soil contents, whatever the soil characteristics and analysis techniques are. In contrast, the relations of Zn and in particular Ni are very different according to analysis methods and soil characteristics.

C. Earthworms as a Biorisk in Food Chains

The heavy metal burden in earthworms from the study areas may pose a risk to earthworm predators.

Laboratory toxicological studies on vertebrates (mouse, rat, or others) provide the basis of assessments of human risks. Most of these risks are expressed as a limit: the maximum daily admissible amount in food. This is expressed as a ratio: weight of ingested toxicant in daily diet/live weight of the consumer. In man[15] for Cd, Cu, Pb, and Zn ratios are 0.0075, 0.275, 0.05, and 0.65 (in mg/kg), respectively.

As toxicologists using rat and mouse as a "human" representative, these results can also be used as bird or mammal representatives in general.

Granval and Aliaga[4] listed 200 vertebrate earthworm predators in Western Europe. They ordered them as using earthworms as occasional, important, or major food. In fact, all the literature gives an underestimated figure both qualitatively and quantitatively because of the erratic and scarce studies. About invertebrate predators we have only scattered observations for vertebrates: only four valid diet estimates are available. Mean live weight of predators and daily ingested amount of earthworms for woodcock are 330 g and 100 g/d;[16] for black headed gull: 290 g and 120 g/d;[17,18] for badger: 10.85 kg and 720 g/d;[19,20] and for pig 29.34 kg and 419 g/d,[21,22] respectively.

Using these two values and assuming that earthworms ingested by such predators have the same heavy metal body burden as observed in our sampling (Table 6), we have calculated the ratio of estimated daily food content/daily admissible diet limit. This ratio gives us a level of risks for the wild fauna in comparison with the admissible limits for men. Table 7 presents these values.

For simplicity, we have assumed that predators consume only earthworms. This may not be the case, especially for omnivores such as badgers and pigs.

Table 6. Number, Weight, and Concentration of Earthworms in Heavy Metals on Studied Areas

Studied areas	Number	Weight (mg)	Cd mg/kg	Cu mg/kg	Ni mg/kg	Pb mg/kg	Zn mg/kg
Frontignan	96	2898.5	2.4	12.9	5.9	10.9	50
Gignac	34	7738.6	2.6	6.7	6.4	11.5	98
Montpellier F[a]	34	777.4	3.0	16.8	3.6	26.9	74
Montpellier T[a]	34	1451.2	2.3	14.2	4.4	29.4	144
Le Lez	63	577.7	2.3	10.4	4.6	13.5	85
Anduze	136	1195.4	4.7	7.6	3.5	180.8	160
Narbonne	114	1222.6	19.2	14.0	3.6	5.2	42
Mean	511	1869.8	6.8	11.4	4.4	57.5	93

[a] F = fields, T = town.

These animals may take in, in addition, heavy metals associated with other components of their diet. It is obvious that for all predators the heavy metal load due to earthworms is too contaminated, especially for Cd and Pb. This contamination reaches a tremendous level for specialized earthworm predators such as gulls and woodcock, with more than 250 times (mean) the admissible maximum food burden and in some areas more than 1000 times!

Obviously, this estimation of risks is only an estimate with some assumptions to compare the burden of food in an ecological web with the quality of food judged by traditional hygienist toxicological tests. This gives an index of the

Table 7. Ratios of Estimated Daily Food Content/Daily Admissible Diet Limit for Four Predators Supposedly Eating in Studied Areas

Areas	Cd	Cu	Pb	Zn	Cd	Cu	Pb	Zn
	Black headed gull				Woodcock			
Frontignan	130	19	90	32	96	14	66	23
Gignac	145	10	95	63	106	7	70	46
Montpellier F[a]	166	25	222	47	121	19	163	34
Montpellier T[a]	124	21	243	92	91	16	178	67
Le Lez	128	16	112	54	94	11	82	40
Anduze	259	12	1496	102	190	8	1096	75
Narbonne	1060	21	43	27	776	15	32	20
Mean	375	17	476	59	274	13	349	43
	Badger				Pig			
Frontignan	21	3	15	5	5	0.7	3	1
Gignac	23	2	15	10	5	0.3	3	2
Montpellier F[a]	27	4	36	8	6	0.9	8	2
Montpellier T[a]	20	3	39	15	4	0.7	8	3
Le Lez	21	3	18	9	4	0.5	4	2
Anduze	42	2	240	16	9	0.4	52	4
Narbonne	170	3	7	4	37	0.7	2	0.9
Mean	60	3	76	10	13	0.6	16	2

[a] F = fields, T = town.

level of risk for the wild fauna and in some extent, an estimation of the potential bioconcentration in this fauna as a food for men (game).

IV. DISCUSSION

To assess biorisks we need to know (1) if organisms are affected by the observed risk, and (2) at which levels the risk in food webs are present. The observed food risk is itself more or less well assessed in soil and we need to know (3) if a given level of heavy metal in soil is a hazard or not and (4) how to measure this level.

Our study has been designed to observe "biocirculation" of heavy metals directly in the field through concentration in earthworm bodies. Earthworms ingest mineral particles, organic matter from dead plants and animals, and microorganisms. They "integrate" by ingestion the most important live and dead ecosystem parts. Earthworms play a key role in the ecosystem soils in the recycling of chemicals, for example heavy metals, and they are an index of the "circulating fraction".

We compared this key position to the total soil content and the so-called available fractions classically used in agronomy to predict a deficiency in future plant needs. Clearly, the "index of the circulatory fraction" (or earthworm bioconcentration) depends more on the total soil concentration than on so-called "assimilable" fractions.

In addition, earthworms themselves are directly an indicator of biorisks of two types: as existential bioindicators and as physiological bioindicators.

During this study, thanks to a rather large sample of 186 locations, an eradication of an animal genus due to an overburden of heavy metals in soils has been demonstrated. While the genus *Scherotheca* accumulates some heavy metals more than other genera, they disappear when the soil concentration is higher than critical. As far as this sampling is representative, this Southern European genus suffers strongly from soil contamination.

Conversely, earthworms resistant to soil contamination accumulate heavy metals. Earthworms are the sole practical physiological bioindicator available as a compartment in soils. Their bioaccumulation plays a key role in estimating the level of contamination of the biocenoses of ecosystems thanks to their key function and can sustain biomonitoring.

This bioconcentration could also be an indicator of the biorisks coming from the earthworms in food webs. The observed earthworm population burdens eaten every day by vertebrate predators were compared with the maximum admissible dose in daily diet. We observed that vertebrates eat more than 10 times, and usually 100 times this maximum daily dose for Cd and Pb. In some areas the level of contamination is in vertebrate food potentially higher than 1000 times the admissible level for humans.

Our method to directly observe field relationships between toxicants and earthworms as survivors, earthworms as "biocontainers", and earthworms as a prey, leads to three conclusions:

1. The direct observation of toxicant burden in a biological compartment could not be substituted by a chemical analysis of soil, and especially by the so-called available fraction estimates.
2. The observation of the total eradication of an earthworm genus in contaminated soil reveals a serious ecological disaster.
3. The level of observed earthworm contamination reveals a nonadmissible burden of the food for wildlife, including game consumed by man.

Obviously, the described method needs to be expanded to more applications and controls to extend our biomonitoring, but it seems that today this method is enough settle to be practically used.

REFERENCES

1. Ramade, F., *Ecotoxicologie*, Masson, Paris, 1977, 205.
2. Bouche, M.B., Discussion d'écologie. III. Transferts d'énergie entre maillons trophiques, *Bull. Ecol.*, 9, 289, 1978.
3. Sillanpaa, M., Micronutrients and the nutrient status of soils: a global study, *FAO Soils Bull.*, 48, 1, 1982.
4. Granval, Ph. and Aliaga, R., Analyse critique des connaissances sur le prédateurs de lombriciens, *Gibier Faune Sauvage*, 5, 71, 1988.
5. Lee, K.E., *Earthworms: Their Ecology and Relationships with Soils and Land Use*, Academic Press, New York, 1985, 411.
6. Bouche, M.B., Discussion d'écologie. I. Introduction. II. Obtention de données écologiques et uniformation spatiale, *Bull. Ecol.*, 6, 23, 1975.
7. Bouche, M.B., *Ecologie Opérationnelle Assistée par Ordinateur*, Masson, Paris, 1990, 572.
8. Abdul Rida, A.M.M., Biosurveillance de la Contamination du Sol: Apport de l'Étude des Lombriciens à l'Évaluation des Risques Liés aux Éléments Traces, Doc. Pédozool., Vol. 1, Fasc. 4, Montpellier, France, 1992, 234.
9. Lindsay, W.L. and Norvell, W.A., Development of a DTPA soil test for zinc, iron, manganese and copper, *Soil Sci. Soc. Am. J.*, 42, 421, 1978.
10. Abdul Rida, A.M.M. and Bouche, M.B., An earthworm genus eradication by heavy metals in Southern France, in preparation.
11. Marchandise, P., Olie, J.L., Robbe, D., and Legret, M., Dosage d'éléments traces dans les sédiments de cours d'eau et les boues de stations d'épuration: Comparaison inter-laboratoires de diverses méthodes de minéralisation, *Environ. Tech. Lett.*, 3, 157, 1982.

12. Arnoux, A., Nienchewski, L.P., and Tatossian, J., Comparaison de quelques méthodes d'attaque des sédiments marins pour l'analyse des métaux lourds, *J. Fr. Hydrol.*, 1, 29, 1981.

13. Welte, B. and Montiel, A., Comparaison de différentes méthodes d'attaque des sédiments et boues de station d'épuration, *Trib. CEBEDEAU*, 503, 3, 1985.

14. Juste, C., Problèmes posés par l'évaluation de la disponibilité pour la plante des éléments-traces du sol et de certains amendements organiques, *Sci. Sol*, 2, 109, 1983.

15. FAO/OMS, Codex alimentarius. Vol. XVII. Contaminants, Commission du Codex alimentarius, Rome, 1984.

16. Granval, Ph., personal communication, 1990.

17. Cuendet, G., Etude du Comportement Alimentaire de la Mouette Rieuse (*Larus ridibundus* L.) et de Son Influence Sur Les Peuplements Lombriciens, Ph.D. thesis, Suisse, 1979, 111.

18. Cuendet, G., Predation on earthworms by the black-headed Gull (*Larus ridibundus* L.), in Satchell, J.E., Ed., *Earthworms Ecology: from Darwin to Vermiculture*, Chapman and Hall, London, 1983, 415.

19. Neal, E.G., *Badgers*, Blandford Press, Poole, Dorste, 1977, 321.

20. Kruuk, H., Spatial organization and territorial behaviour of the European badger *Meles meles*, *J. Zool. London*, 184, 1, 1978.

21. Rose, C.J., Preliminary observations on the performance of village pigs (*Sus scrofa papuesis*) under intensive outdoor management. Part I. Dietary intake and live weight gain, *Sci. in New Guinea*, 8, 132, 1981.

22. Rose, C.J., Preliminary observations on the performance of village pigs (*Sus scrofa papuensis*) under intensive outdoor management. Part II. Feed conversion efficiency, carcase composition and gastro-intestinal parasites, *Sci. in New Guinea*, 8, 156, 1981.

SECTION V

Ecotoxicological Assessment Procedures

CHAPTER 29

Progressing Limits For Soil Ecotoxicological Risk Assessment

N.M. Van Straalen, P. Leeuwangh, and P.B.M. Stortelder

TABLE OF CONTENTS

ABSTRACT

Recent years have seen the development of various ecotoxicological as-sessment procedures for polluted soils and sediments. In several European coun-tries, there is an increased interest in the results of ecotoxicological experiments designed to quantitatively estimate the relationship between exposure and effect. The no-effect-level obtained from these experiments provides the best available basis to derive environmental quality criteria and to estimate the *in situ* risk of polluted sites. This chapter reviews five approaches currently adopted by U.S. EPA, the Dutch Ministries of Environment and Waterways, and the German Federal States. The approaches are evaluated on the basis of the way they deal with (1) scarcity of data, (2) variability among species, (3) food-chain effects, (4) combination toxicity, and (5) laboratory-field extrapolation. None of the approaches is considered satisfactory from a scientific point of view. Gaps of knowledge and recent developments are discussed for each of the issues.

I. INTRODUCTION

Maximum acceptable concentrations for potentially toxic substances in the environment are often established on the basis of laboratory toxicity experiments. Traditionally this approach has been directed towards taking the most sensitive species, but recent developments demonstrate a tendency to use all available toxicity data from various species investigated, and to take account of ecological information on exposure routes, body plan, and ecological function when se-lecting test species. Extrapolation methodologies have been proposed to derive maximum acceptable concentrations from a set of laboratory data.

Soils and sediments are effective sinks for a long suit of environmental contaminants. Modern analytical techniques are able to demonstrate the presence of various substances in very low concentrations. Whether or not these actually constitute an ecological risk depends on the ratio of exposure to effect concen-trations. In addition to chemical analysis, toxicity tests and bioassays are needed to evaluate the probability of undesirable effects.

When developing a system of soil ecotoxicological risk assessment, it may be expedient to make an inventory of approaches already developed for the aquatic field. From a scientific point of view, there is no reason to make a distinction between these two fields, since the science of ecotoxicology is the same. Indeed, the assessment of (dredged) contaminated sediments involves both aquatic and soil organisms and forces us to develop an integrated evaluation. This paper will therefore consider approaches both from the aquatic and the terrestrial fields.

A common feature of most of the recently developed approaches is that they are based on the no observed effect concentration (NOEC), i.e., the highest concentration applied in an experiment that does not yet cause an adverse effect to the experimental organisms. All approaches assume that such a concentration exists and that it is possible to protect ecosystems by ensuring that none of the species present in the ecosystem is exposed to a level higher than its NOEC. The validity of this assumption can hardly be tested actually.

The present chapter reviews some of the extrapolation methodologies proposed by authorities in the U.S., Germany, and the Netherlands. The approaches are evaluated on the basis of five criteria that address the way in which each approach takes account of (1) information deficiency, (2) variability between species, (3) biomagnification, (4) chemical interactions, and (5) extrapolation to the field. For each criterion an analysis is made of the scientific knowledge that would have to be developed to provide a better base for risk assessment.

II. THE NO EFFECT LEVEL AS A BASIS FOR RISK ASSESSMENT

Chronic toxicity experiments allow the relationship between exposure concentration and the performance of the test organism to be established quantitatively. The response is often characterized by a threshold level, such that the organism is able to maintain its performance at all exposure concentrations below the threshold. Since the no effect level plays such an important role in risk assessment procedures, it may be relevant to consider some of the scientific arguments for the existence of thresholds in toxicological responses and the way these thresholds determine the range of ecological performance of a species.

It is evident that NOEC values derived from chronic experiments will be lower than LC_{50} values, especially when the latter are estimated from short-term experiments. However, the LC_{50} is not only higher, it also does not necessarily provide a good measure of responses at higher levels of integration. This is due to the fact that the capacity for population increase for most organisms, especially those with high population turnover, is very much dependent on the reproductive rate; any chemical that inhibits reproduction will directly affect population growth, even at concentrations far below the LC_{50}. The importance of reproduction and survival for the maintenance of population growth capacity is determined by the life cycle of the species.[1]

From a physiological point of view, the existence of a concentration range causing no adverse effects may be expected on the basis of the various mechanisms aimed at detoxification, excretion, and repair. Numerous feedback controls operate to maintain homeostasis of the "internal environment". The no effect level marks the point above which control mechanisms are no longer able to prevent the toxicant from causing an adverse physiological disturbance.[2]

The operation of feedback control mechanisms in an organism may also provide an explanation for the stimulatory effect of low doses, an often observed phenomenon in toxicity experiments called "hormesis". Many toxicants that inhibit the growth of an organism at high exposure levels stimulate growth at low levels.[3] The widespread occurrence of hormetic effects (many organisms, all kinds of substances) suggests that a very basic mechanism is involved. It has been argued that physiological control mechanisms in organisms have evolved in such a way that they often overreact to small deviations of the physiological norm.[3] This would explain why hormesis is such a common phenomenon.

When the dose-response curve includes a hormetic effect, the definition of a "no effect" concentration poses a problem. There is a tendency to consider the hormetic effect as "not adverse" and to redefine the no effect level as the "zero equivalent concentration", that is, the lowest concentration above the stimulatory effect where performance is equal to the blanc. Whether or not this is justified, and how to evaluate the subtle physiological changes detectable before an adverse effect on growth or reproduction occurs, cannot be said at the moment. Research in the field of "biomarkers" may shed a new light on this problem.[4]

Another issue under discussion is how to statistically estimate the no effect level from a set of observations in a toxicological experiment. The usual procedure is to apply Williams' test,[5] which will identify one of the experimental exposure levels as the NOEC. A major objection to this approach is that increased variability among replicates within each exposure level may lead to a higher estimate of NOEC. Another objection is that the no effect level is equalled to one of the concentrations applied in the experiment, while in practice it will fall somewhere in between two dose levels. Parametric models including the no effect level as one of the parameters to be estimated from the data may remove these objections.[6,7]

It can be concluded from the discussion given above that the no effect concentration based on ecologically relevant responses, such as reproduction, is thought to provide the best toxicological criterion for risk assessment. Yet there are some unsolved problems with respect to the interpretation and the statistical analysis of responses in the low exposure range, e.g., due to hormesis. Hopefully, these problems may be settled by new physiological and modeling research.

III. SOME APPROACHES IN RISK ASSESSMENT

Different countries and agencies have adopted different risk assessment schemes based on toxicological criteria. The variety of approaches and their common features are illustrated below.

A. Concern Levels

In the U.S., the Environmental Protection Agency (EPA) has adopted two different approaches, one for screening chemicals in a first-tier evaluation and one for deriving standards of water quality. The first-mentioned approach aims at deriving "concern levels" from simple laboratory toxicity data;[8] this procedure has, as yet, only been applied for the aquatic environment.

- If only one LC_{50} value is available, the concern level equals 0.001 times this value.
- If three LC_{50} values are available for fish, crustaceans, and algae, the concern level equals 0.01 times the lowest of these three values.
- If NOEC values are available, based on chronic toxicity experiments, the concern level equals 0.1 of the lowest value.

The concern level is interpreted as a concentration such that populations of aquatic organisms exposed to this concentration may be affected adversely. The method aims to identify substances for further research by comparing the concern level with a predicted environmental concentration.

B. National Water Quality Criteria

Another approach adopted by the U.S. EPA is focused on national water quality criteria.[9] These are derived from acute toxicity data obtained for at least eight different species from different families of aquatic organisms; in addition, chronic data are required for at least three families. By assuming that the data are distributed according to a triangular function, the 5-percentile is estimated, both for the acute and the chronic data. These are interpreted as maximum acceptable concentrations for the 1-h mean (in the case of acute data), and for the 4-d mean (in the case of chronic data). The criteria may be exceeded once in 3 years. The method aims to derive a concentration range such that adverse effects are prevented for 95% of the families in the aquatic ecosystem.

C. Ecotoxicological Values

The Dutch Institute for Inland Water Management and Waste Water Treatment (RIZA) derived quality criteria ("ecotoxicological values") for surface waters and bottom sediments according to the following scheme:[10]

- Chronic NOEC values obtained for unicellular algae, molluscs, crustaceans, and fish are necessary; these are adjusted, if applicable, for differences between nominal and actual exposure. In case of lack of data, chronic NOECs are estimated from LC_{50}s or from QSARs.
- The lowest of the available NOECs for each substance is divided by a number equal to the number of substances in the chemical group to which it belongs,

and which are present simultaneously; this allows for possible additive interactions among chemicals. If applicable, a further factor is applied to allow for combination toxicity among different groups.

For chemicals with suspected potential for biomagnification (e.g., organochlorines), quality criteria are derived from the maximum acceptable levels for birds or mammals. The "ecotoxicological value" for sediment is obtained from the one for surface water by applying substance-specific factors, based on the equilibrium distribution between water and sediment.[11]

The "ecotoxicological value" is interpreted as a maximum acceptable concentration for the 90-percentile of concentrations in the water measured over a year, or for the mean of concentrations in bottom sediments, measured for a spatially homogeneous water body.

D. Qualitätsziele

The German Federal Environmental Agency and the Federal States Working Group "Quality Objectives" (abbreviated name: BLAK QZ) have derived quality criteria ("Qualitätsziele") for surface waters. The approach has been laid out in a draft paper developed by BLAK QZ, and can be summarized as follows:[12]

- NOEC values obtained for four trophic levels (primary producers, primary consumers, secondary consumers, and detritivores) are necessary.
- The lowest from the four NOECs is divided by 10; if only two or three NOECs are available, the lowest of these is divided by 100. These factors are used to account for uncertainty associated with the possibility of susceptible species living under field conditions.
- In the case of additional risk factors such as persistent chemicals, and in the case of possibly hazardous products obtained by metabolism or degradation, and in the case of susceptible ecological target organisms, the result of the preceding step is again divided by 10.

The quality criteria obtained in this way are under discussion in international consultations on the Rhine River Basin. The interpretation is similar to the ecotoxicological values derived by RIZA.

E. Maximum Risk Levels

The Dutch Ministry for the Environment (VROM) has proposed a system of "maximum acceptable risk levels" and "negligible risk levels", where the latter equals 0.01 times the former.[13] The RIVM Institute has implemented the approach in environmental consultancies.[14] The maximum acceptable risk level is derived in the following way.

- NOEC values for a set of organisms covering different ecological functions, different body plans, and different exposure routes are required.
- Assuming a symmetric distribution of these values on a logarithmic concentration scale, the 5-percentile of the distribution is estimated.

- An uncertainty margin is applied to the 5-percentile depending on the number of NOEC values used in the estimation. The result is called "hazardous concentration for 5% of the species", abbreviated to "HC5".[15]

The quality criterion will be low if few data are available, and may increase with more information. The way the uncertainty margin is estimated, the assumed distribution of NOEC (logistic or normal) and the incorporation of biomagnification is presently under discussion.[16,17]

IV. EVALUATION OF THE APPROACHES

The above given summary of environmental risk assessment schemes adopted by different authorities have in common that they are all largely based on NOEC values. Only in a first-tier evaluation $LC_{50}s$ are being used, with application of large safety factors (cf. the EPA approach). The use of NOECs can be seen as a recognition of the fact mentioned as a conclusion of Section II, i.e., no effect levels provide up to now the best starting point for environmental protection. To compare the approaches further, we identify five questions to be addressed in each case.

1. How is scarcity of information taken into account? Has the quantity of ecotoxicological information any consequences for the derivation of quality criteria?
2. Does the approach allow for the variability between species? Does it explicitly protect sensitive species outside the set of test organisms?
3. Is it recognized that effects may occur outside the system under consideration? Does the approach take biomagnification into account?
4. Are interactions between chemicals taken into account? Does the approach allow for mixtures?
5. What is the relevance of derived criteria for systems where different species, in mutual interaction, are exposed simultaneously? Does the approach allow for extrapolation to the field, in other words, has it "ecosystem realism"?

In Table 1 we have indicated the evaluation results according to the above identified five questions. A " + " marks a positive answer to the question, i.e., the issue is explicitly taken into account; a " − " marks a negative or a partly

Table 1. Comparison of Five Assessment Methods For Environmental Chemicals, Evaluated According to Five Criteria

	EPA concern levels	EPA national criteria	RIZA	BLAK QZ	VROM/RIVM
Information uncertainty	+	+	−	+	+
Variability between species	−	+	−	−	+
Biomagnification	−	−	+	−	−
Toxicity of mixtures	−	−	+	−	−
Ecosystem realism	−	−	−	−	−

Note: See text for explanation.

positive answer, i.e., the issue is not taken into account, or is hidden in an application factor. For example, U.S. EPA and BLAK QZ derive quality criteria by multiplying NOECs by a factor of 0.1.

Not all questions in Table 1 can be answered positively. In the remainder of this section we aim to address each of the issues in more detail, to identify important gaps of knowledge and needs for a better risk assessment system.

A. Information Deficiency

A major problem with all risk assessment schemes is the scarcity of pertinent ecotoxicity data. This holds in particular for the soil environment, where only a few standardized toxicity experiments have been described.[18] One of the ways to deal with lack of information is to make sure that the maximum acceptable standards are lower when there are less data. This is especially recognized in the HC5 approach, which considers the lower limit of the confidence interval for the 5-percentile, estimated from a set of observations. In the EPA approach for concern levels, this is also recognized, since the application factors (10, 100, 1000) depend on the information present.

In the estimation of HC5, with recent statistical modifications,[15-17] the influence of the uncertainty margin becomes very large when the number of species tested is less than four.[19] Therefore, in the present implementation of this method it has been established that at least four different NOEC values are required.[14] In fact, the number of data has to increase above ten to reduce the uncertainty margin to a range comparable to the repeatability of toxicity experiments.

From a scientific point of view, not only the number of data is important, but also their quality, i.e., their "representativeness" to the ecosystem. For the soil environment, criteria for selecting test species have recently been reviewed and a list of possible test systems has been established.[20] One of the problems is the fact that the HC5 model considers the species as a random sample from an infinite community, while in practice the species are deliberately chosen on the basis of experimental manageability. This type of information uncertainty can only be dealt with by regular adjustment of the methods, using additional research to validate or invalidate the outcomes of theoretical calculations.

B. Variability Between Species

A single indicator organism for all environmental pollutants does not exist. Every risk assessment scheme has to deal with biological variation. This is implemented in some of the approaches by ensuring that at least certain ecological groups are represented in the set of test organisms (cf. BLAK QZ). The test organisms would also have to be selected from different taxonomic groups, since each taxon represents a certain body plan and a certain physiological equipment to deal with toxicants.

Comparative toxicological research on 22 aquatic organisms[21] has shown that there are no great differences between species in their susceptibility when for each species the no effect level is averaged over various chemicals. The average sensitivity to 15 chemical compounds differed by a maximal factor of

6.3. However, for separate compounds much larger species differences appeared: the most sensitive organism was 9000 times more sensitive than the least sensitive one. On a substance-specific basis, differences between taxonomic groups are often larger than differences within each group, but many exceptions exist.[22]

For soil organisms, no specific rules can be given on susceptibility patterns between species. In a few cases, certain groups of invertebrates have been identified that seem to be particularly sensitive to certain toxicants, for example:

- rhodacarid mites to DDT[23]
- oribatid mites to cadmium[1]
- snails and earthworms to copper[24]
- linyphiid and erigonid spiders to pyrethroids[25]

These apparently intrinsic susceptibilities probably reside in the presence of physiological targets and the inability to effectively detoxify the chemicals. Among the vertebrates it has been shown that the monooxygenase enzyme system, responsible for many biotransformation reactions with xenobiotics, is not equally distributed among the various taxa.[26] Within the Lepidoptera, polyphagous species seem to be better equipped in detoxifying certain insecticides compared to monophagous herbivores, which can be explained by the fact that species with a broad feeding habit, being exposed to a variety of plant allelochemicals, must have evolved a flexible detoxification system.[27] Comparative physiological research may considerably improve our understanding of possible species differences within the soil community.

C. Biomagnification

Persistent chemicals may be transferred in food chains and sometimes are concentrated by top predators. A pollutant present in soil may not be hazardous to organisms living in the soil itself, but it may reach sensitive targets when transferred in the terrestrial ecosystem. In the approach formulated by RIZA the phenomenon of biomagnification is explicitly recognized, for example, by deriving a critical level for polychlorinated biphenyls in bottom sediments not from toxicity data on benthic organisms, but from data on mustelids, which are exposed through the food chain.

For soil organisms, biomagnification seems to follow a limited number of critical pathways. Within the diverse community of invertebrates, no relationship has been found between residues of metals and trophic level; the same holds for polycyclic aromatic hydrocarbons (PAH).[28] However, some small mammals with a specialized diet, such as the mole, *Talpa europea,* and the common shrew, *Sorex araneus,* may accumulate soil pollutants such as metals to a very high degree.[29]

A recent discussion on biomagnification in the terrestrial environment has demonstrated the difficulties of predicting the behavior of a persistent chemical in an ecosystem.[30-32] This is due to the fact that the residue depends on a complex interaction between various processes such as ingestion, defecation, excretion, metabolism, etc. Experiments aiming at the kinetics of uptake and elimination, using compartment models to analyze the data, seem to be the right approach.[33]

D. Toxicity of Mixtures

The fourth factor identified to evaluate the assessment approaches is the issue of combined toxicity. This is explicitly accounted for in the RIZA approach by dividing each NOEC by the number of chemicals from the same group that might be present simultaneously. This procedure is inspired by the often applied model for relative addition of concentrations, also known as the "toxic unit approach".[34]

According to the toxic unit model, chemicals with the same mode of action will differ in their LC_{50} only because of differences in the bioconcentration factor; consequently, their dose-response curves become identical when the concentration axis is rescaled by expressing concentrations relative to the LC_{50} (toxic units). By addition of toxic units, the effect of a mixture can be predicted from the effects of its component parts.

The validity of the concentration-addition model has been shown for aquatic organisms using various mixtures of nonreactive organic chemicals.[35] For metals the picture is less clear. For example, in toxicity experiments using zebra mussels, Zn and Cd were concentration-additive, but Zn and Cu were less than additive, while Cu and CD were more than additive.[36] The effects of mixtures of metals cannot be predicted simply from their effects in single-metal experiments, because their modes of action are different.

How the toxicity of mixtures must be incorporated in risk assessment schemes is not yet clear. This especially relates to the NOEC as a possible base for expressing toxic units. In soil, several extremely complex mixtures of chemicals are present, for example PAHs, that cannot be assessed adequately at the moment.

E. Laboratory-Field Extrapolation

The largest source of uncertainty in risk assessment of chemicals is due to extrapolation of laboratory results to the field. Such an extrapolation seems theoretically impossible, as the essential characteristics of an ecosystem are not incorporated in a single-species laboratory test.

For the aquatic environment, several programs have been conducted to assess toxicity under field conditions in ponds, experimental ditches, etc. For the terrestrial environment, studies like these refer mainly to pesticides applied to experimental plots. The impression from these studies is that effects of chemicals on certain species can be predicted in the field, provided that the same species have been used in the laboratory tests, and the exposure concentration is accurately known.

In a review of laboratory and field data on the toxicity of benomyl, car-bendazim, carbofuran, and carbaryl to earthworms, a quantitative comparison was made of the effect levels.[37,38] Under field conditions, exposure of earthworms is strongly dependent on the rate of deposition of pesticides on the soil surface, on the behavior of pesticides in the soil column, and on the vertical distribution of the earthworms. Still, the results of field studies were in reasonable agreement with those from laboratory tests, when a homogeneous distribution of pesticides over the top 2.5 cm soil layer was assumed.[37,38]

To validate the assessment schemes with respect to their ecological relevance, a more fundamental ecological approach is needed. This requires knowledge of not only the properties of chemicals that confer environmental hazards, but also the properties of ecological receptors that confer environmental vulnerability.

V. CONCLUSIONS

No effect levels obtained from chronic reproduction experiments presently provide the best available base for risk assessment and are usually adopted. However, none of the approaches applied to derive environmental quality standards seems to be completely satisfactory from a scientific point of view. To establish criteria, various nonscientific and pragmatic arguments have to be applied. This should, however, not detract attention from strengthening the science of ecotoxicology which can improve the base for a better protection of ecosystems only by progressing the limits of knowledge.

ACKNOWLEDGMENTS

The authors are indebted to Dr. H. Eijsackers (NISRP, Wageningen) for encouragement to compile this paper and to Ms. Désirée Hoonhout for word processing.

REFERENCES

1. Van Straalen, N.M., Schobben, J.H.M., and De Goede, R.G.M., Population consequences of cadmium toxicity in soil microarthropods, *Ecotoxicol. Environ. Saf.*, 17, 190, 1989.
2. Depledge, M., The rational basis for detection of the early effects of marine pollutants using physiological indicators, *Ambio*, 18, 301, 1989.
3. Stebbing, A.R.D., Hormesis — the stimulation of growth by low levels of growth inhibitors, *Sci. Total Environ.*, 22, 213, 1982.

4. McCarthy, J.F. and Shugart, L.R., Eds., *Biomarkers of Environmental Contamination,* Lewis Publishers, Chelsea, MI, 1990.
5. Williams, D.A., A test for differences between treatment means when several dose levels are compared with a zero dose control, *Biometrics,* 27, 103, 1971.
6. Cox, C., Threshold dose-response models in toxicology, *Biometrics,* 43, 511, 1987.
7. Hoekstra, J.A. and Van Ewijk, P.H., Alternatives for the no effect level, *Environ. Toxicol. Chem.,* 12, 187, 1993.
8. EPA, Estimating "Concern Levels" for Concentrations of Chemical Substances in the Environment, Environmental Effects Branch, Health and Environment Division, U.S. Environmental Protection Agency, Washington, D.C.
9. Stephan, C.E., Proposed goal of applied aquatic toxicology, in *Aquatic Toxicology and Environmental Fate,* Poston, T.M. and Purdy, R., Eds., American Society for Testing and Materials, Philadelphia, 3, 1986.
10. Van der Gaag, M.A., Stortelder, P.B.M., Van der Kooij, L.A., and Bruggeman, W.A., Setting environmental quality criteria for water and sediment in the Netherlands: a pragmatic ecotoxicological approach, *Eur. Water Pollut. Control,* 1, 13, 1991.
11. Van der Kooij, L.A., Van de Meent, D., Van Leeuwen, C.J., and Bruggeman, W.A., Deriving quality criteria for water and sediment from the results of aquatic toxicity tests and product standards: application of the equilibrium partitioning methods, *Water Res.,* 25, 697, 1991.
12. Gottschalk, C. and Stix, E., *Quality Objectives for Hazardous Substances in Surface Waters,* Report Umweltbundesamt, Berlin, 1991.
13. Van Leeuwen, K., Ecotoxicological effects assessment in the Netherlands: recent developments, *Environ. Manage.,* 14, 779, 1990.
14. Slooff, W., Ecotoxicological Effect Assessment: Deriving Maximum Tolerable Concentrations (MTC) from Single-Species Toxicity Data, RIVM Guidance Document, Report No. 719102018, RIVM, Bilthoven, 1992.
15. Van Straalen, N.M. and Denneman, C.A.J., Ecotoxicological evaluation of soil quality criteria, *Ecotoxicol. Environ. Saf.,* 18, 241, 1989.
16. Aldenberg, T. and Slob, W., Confidence Limits for Hazardous Concentrations Based on Logistically Distributed NOEC Data, RIVM Report No. 719102002, RIVM, Bilthoven, 1991.
17. Wagner, C. and Løkke, H., Estimation of ecotoxicological protection levels from NOEC toxicity data, *Water Res.,* 25, 1237, 1991.
18. Van Gestel, C.A.M. and Van Straalen, N.M., Ecotoxicological test methods using terrestrial invertebrates, in *Ecotoxicology of Soil Organisms,* Donker, M.H., Eijsackers, H., and Heimbach, F., Eds., Lewis Publishers, Chelsea, 1993.
19. Okkerman, P.C., Van de Plassche, E.J., Slooff, W., Van Leeuwen, C.J., and Canton, J.H., Ecotoxicological effects assessment: a comparison of several extrapolation procedures, *Ecotoxicol. Environ. Saf.,* 21, 182, 1991.
20. Eijsackers, H. and Løkke, H., SERAS, Soil Ecotoxicological Risk Assessment System, Report, National Environmental Research Institute, Silkeborg, 1992.
21. Slooff, W., Canton, J.H., and Hermens, J.L.M., Comparison of the susceptibility of 22 freshwater species to 15 chemical compounds. I. (Sub)acute toxicity tests, *Aquat. Toxicol.,* 4, 113, 1983.

22. LeBlanc, G.A., Interspecies relationships in acute toxicity of chemicals to aquatic organisms, *Environ. Toxicol. Chem.*, 3, 47, 1984.

23. Edwards, C.A. and Thompson, A.R., Pesticides and the soil fauna, *Res. Rev.*, 45, 1, 1973.

24. Hopkin, S.P., *Ecophysiology of Metals in Terrestrial Invertebrates*, Elsevier Applied Science, London, 1989.

25. Everts, J.W., Aukema, B., Hengeveld, R., and Koeman, J.H., Side-effects of pesticides on ground-dwelling predatory arthropods in arable ecosystems, *Environ. Pollut.*, 59, 203, 1989.

26. Walker, C.H., Species variations in some hepatic microsomal enzymes that metabolize xenobiotics, *Progr. Drug Metab.*, 5, 113, 1980.

27. Brattsten, L.B., Ecological significance of mixed-function oxidations, *Drug Metab. Rev.*, 10, 35, 1979.

28. Janssen, M.P.M., Ma, W.-C., and Van Straalen, N.M., Biomagnification in terrestrial ecosystems, *Sci. Total Environ.*, in press.

29. Ma, W.-C., Denneman, W., and Faber, J., Hazardous exposure of ground-living small mammals to cadmium and lead in contaminated terrestrial ecosystems, *Arch. Environ. Contam. Toxicol.*, 20, 266, 1991.

30. Laskowski, R., Are the top carnivores endangered by heavy metal biomagnification?, *Oikos*, 60, 387, 1991.

31. Van Straalen, N.M. and Ernst, W.H.O., Metal biomagnification may endanger species in critical pathways, *Oikos*, 62, 255, 1991.

32. Fagerström, T., Biomagnification in food chains and related concepts, *Oikos*, 62, 257, 1991.

33. Moriarty, F. and Walker, C.H., Bioaccumulation in food chains — a rational approach, *Ecotoxicol. Environ. Saf.*, 13, 208, 1987.

34. Sprague, J.B., Measurement of pollutant toxicity to fish. II. Utilizing and applying bioassay results, *Water Res.*, 4, 3, 1970.

35. Hermens, J., Canton, H., Steyger, N., and Wegman, R., Joint effects of a mixture of 14 chemicals on mortality and inhibition of reproduction of *Daphnia magna*, *Aquat. Toxicol.*, 5, 315, 1984.

36. Kraak, M.H.S., Lavy, D., Schoon, H., Toussaint, M., Peeters, W.H.M., and Van Straalen, N.M., Evaluation of the ecotoxicity of mixtures of metals to the zebra mussel *Dreissena polymorpha*, using the toxic unit concept, *Environ. Toxicol. Chem.*, in press.

37. Van Gestel, C.A.M., Validation of earthworm toxicity tests by comparison with field studies: a review of benomyl, carbendazim, carbofuran and carbaryl, *Ecotoxicol. Environ. Saf.*, 23, 221, 1992.

38. Heimbach, F., Effects of pesticides on earthworm populations: comparison of results from laboratory and field tests, in *Ecotoxicology of Earthworms*, Greig-Smith, P.W., Becker, H., Edwards, P.J., and Heimbach, F., Eds., Intercept Ltd., Andover, 100, 1992.

CHAPTER 30

Ecotoxicological Extrapolation: Tool or Toy?

H. Løkke

TABLE OF CONTENTS

ABSTRACT

A short review on the existing tools for statistically-based ecotoxicological extrapolation is given. A calculation method is presented for the inclusion of aquatic data in these cases where data on endogeic and epigeic terrestrial organisms are lacking. Further, data on endogeic and epigeic terrestrial organisms are combined by a proposed calculation method. The use of different parameters of the method is discussed. The chapter demonstrates how an extrapolation method can be used for setting a safe, no-effect level for soil using the insecticide dimethoate as an example. By use of an aquatic data set it was found that at 0.25 g ha^{-1} of dimethoate and, with the probability of 95%, no more than 5% of the species in the agroecosystem might be sublethally affected. By use of the available limited data set for terrestrial species the similar value was found at 0.006 g ha^{-1}. By combining the terrestrial and the aquatic data sets the value was estimated at 0.30 g ha^{-1}.

I. INTRODUCTION

An important objective of laboratory testing of the effects of chemical compounds or of complex mixtures of compounds is to assess the impact on natural systems and to calculate risk factors. Environmental risk assessment combines exposure assessment and effects assessment, respectively.

In most cases the effects assessment is performed by use of single species tests. However, regulatory guidelines for handling of polluted soil need the development of ecotoxicity methods and risk assessment procedures. At this moment the assessment of ecotoxicological effects on soil organisms is hardly possible, as an adequate number of soil ecotoxicity guidelines is not available.[1] Very limited toxicity data sets are available for existing or new industrial chemicals, although for pesticides larger data sets may be present. For the generation of data, quantitative structure-activity relationship (QSAR) procedures have been developed for the extrapolation of single species acute data interspecies or for extrapolation from single species LC$_{50}$ to chronic toxicity effects (no observed effect concentration, NOEC). The application of QSAR estimates for narcotic chemicals in the aquatic environment has been demonstrated by van Leeuwen et al.[2]

The main possibilities of extrapolation are illustrated in Figure 1. Single species chronic tests may as well be extrapolated interspecies. The chronic NOEC may be used to extrapolate from the population level of species in the ecosystem by use of statistical based extrapolation methods resulting in hazardous or safe concentrations. Alternatively, hazardous or safe concentrations may be elaborated by multispecies tests.

The aim of the present work is to give a short review on the existing tools for ecotoxicological extrapolation. A calculation method is presented for the

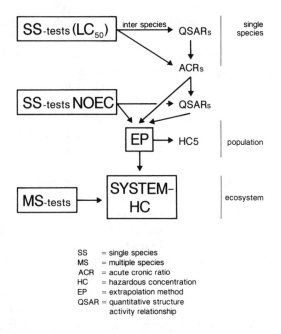

SS = single species
MS = multiple species
ACR = acute cronic ratio
HC = hazardous concentration
EP = extrapolation method
QSAR = quantitative structure
         activity relationship

FIGURE 1. Outline of the possibilities of extrapolation in the terrestrial environment.

inclusion of aquatic data in these cases where data on terrestrial organisms are lacking. Further, data on endogeic and epigeic terrestrial organisms are combined by a proposed calculation method. The insecticide dimethoate is used for a calculation example, and preliminary parameters for combining aquatic and endogeic and epigeic data are presented. The paper demonstrates how an extrapolation method can be used for setting a safe no effect level for soil.

II. TOOLS FOR EXTRAPOLATION FROM SINGLE SPECIES TESTS

Several extrapolation methods have been developed that are based on species sensitivity distributions. The U.S. EPA[3] was first to present an extrapolation method aiming at protecting 95% of aquatic genera. This procedure assumed that the distribution curve of the sensitivities of the species is a log-triangular form. Kooijman[4] developed a statistical approach to show how the difference in the sensitivities of species for a chemical can be used to predict the range of sensitivities for all other untested species. This method assumes that the sensitivities of the species are described by a log-logistic distribution. A concentration is determined at which the LC_{50} of the most sensitive of the species present in the community exceeds that concentration by a specified probability. The method of Kooijman was adapted by Van Straalen and Denneman[5] for the calculation of a protection level for 95% of the species in a soil ecosystem. Two modifications of this approach were suggested by Aldenberg and Slob[6] and by Wagner and Løkke.[7]

The selection of test species involves ecological and statistical considerations. Ideally, by using extrapolation methods which are based on sensitivity distributions, test species should be selected at random. In statistical terms the sampling of test species should represent all species in an ecosystem. For practical reasons, only those species which may be reared in the laboratory or collected in sufficient numbers in the field are used for testing. Therefore, the sampling of test species is normally not respecting the ideal requirement of random sampling.

By selection of test species the ecological background of the species should be considered to perform an adequate selection of various biological and physiological aspects which may be important to the ecology of terrestrial biota. However, the extrapolation methods do not take into account the interactions between species. Species can be selected on the basis of their ecological function (trophic level), their morphological structure, and their route of exposure. Social, economic, and recreational factors may also play a role. Because ecosystems can tolerate some stress and occasional adverse effects, protection of all species at all times and places is not necessary. With data available for a large number of appropriate taxa from an appropriate variety of taxonomic and functional groups, a reasonable level of protection can probably be provided if all except a small fraction of the taxa are protected. In the U.S. EPA[3] method there is a requirement for at least eight different toxicity data. In the Wagner and Løkke[7] method it is recommended at least five test species be used to be reasonably sure of determining limits which are not too high. By using the distributions a concentration is calculated with a certain percentage of the species in the community or ecosystem unlikely to be affected, if the concentration is not exceeded.

The method of Aldenberg and Slob[6] assumes that all toxicity samples come from one log-logistic distribution. The method of Wagner and Løkke[7] assumes a log-normal instead of a log-logistic distribution for the toxicity data. The lower left fractile of the log-normal distribution for the probability p is shown in Figure 2. The value of p is the probability that a species chosen at random has a NOEC lower than the environmental concentration. Arbitrarily, the value $p = 0.05$ has been chosen corresponding to the protection of $(1 - p) \cdot 100\% = 95\%$ of all species in the community. However, this fractile is determined with a confidence of 50%. To be reasonably sure that the $p \cdot 100\%$ of the species are protected, a lower statistic tolerance limit K_p is determined as illustrated in Figure 2 so that with the probability $(1 - \delta)$ no more than $p \cdot 100\%$ of all species in the community have a NOEC value smaller than K_p.

At an Organization for Economic Cooperation and Development (OECD) workshop held in Washington 1990[8] the U.S. EPA,[3] Aldenberg and Slob,[6] and the Wagner and Løkke[7] methods were compared by application to a set of data for 11 species and eight chemicals. The chronic data sets were taken from Slooff and Canton.[9] It was found that the log-logistic and the log-normal extrapolation methods hardly differed in cases where the levels of confidence are identical.

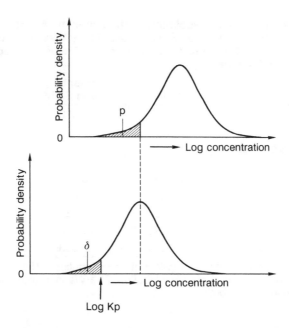

FIGURE 2. Hazardous concentration, Kp, for species in the log-normal concentration. The lower statistic tolerance limit Kp is determined by use of the log-normal distribution so that with the probability (1-δ) no more than $p \cdot 100\%$ of all species in a community have a NOEC value smaller than Kp.

However, the U.S. EPA method[3] currently has no provision for specifying the uncertainty on the fifth percentile estimate. From Table 1 it is seen that all three methods yield similar results with a 50% confidence level.

It can be concluded that the existing extrapolation methods are feasible for further consideration as useful tools for risk assessment and for setting maximum acceptable concentrations of single compounds in soil. The numerical values of p and δ should be chosen on the basis of investigations on several data bases. In general, the methods need validation at the ecosystem level before they are implemented for regulatory purposes.

III. COMBINATION OF AQUATIC AND TERRESTRIAL TOXICITY DATA IN EXTRAPOLATION METHODS

A. Background

Due to lack of data on terrestrial organisms the environmental hazard and risk assessments of soil ecosystems may include data of aquatic organisms or on mammals.

Table 1. Comparison of Extrapolation Methods For the Aquatic Environment Based on Species Sensitivity Distributions for 11 Aquatic Species;[8] Protection Levels for 95% of All Species With 50% Confidence [mg l⁻¹]

	U.S. EPA[4]	Aldenberg and Slob[6]	Wagner and Løkke[7]
K_2CrO_4	0.086	0.070	0.062
NaBr	5.5	4.1	3.3
Tetrapropylene benzene sulphonate	0.27	0.27	0.23
2,4-Dichloroaniline	0.018	0.063	0.055
4-Nitroaniline	0.19	0.44	0.40
DNOC	0.019	0.031	0.026
Dimethoate	0.019	0.018	0.014
Pentachlorophenol	0.0023	0.0039	0.0035

The U.S. EPA method uses the log-triangular distribution.[4] The Aldenberg and Slob[6] method uses the log-logistic distribution. The Wagner and Løkke[7] method uses the log-normal distribution.

In terrestrial ecotoxicology chemical compounds are assumed to be partly adsorbed to soil particles, partly in solution in the soil pore water. Normally, the effects concentrations are given on a dry matter basis, e.g., in mg kg⁻¹. In aquatic ecotoxicology the corresponding effects concentrations are related to the water phase and given, e.g., in mg l⁻¹. Van Gestel and Ma[10] and van Gestel et al.[11] has shown that the soil pore water concentration is of primary importance for the exposure of earthworms to organic chemicals in the soil. Their approach renders QSAR studies for organic chemicals in soil a feasible option in terrestrial ecotoxicology.

In the present chapter the use of combined aquatic and terrestrial single species toxicity data is proposed for extrapolation. The concept of the procedure is depicted in Figure 3. By assuming that only the water-soluble fraction of any compound in question is bioavailable, it is possible to normalize data of toxic effects on aquatic and terrestrial organisms to the water phase common for both groups of organisms. The data are further normalized on an area-based unit. The approach is extended to include epigeic species in the risk assessment. It is assumed that the NOEC values of all species in a given ecosystem is log-normal distributed. The calculations of safe concentrations for terrestrial ecosystems are performed with the log-normal extrapolation method by Wagner and Løkke.[7]

B. Outline of the Calculation Method

The concentration of a chemical compound in soil pore water is dependent on the specific adsorption coefficient in the soil, K_d, and of the actual soil water fraction w. The soil/water partition coefficient K_d is measured or calculated from

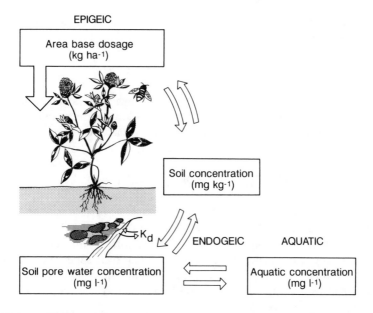

EPIGEIC

Area base dosage
(kg ha⁻¹)

Soil concentration
(mg kg⁻¹)

K_d ENDOGEIC AQUATIC

Soil pore water concentration
(mg l⁻¹)

Aquatic concentration
(mg l⁻¹)

FIGURE 3. Outline of the normalization and combination of epigeic, endogeic, and aquatic toxicity data for extrapolation and risk assessment.

the K_{OC} value. The concentration of the chemical compound adsorbed on the soil fraction, c_s, is determined by the definition relation of K_d,

$$c_s = K_d \cdot c_w \tag{1}$$

where c_w is the concentration in the soil pore water fraction. The total concentration in the soil, c_t, is determined by

$$c_t = c_s + w \cdot c_w \tag{2}$$

where w is the soil pore water fraction (volume/soil dry weight).
By inserting Equation 1 into Equation 2, c_t is determined as

$$c_t = c_w \cdot (K_d + w) \tag{3}$$

In a given soil, the water concentration, c_w, increases with the soil concentration c_t and it decreases with increasing concentrations of organic carbon in the soil. By this formula, toxicity concentrations for aquatic species may be transformed into a set of terrestrial toxicity concentration data or vice versa. Based on the assumption described above that only the water-soluble fraction of any compound in question is bioavailable, it is proposed that for an aquatic species i the $NOEC_{w,i}$ can be transformed into the terrestrial toxicity data set value $NOEC_{t,i}$ by the formula

$$NOEC_{t,i} = NOEC_{w,i} \cdot (K_d + w) \tag{4}$$

However, the large differences of exposure between species should be taken into account. The exposure is dependent on the route which may be through the surface of the organism or through the alimentary canal or combined uptake. To each species a correction factor f_i should be applied:

$$NOEC_{t,i} = f_i \cdot NOEC_{w,i} \cdot (K_d + w) \tag{5}$$

In the terrestrial ecosystems the uptake through the surface may be low due to low direct contact with the soil pore water. On the other hand, the uptake by food items containing adsorbed amounts of the chemical compound in question may be important. For endogeic organisms the uptake is supposed to be from the soil pore water solution by combined penetration through the surface of the organism and by absorption in the alimentary canal. Therefore, as a starting point the factor f_i may be set as unity.

NOEC values of epigeic organisms which are exposed to pesticides are normally related to the rate of application of the active ingredient. The exposure is calculated on basis of either amount per area, D [kg ha^{-1}], or by mean concentration in food items [mg kg^{-1}], e.g., by following a spraying event. Similar area-based exposure values may be found for pollutants. It is assumed that endogeic organisms are exposed to chemical compounds which are uniformly distributed in the topsoil layer. The depth of penetration into the soil, L [dm], depends on the preceding history and the mobility of the compound in the particular soil. In a soil with dry weight density d [kg dm^{-3}] the area-based normalized exposure value D [kg ha^{-1}] can be calculated from the total concentration in the soil c_t [mg kg^{-1}] by the equation,

$$D = c_t \cdot L \cdot d \tag{6}$$

Equation 6 may also be used on soil-dwelling species feeding on contaminated food items on the ground surface, e.g., earthworms eating pesticide-treated leaves from the soil surface. The exposure is assessed from the application rate on the area followed by a normalization to the standard soil concentration for the pesticide and soil in question. Equation 6 describes the no-effect field dosage for the species, i, $D_{0,i}$, when c_t is substituted with $NOEC_{t,i}$:

$$D_{0,i} = NOEC_{t,i} \cdot L \cdot d \tag{7}$$

The use of this equation for aquatic and epigeic species should be regarded as tentative.

C. Calculation Example: Dimethoate

The organophosphorous insecticide dimethoate is chosen as an interesting chemical compound for demonstration of the procedure. Due to a high water solubility (25 g l^{-1})[12] the compound may penetrate into the soil. Further, dimethoate is systemic and is taken up in plants from the roots or from leaf surfaces.

The extrapolation models are based on the assumption that the species in ecosystems have some distribution of sensitivity with few very sensitive species and many moderately sensitive species. However, it may be expected that dimethoate does not fit the log-normal distribution due to the specific acetylcholinesterase inhibition of this compound. Thus, relatively many species, especially including the insects, are very sensitive to the insecticide dimethoate. Correspondingly, relatively many species, especially plants, may be very insensitive to the compound so that the data may belong to a distribution with two maxima. The present calculations include abiotic data as shown in Table 2, and toxicity data of aquatic species, and of endogeic and epigeic species, respectively, as shown in Table 3.

1. Aquatic Data

Aquatic data from Slooff and Canton given by Okkerman et al.[13] were used for the comparison. Eleven different species were used, bacteria (*Pseudomonas fluorescens* and *Microcystis aeruginosa*), algae (*Scenedesmus pannonicus*), plants (*Lemna minor*), crustaceans (*Daphnia magna*), insects (*Culex pipiens*), hydrozoans (*Hydra oligactis*), molluscs (*Lymnaea stagnalis*), fish (*Poecilia reticulata* and *Oryzias latipes*), and amphibians (*Xenopus laevis*). The end points were survival, growth, and reproduction. The NOEC data are shown in Table 3. They were normalized to the area-based values, D, by use of Equations 4 and 7 and by inserting the abiotic data of Table 2.

The data were tested for normality by the UNIVARIATE procedure.[14] The hypothesis was accepted that the available toxicity data resemble the log-normal distribution. By using the log-normal extrapolation formula given by Wagner and Løkke,[7] the lower statistic tolerance limit K_p is determined so that with the probability of 95%, no more than 5% of all species in the terrestrial system have a NOEC value lower than K_p,

$$K_p = \exp(\bar{x}_m - s_m \cdot k) \tag{8}$$

Table 2. Abiotic Data of Dimethoate Used For Calculations

Soil depth	L = 0.5 dm
Soil density	d = 1.44 kg l^{-1}
Soil pore water fraction	w = 0.15
Adsorption coefficient	K_d = 0.46[21]
Soil organic carbon	OC = 1.7%

Table 3. Toxicity Data For Selected Test Species Exposed to Dimethoate in Water (NOEC$_w$) or in the Terrestrial Environment (NOEC$_s$), Respectively; the Data Are Normalized to an Area-Based Unit (g ha^{-1})

Taxonomical group	Species	Endpoint	NOEC$_w$ mg l^{-1}	NOEC$_s$ mg kg^{-1}	Normalized data, g ha^{-1}
Bacteria	*Pseudomonas fluorescens*	Specific growth rate	320		140,000
	Microcystis aeruginosa	Specific growth rate	32		14,000
Fungi	*Saccharomyces cerevisiae*	Growth (biomass)	10		4,400
Algae	*Scenedesmus pannonicus*	Growth (biomass)	100		44,000
Protozoa	*Tetrahymena pyriformis*	Growth (biomass)	<1		220
Pteridophyta	*Lemna minor*	Specific growth rate	32		14,000
Crustacea	*Daphnia magna*	Mortality	0.032		14
Collembola	*Folsomia fimetaria*	Reproduction		0.2	144
Insecta	*Culex pipiens*	Mortality	0.32		141
	Apis mellifera	Pupation success		0.156	9
	Drosophila melanogaster	Pupation success		0.5	28
Coelenterata	*Hydra oligactis*	Specific growth rate	100		44,000
Mollusca	*Lymnaea stagnalis*	Reproduction	10		4,400
Pisces	*Poecilia reticulata*	Growth	0.1		44
	Oryzias latipes	Mortality	0.32		141
Amphibia	*Xenopus laevis*	Mortality	1		440

where $\bar{x}_m$ is the mean of the natural logarithm of the D values, s_m is the standard deviation, and k a factor dependent on the number of species, N, and the degree of confidence. By use of the values $\bar{x}_m$ = 7.557, s_m = 3.176, and k = 2.815 for N = 11 and confidence level 95%, the value of K_p is found to be 0.25 g ha^{-1}.

2. Terrestrial Data

The NOEC value for an endogeic species was obtained for collembola (*Folsomia fimetaria*)[15] and transformed by Equation 7 to the D value. Further, NOEC values of two insects were included, representing different epigeic species, namely the honey bee *Apis mellifera* and the fruit fly *Drosophila melanogaster*.

Exposure of insects following the use of insecticides may be due to direct contact with the spray and/or with residues left on the surfaces of plants, or even direct contamination of pollen. However, systemic insecticides may add hazards resulting from their translocation into nectar and pollen, which may be rendered toxic to insects collecting nectar and pollen, or could affect both the brood and honey stores of the honey bee. Lord et al.[16] detected dimethoate at 0.5 mg kg^{-1} in field bean nectaries when applied at 400 g ha^{-1} to the soil in pots. The maximum concentration was reached a few days after treatment and the insecticide seemed to persist for a longer time with smaller than with larger doses.

For honey bees, a NOEC value at 0.156 mg kg^{-1} is given by Davis et al.[17] for the pupation success of larvae which were placed on diets containing dimethoate. For fruit flies, Dhingra and Vijayakumar[18] found a NOEC value at 0.5 mg kg^{-1} by exposure of eggs, third-instar larvae, and adults to dimethoate in standard corn meal-yeast-agar medium. It is very difficult to relate these NOEC values to the field application rate. In the present calculations a conservative estimate was applied. The mean residue concentrations in the plant material was set at 1 mg kg^{-1} wet weight as caused by an application rate at 55 g ha^{-1}.

The terrestrial data set was extended with NOEC values on yeast (*Saccharomyces cerevisiae*)[19] and on a holotrichous ciliate, *Tetrahymena pyriformis*.[20] These data are normalized in a similar way as the aquatic data to the area-based values, D, by use of Equations 4 and 7 and by inserting the abiotic data of Table 2.

The terrestrial data set was also tested for normality by the UNIVARIATE procedure,[14] and a reasonable fit to the log-normal distribution was found. By log-normal extrapolation the lower statistic tolerance limit K_p is determined so that with the probability of 95% no more than 5% of all species in the terrestrial system have a NOEC value lower than K_p. By use of the values $\bar{x}_m$ = 4.856, s_m = 2.355, and k = 4.210 for N = 5 and confidence level 95%, the value of K_p is found to be 0.006 g ha^{-1}.

The combined aquatic and terrestrial data set was also tested for normality by the UNIVARIATE procedure,[14] and it was found that the data set resembles the log-normal distribution. Further, the values which are depicted in Figure 4 do not indicate that the distribution curve of the log-normal data of dimethoate has two maxima. By log-normal extrapolation the following values were used: $\bar{x}_m = 6.713$, $s_m = 3.142$, and $k = 2.523$ for $N = 16$ and confidence level 95%, the value of K_p is found to be 0.30 g ha^{-1}.

IV. DISCUSSION

The example of dimethoate shows that the combined data obviously belong to a log-normal distribution. Although many assumptions are involved, the values found by use of combined aquatic and terrestrial data (0.30 g ha^{-1}) is similar to the value found by use of the aquatic data set (0.25 g ha^{-1}). The value found

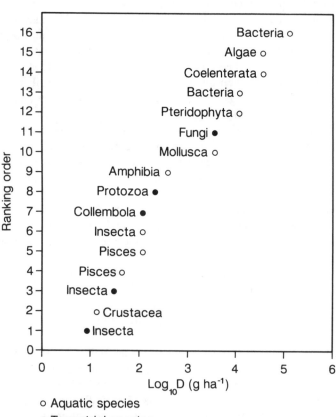

FIGURE 4. Combined aquatic and terrestrial toxicity data on dimethoate depicted in ranking order with respect to the logarithm of the calculated field No Observed Chronic Effects level, log$_{10}$D (g ha^{-1}).

by use of terrestrial data only is considerably lower (0.006 g ha^{-1}) than the value derived from aquatic data. This may be due to the lower number of data and to a poor fit to the log-normal distribution curve.

The tentative use of K_d, L values, and other parameters for assessing the exposure concentration should be refined. It is not justified that only the water-soluble fraction of the compound in question is bioavailable. This assumption is preliminary and may certainly only be true for soft-body organisms. At present, scarce data only are available on the bioavailability of adsorbed chemical compounds for soil-dwelling organisms. Further research is highly needed.

The use of extrapolation methods on combined aquatic and terrestrial data sets may be regarded as dubious. However, a broader range of phylla are covered. On the other hand it is not justified to use large taxonomic distances ranging from bacteria and plants to insects and fishes. A large variation of end points and taxa in combination with a limited number of test species may cause serious errors in the extrapolation procedure because of wrong assumptions of the distribution model or the lack of random sampling of the taxa. Therefore, when only limited data are available it is recommended to use homogeneous end points. Additionally, the distribution function of the test data should always be controlled. The statistical implications of the fact that random sampling of test species hardly may be fulfilled, should be further regarded. More work should be performed on normalizing and standardizing the transformation calculations before this practice should be regarded as more than a toy.

Uncertainties in data may arise either due to the data selection procedure being biased relative to desired levels of concern (accuracy), or to variability in the data (precision).[8] The former point concerns whether or not the test end point, biological property, is adequately protected, whether or not the set of test species is adequately representative of the communities to be protected, whether or not the fifth percentile of this set establishes an appropriate level of protection for diverse ecosystems, and whether or not the test conditions are representative of field exposures. The variability in the data can come from either differences among the species or from variation among tests.

Of particular concern is the possibility that an extreme value in the upper tail of the data set causes the lower percentiles of the distribution to be lower. It may seem somewhat paradoxical that a very tolerant species should lead to a more extreme estimate for sensitive species. However, it may be argued that in this case of high variance of the distribution also extremely sensitive species might be expected. Alternatively, following statistical tests for log-normality of the data set, data for extremely tolerant species may be censored if they cause a skewed log-normal distribution. More work should be done on the evaluation of criteria for the elimination of outliers.

It may be argued that toxicity data obtained from laboratory experiments are not valid under field conditions. The extrapolation models do not account for other aspects such as the interaction between species or the modifying influence of the nonbiological components of the environment on the exposure and on the biological effect. Chemicals may affect organisms by stimulating or inhibiting their functioning. The chemical concentration and the duration of exposure are important determinants of these effects. Furthermore, the toxicological end points used for calculations are heterogeneous, e.g., lethality, growth rate, and hatching. However, it may be verified that the distribution of sensitivity of laboratory species approximates that distribution of species in the field. More research is needed to verify if the species sensitivity in isolation under laboratory conditions is equal to the sensitivity of the species as a member of a complex community.

The extrapolation methods can be employed in the environmental setting of soil quality standards and for setting maximum acceptable concentrations of single compounds in polluted soil. In some cases the combined effect of all compounds present in the environment should be accounted for. This may be done by an addition model, summing the critical concentrations of the single compounds.

Protection of 5% of all species against any significant impact may be reasonable. The confidence level of this calculation is arbitrarily set at 95%. This confidence may be regarded as a "politically" fixed value. It must be recognized that a 95% lower confidence limit for the fifth percentile may be overly conservative. Effort should be addressed to reduce this uncertainty. The actual confidence level may be assessed on the basis of field trials validating the extrapolated values, although this task is very complicated and expensive.

Because the extrapolation models have reasonable underlying assumptions and have been accepted for regulatory use, there are good bases for their application in assessing hazards of chemicals. The main problem is still the scarcity of reliable data. The proposed strategy to use combined data for aquatic and terrestrial species may be useful as a first step until sufficient and reliable data are available for the chemical compounds in question.

REFERENCES

1. OECD (Organisation for Economic Co-operation and Development), 1989. Report of the OECD Workshop on Ecological Effects Assessment, OECD Environment Monogr. No. 26, OECD, Paris, 65 pp.
2. Van Leeuwen, C.J., Van der Zandt, P.T.J., Aldenberg, T., Verhaar, H.J.M., and Hermens, J.L.M., Application of QSARs, extrapolation, and equilibrium partitioning in aquatic effects assessment. I. Narcotic industrial pollutants, *Environ. Toxicol. Chem.*, 11, 267, 1992.
3. U.S. EPA (U.S. Environmental Protection Agency), Water Quality Criteria; availability of documents, Fed. Regist. 50:30784–30796, 1985.
4. Kooijman, S.A.L.M., A safety factor for LC_{50} values allowing for differences in sensitivity among species, *Water Res.*, 21, 269, 1987.
5. Van Straalen, N.M. and Denneman, Ecotoxicological evaluation of soil quality criteria, *Ecotoxicol. Environ. Saf.*, 18, 241, 1989.

6. Aldenberg, T. and Slob, W., Confidence Limits for Hazardous Concentrations Based on Logistically Distributed NOEC Toxicity Data, Report No. 719102002, National Institute of Public Health and Environmental Protection, Bilthoven, The Netherlands, 1991.

7. Wagner, C. and Løkke, H., Estimation of ecotoxicological protection levels from NOEC toxicity data, *Water Res.*, 25, 1237, 1991.

8. OECD, Report of the OECD Workshop on the Extrapolation of Laboratory Aquatic Toxicity Data to the Real Environment, OECD Environment Monogr. No. 59, OECD, Paris, 1992, 43 pp.

9. Slooff, W. and Canton, J.H., Comparison of the susceptibility of 11 freshwater species to eight chemical compounds. II. (Semi)chronic toxicity tests, *Aquat. Toxicol.*, 4, 271, 1983.

10. Van Gestel, C.A.M. and Ma, W.-C., An approach to quantitative structure-activity relationships (QSARs) in earthworm toxicity studies, *Chemosphere*, 21(8), 1023, 1990.

11. Van Gestel, C.A.M., Ma, W.-C., and Smit, C.E., Development of QSARs in terrestrial ecotoxicology: earthworm toxicity and soil sorption of chlorophenols, chlorobenzenes and dichloroaniline, *Sci. Total Environ.*, 109/110, 589, 1991.

12. Worthing, C.R., *The Pesticide Manual. A World Compendium,* 8th ed., The British Crop Protection Council, 298, 1987.

13. Okkerman, P.C., Plassche, E.J., v.d., Slooff, W., Van Leeuwen, C.J., and Canton, J.H., Ecotoxicological effects assessment: a comparison of several extrapolation procedures, *Ecotoxicol. Environ. Saf.*, 21, 182, 1991.

14. SAS Institute Inc., *SAS® Procedures Guide,* Release 6.03 Edition, SAS Institute, Cary, NC, 1988, chap. 34.

15. Krogh, P.H., Unpublished results, National Environmental Research Institute, Department of Terrestrial Ecology, Silkeborg, Denmark, 1991.

16. Lord, K.A., Margaret, A.M., and Stevenson, J.H., The secretion of the systemic insecticides dimethoate and phorate into nectar, *Ann. Appl. Biol.*, 61, 19, 1968.

17. Davis, A.R., Solomon, K.R., and Shuel, R.W., Laboratory studies of honeybee larval growth and development as affected by systemic insecticides at adult-sublethal levels, *J. Agric. Res.*, 27, 146, 1988.

18. Dhingra, G. and Vijayakumar, N.K., Toxicity in *Drosophila* exposed to organophosphorus pesticides, *Ann. Biol.*, 5, 9, 1989.

19. Kumar, S., Dnanaraj, P.S., and Bhatnagar, P., Bioconcentration and effects of dieldrin, dimethoate, and permethrin on *Saccharomyces cerevisiae, Bull. Environ. Contam. Toxicol.*, 43, 246, 1989.

20. Kumar, S. and Lal, R., The effects of dieldrin, dimethoate and permethrin on *Tetrahymena pyriformis, Environ. Pollut.*, 57, 275, 1989.

21. Lystbæk, K., Personal communication. Mean value estimated from K_{OC}-values of six soils, Cheminova Ltd.

CHAPTER 31

A Case Study on Bioindication and Its Use For the Assessment of Ecological Impact

H.H. Koehler

TABLE OF CONTENTS

ABSTRACT

Results from a four-year field study on the succession of soil mesofauna after a single, initial Aldicarb application are presented, with special reference to the predacious mesostigmatid mites (Gamasina). The observed consequences of the pesticide application varied according to animal groups, soil stratum, type of vegetation, successional status and parameters (importance values) measured. Responses range from total inhibition to promotion. True assessment of ecological impact goes beyond the simple documentation of the effects of some disturbance. For a prospective judgment, assessment of influences on ecosystem functions, like microbial activity, nutrient cycling or energy flow are required. The reported study illustrates the problems experienced when attempts are made to relate the results of observations on ecosystem structure to those on ecosystem functions within the context of a conceptual model. The most prevalent problems are the practical limitation of the research to some selected ecosystem compartments and the insufficient knowledge of the biology of the organisms concerned.

I. INTRODUCTION

Field research, in addition to laboratory tests and microcosm studies, is indispensable for the assessment of man's impact on his environment, particularly with respect to his use of chemicals. In this context, soil organisms have been extensively studied (e.g., see References 1 to 6) and soil mesofauna, particularly predacious mites are considered by many authors to be good bioindicators of environmental impact.[7-11]

A proper ecological judgment goes beyond the simple documentation of observed population changes. It requires not only investigations on the structural compartments of a specific ecosystem, but also assessment of functional and integrative parameters, such as microbial activity, nutrient cycling, or energy flow, including the consideration of ecosystem properties. In the experiment to be described, the development of soil mesofauna populations and microbiological parameters, including enzyme activities and N mineralization, were studied after a single initial application of the systemic insecticide Aldicarb. The experimental design intended synchronous sampling of faunal populations and functional parameters on the same sites for 1 year.[12] Additional samples were taken after 3 and 4 years to study longer term effects.

Soil mesofauna taxa were aggregated to form guilds (predators, fungal feeders, microbial feeders), whose incidence could be related to the microbial data collected. The results are used to assess the possibility of making a wider ecological judgment of environmental impact with an ecosystem context.

II. SITE, MATERIALS, AND METHODS

A. Site Description

The experiment was established on an area of 30 × 20 m² in the vicinity of Bremen, northwestern Germany, on top of a rubble and debris dump. The underlying material on the site is not known to be toxic. The site has no capillary contact with ground water. After excavation to a depth of 80 cm, the site was refilled with mixed soils to achieve a relatively homogeneous substrate (April 1985). The soil type was loamy sand, pH = 6.9, C = 1.6%, and N = 0.09%. The soil was allowed to settle for 2 weeks. Then it was rotary tilled and carefully leveled. On May 22, 1985, half of the area was sown with grass and subsequently mown regularly (grassland or recultivation site, REC), while the remainder was left to allow undisturbed succession (ruderal or successional site, SUC, Figure 1). Clippings were removed from the first mowing only (July 20). Because of the incompatibility of sampling methods, subareas of 10 × 20 m² each were reserved for microbiological, zoological, and botanical investigations. At one

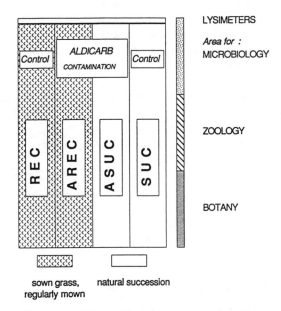

FIGURE 1. The design of the research site on top of a rubble and debris dump: an area of 20 * 30 m was excavated to a depth of 80 cm and refilled. One half of the site was left for undisturbed succession, the other was sown with grass. The respective control plots are named SUC (successional site) and REC (recultivated site), the prefix A means application of Aldicarb (2.5 g/m² ai). Specific areas were assigned for microbiological, zoological, and botanical investigations.

end of the experimental site, a row of 12 lysimeters was embedded in the ground of the subplot for microbiological studies (Figure 1).

B. Pesticide Application

On May 22, 1985, immediately after sowing, Temik® 5G granules (2.5 g ai Aldicarb/m²) were applied to two adjacent subplots measuring 5 × 30 m² (pesticide-treated experimental areas, AREC and ASUC), within the untreated REC and SUC halves of the site, which served as controls (Figure 1). After sowing and pesticide application, the topsoil of the whole area was raked to a depth of 5 cm. Finally, the site was treated with a roller.

C. Sampling Methodologies

1. Soil Fauna

A first set of ten soil cores was taken on May 22, 1985, before sowing and pesticide treatment. From June 1985 to June 1986, nine soil cores per plot were taken on each sampling occasion along a transect across the four equal sized subplots REC-AREC-ASUC-SUC for the estimation of microarthropod populations. Sample size for the estimation of enchytraeid populations was ten cores. With each sampling occasion, the transect was relocated for 0.7 m. Additional sets were taken in June 1988 and in June 1989 to study longer term effects on microarthropods. In all cases, the soil cores had a surface area of 25 cm² and were divided into two depths of 0 to 4 cm and 4 to 8 cm to produce subsamples of a volume of 100 cm³. Sampling frequency from June 1985 to June 1986 was biweekly to monthly. For the first year, the time scale is given in succession weeks. New year 1986 was in the 32nd week of succession. A frost period was between succession week 33 and 42.

Microarthropods were recovered from the cores with a Macfadyen canister-type extractor over a 10-d period. Temperature was raised in 5°C increments per day to 60°C on the upper soil core surface and more than 45°C on the soil core bottom. Enchytraeidae were extracted with an O'Connor-type apparatus with the application of heat for 4 h (45°C) after a period of 24 h of watering at room temperature.

2. Microbiological and Chemical Analyses

Samples for microbiological investigations were taken synchronously with those for soil mesofauna. For the study of enzyme activity, 25 presamples per site, 0 to 4 and 4 to 8 cm depth, respectively, were combined to two mixed samples. Dehydrogenase enzyme activity was determined by the reduction of triphenyltetrazoliumchloride within 24 h.[13] Protease activity was analyzed according to Beck.[14] Net N mineralization was determined after Gerlach[15] with a 3-week on-site incubation period in six soil-filled polyethylene bags.

For residue analyses, six presamples were taken from each subplot, from 0 to 4 and 4 to 8 cm depth, respectively. They were mixed according to site and soil depth. The samples were taken in the subarea for microbiological investigations. Sampling occasions are synchronous to those from the other studies. The soil was deep frozen for storage ($-19°C$). After extraction from the soil with acetone and dichlormethane, the solution was buffered with acetonitrile and analyzed with high-pressure liquid chromatography for residues of Aldicarb and its main metabolites Aldicarb sulfoxide and Aldicarb sulfon (Perkin-Elmer® two-pump system, identification at 205 μm).[16] Tests with standards revealed a recovery rate of 90% for Aldicarb, and of 80% for Aldicarb sulfoxide and Aldicarb sulfon. The practically established resolution was 0.5 μg Aldicarb/g soil (dry weight). In the paper, the combined residues of Aldicarb, Aldicarb sulfoxide, and Aldicarb sulfon are referred to as "total toxic residues" (TTR). The toxicity of these substances is well documented.[17,18]

3. Evaluation and Statistical Analysis

Extracted Gamasina were identified to species level mainly after Karg,[19] Collembola after Gisin[20] and Fjellberg,[21] the Enchytraeids after Nielsen and Christensen.[22] Although it is acknowledged that the extraction methods used have varying efficiencies for different taxa, the fauna obtained were grouped into the following guilds for analysis and comparison with microbiological parameters: Predators = Gamasina + 20% (Prostigmata + Astigmata) + larvae of Coleoptera; fungal feeders = Collembola + Uropodina + Oribatei + 30% (Prostigmata + Astigmata) + 20% Enchytraeidae + 50% Protura; microbial feeders = 80% of Enchytraeidae + 50% (Prostigmata + Astigmata) + 50% Protura + larvae of Diptera. From the Prostigmata + Astigmata, Nanorchestidae and Anoetidae were regarded as microbial feeders and Tyroglyphidae (Acarida) as fungal feeders. Results are presented separately for soil depths of 0 to 4 and 0 to 8 cm. Differences between the abundances from the Aldicarb-treated site and the untreated control were tested for significance (5% level) by using the Mann-Whitney U-test.

III. SOIL MESOFAUNA AND MICROBIOLOGICAL PARAMETERS IN SECONDARY SUCCESSION

Residue analysis of soil from the pesticide treated areas shows that the degradation of Aldicarb followed closely that observed by previous workers.[23,24] Aldicarb is degraded to its two main metabolites by more than 66% within 1 week and below detection limit within 10 weeks. The two metabolites Aldicarb sulfoxide and Aldicarb sulfone were found until succession week 37. After succession week 10, no major differences of TTR are found between the grassland (AREC) and the successional site (ASUC) or between the sample depths 0 to 4 cm and 4 to 8 cm. Data are given as averages of AREC and ASUC and for a

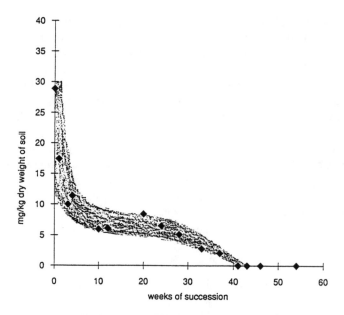

FIGURE 2. The degradation of Aldicarb after a single initial application on May 22, 1985 (total toxic residues = Aldicarb + Aldicarb sulfoxide + Aldicarb sulfone; averages for AREC and ASUC of 0 to 8 cm depth).

depth of 0 to 8 cm (Figure 2). The development of vegetation was rapid and 100% cover was attained within less than 2 months (Müller in Reference 12). The ruderal vegetation on SUC/ASUC produces, compared to the grassland (REC/AREC), a cooler and drier microclimate and specific conditions in the root zone, which both influence the succession of the soil organisms.[25] In the field, the Aldicarb application has no effect on development of cover or plant species composition. On the successional site, however, the dominant species *Chenopodium album* (L.) shows an earlier senescence on the treated subplot (ASUC) than on the control (SUC). We recovered 31 mg/kg dry weight Aldicarb sulfone from plant material collected from ASUC 2 months after the treatment. In September 1985, the respective concentration was at most 3.5 mg/kg dry weight and in October 1985 slightly more than 1 mg/kg dry weight.

With the establishment of the site in May 1985, succession of mesofauna started. For the taxa studied, successional trends are varied and, for some, very marked population developments were observed. From data sets for various taxa, a typical pattern of population development was extracted, which involves a strong initial increase in abundance until the beginning of autumn 1985, when population increase slowed down or, in some instances, numbers decreased during the winter before a second period of increase during the following spring. The effects of frost early in 1986 had no consistent effect on the population

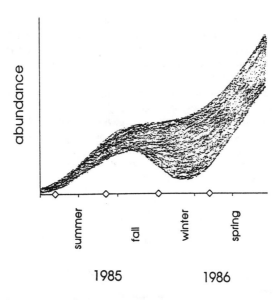

FIGURE 3. Typical generalized trend of abundances of soil mesofauna in a secondary succession from May 1985 to June 1986 (compiled from data sets for various taxa).

levels. This generalized pattern of population changes is summarized graphically in Figure 3.

The effects of Aldicarb application on the succession of soil mesofauna were generally striking, long lasting, and sometimes contrasting. For the first 10 weeks, no predatory mites (Gamasina) colonized the Aldicarb plots at all. Thereafter, gamasine mite numbers fluctuated, being sometimes less and other times equal to numbers on untreated areas. The effect of the chemical impact on the Collembola is clearer: their abundance was consistently lower on the contaminated subplot than on the control. However, in terms of numbers of species collected, the gamasine mites were more seriously reduced by Aldicarb application than were the Collembola.

Not only within taxonomic groupings, such as Gamasina and Collembola, but even more on a species level, a wide range of reactions to pesticide application was observed during the first year of the study. These can be generalized into four major categories (Figure 4), for which some typical examples can be cited (a to d refer to the annotations in the figure):

    a.   Taxa which remained significantly less abundant on treated areas compared with control areas throughout the investigation period: i.e., an almost total inhibition of population development for at least 1 year, e.g., *Rhodacarellus silesiacus* Willmann, *Rhodacarus* spp., small *Pergamasus* spp.; or a strong initial inhibition with a late and still significantly reduced development towards the end of the first year, e.g., *Alliphis siculus* (Oudemans).

b. Taxa whose reaction to Aldicarb varied according to vegetation type; i.e., taxa showing on the two successional subplots similar abundance after initial depression, but sustained depression of abundance on the treated grassland as compared to the grassland control, e.g., Enchytraeidae, Gamasina, *Arctoseius cetratus* (Sellnick).
c. Taxa showing initial, but temporary depression of abundance on both vegetation types with eventual similarity of population size on contaminated and control areas, e.g., *Cheiroseius borealis* (Berlese).
d. Taxa showing an initial depression on pesticide-treated areas, but later reaching population levels significantly higher on Aldicarb-treated areas compared with controls, e.g., *Parasitus eta* Oudemans and Voigts, *Mesophorura* Börner.

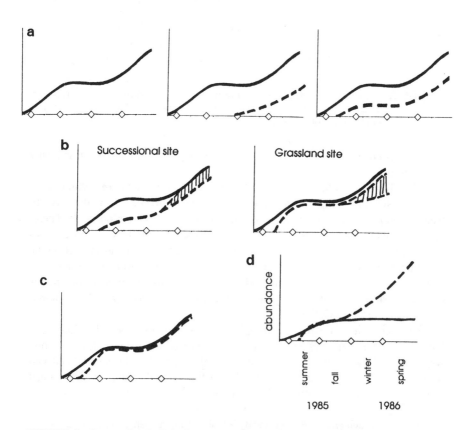

FIGURE 4. Generalized abundance dynamics of soil mesofauna after a single Aldicarb application in a secondary succession from May 1985 to June 1986 (compiled from various data sets; for explanations of a through d, see text). The seasons are delimited by open diamonds. Solid lines indicate the development of abundances of the controls, broken lines those of the contaminated sites. Shaded areas in b describe possible ranges of abundances.

Analysis of the vertical distribution of soil mesofauna taxa suggests an increasing detrimental effect of Aldicarb at greater soil depth. On the soil surface, the large epigeic species *P. eta* and *C. borealis* were not negatively affected by the pesticide application. Hemedaphic species, such as *A. cetratus,* showed an inconsistent reaction, while the deep-dwelling Rhodacaridae could not develop any populations on the Aldicarb-treated sites during the first year of the investigation. Exceptions to this general pattern occurred, however. *A. siculus,* a species of epi- to hemedaphic distribution was severely affected by the Aldicarb application. This species is a specialized nematode feeder and may have suffered from lack of food, its prey being killed by nematicidal Aldicarb.

Gamasina taxocenoses from the untreated controls and Aldicarb-treated areas are compared using Sørensen's quotient of community similarity based on the common occurrence of species and Renkonen's index of similarity based on the relative abundance of species (Figure 5). Both indices show a rapidly increasing faunal similarity within the first 5 months of the experiment. Thereafter, the two indices decline for the following 4 months, which is very striking for the similarities of the gamasine mite communities of the Aldicarb-treated and untreated successional sites. For the gamasine mite taxocenoses of the two grassland subplots, however, the two indices provide contrasting evidence of community similarity; the Sørensen index indicate continued slow increase of similarity, while the Renkonen numbers have a clear decreasing trend documenting a considerable decrease in similarity of the gamasine mite fauna of treated and untreated plots.

On a guild level, the effect of the contamination is generally less obvious after aggregation of the abundances of respective taxa. Fungal feeders as a guild are affected by Aldicarb to a greater and more persistant extent than microbial feeders (Figure 6). The abundance dynamic of predators is very similar to that of their potential prey, the microbial and fungal feeders. An even closer correlation is achieved when the Enchytraeids are excluded. On the Aldicarb-treated successional site (ASUC), predators show, compared to the untreated control (SUC), a more marked initial depression than on the grassland site, where the depressing effect of Aldicarb is more noticeable towards the end of the study period (cf. Figure 4b).

In the following, the development of soil mesofauna abundances on the four subsites are related to data from microbiological investigations. Trends in dehydrogenase enzyme activity match with the development of combined microbial and fungal feeders reasonably well for the first 6 months, particularly on the grassland subplots (Figure 7). After 6 months, towards the faunal development, retarded decline of enzyme activity is observed on the two succession sites, whereas now no relations of the dynamics of the two parameters can be established on the grassland sites. Protease activity (not shown here) behaves very much alike. Net N mineralization from both the treated and untreated successional and the grassland subplots show different trends, which do not match with the abundances of combined microbial and fungal feeders (Figure 8). Very striking is the effect of Aldicarb on net N mineralization in the early weeks of succession,

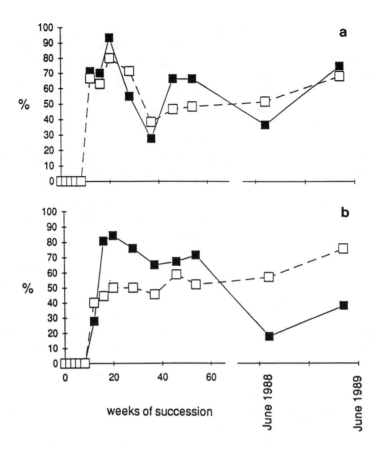

FIGURE 5. Similarity between Gamasina taxocenoses from controls and Aldicarb-treated sites. For the first year, from May 1985 till June 1986, the time scale is in weeks of succession. Additionally, the similarities encountered 3 and 4 years after the Aldicarb application are shown. (a) Successional site; (b) grassland site. ■ = Renkonen index, □ = Sørensen index.

which becomes not so evident from the abundances: immediately after application of the chemical, net N mineralization on both subplots is substantially inhibited. The chemical impact seems to provide easily available organic matter in the form of dead microflora, which is rapidly mineralized after 6 weeks, probably by a microflora tolerant to the chemical and its metabolites.

IV. DISCUSSION

The application of 2.5 g/m² ai Aldicarb to the top 0 to 5 cm of soil produced an immediate toxic residue equivalent to almost 40 mg/kg dry weight soil. This

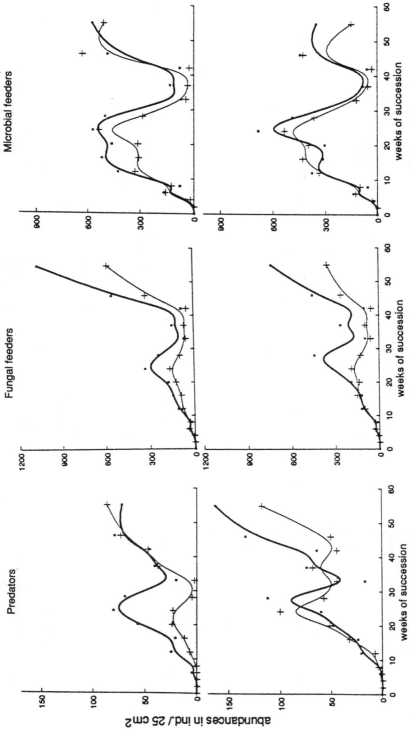

FIGURE 6. Abundance of predators, fungal feeders, and microbial feeders on successional and grassland sites. Bold lines with squared symbols = controls, thin lines with crosses = Aldicarb-treated sites.

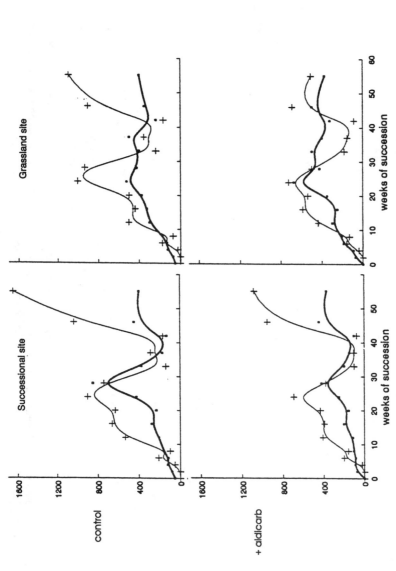

FIGURE 7. Abundances of the combined microbial and fungal feeding guild of soil mesofauna and corresponding synchronous values of dehydrogenase enzyme activity. Bold line with squared symbols = DHA, thin line with crosses = abundance of soil mesofauna, number on the y axes = abundances of soil mesofauna (ind./25 cm², 0 to 4 cm sampling depth) and DHA (µg TPF/g dry soil). For the purpose of better graphical comparison, values of DHA are multiplied by 25.

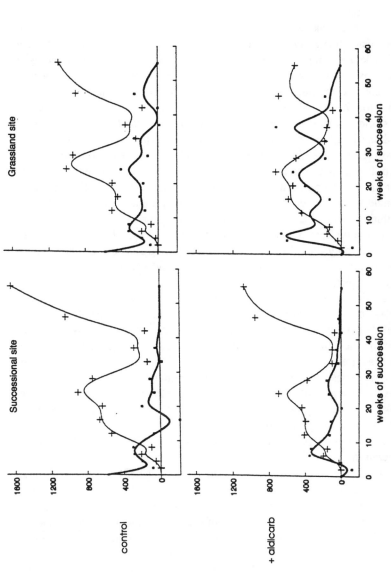

FIGURE 8. Abundances of the combined microbial and fungal feeding guild of soil mesofauna and corresponding synchronous values of N-nettomineralization (NNM). Bold line with squared symbols = NNM, thin line with crosses = soil mesofauna abundance, numbers on the y axes = abundances of soil mesofauna (ind./25 cm², 0 to 4 cm sampling depth), and NNM (µg N/100 g dry soil). For the purpose of better graphical comparison, values of NNM are multiplied by 5.

is substantially towards the upper limit of residues found in other studies or in agricultural practice.[5,26-28] The high application rate in the current study is justified because the purpose of the study is not a practical environmental test of the use of Aldicarb, but an ecological investigation of long-term effects of a single substantial chemical impact.

Repeated experiments in the years since 1980 on the same ruderal site led to the conclusion that early successional phases of ecosystem development show quite a high degree of reproducibility,[29] which is claimed to be a basic prerequisite for any ecotoxicological protocol. This is due to the rather narrow and specific nature of pioneer communities. Older stages of succession tend to be less reproducible for theoretical reasons outlined by Trepl,[30] who concluded that in nature historical processes show a low degree of predictability. The succession of soil animals in the current study was significantly influenced by the application of Aldicarb. Effects are most marked at the single-species level and become less obvious in larger species groupings, culminating in broadly aggregated trophic guilds. It must be borne in mind that there is a wide array of reactions possible at the single-species level, ranging from positive promotion over no effect to total inhibition. When the abundances of groups comprising several species are considered, these effects may cancel out. In communities, which are dominated by few species only, the observed effect is strongly influenced by the reaction of the most dominant species, which will prevail and may obscure reactions of subdominant species.

The sensitivity of the soil animals towards Aldicarb is obviously influenced by complex ecological factors such as successional status and type of vegetation. The results of the current study suggest that indication in the sense of documentation of the effect of any external impact on some specified biological parameter requires time series and synchronous sampling in different phytocenoses. A time series of samples not only gives better information because of a higher overall sample size, but also because of sampling different ecological situations, e.g., development of vegetation or weather. Time lags, which are due, for example, to generation cycles, may be taken into account by such sampling, assuming appropriate timing between the collection of samples and with generation cycles.

The wide range of reactions of soil animal abundances to the single initial Aldicarb application may be due to different exposure to the chemical, different biologies of species, or various indirect effects. The exposure of soil animals is difficult to study, since direct observation of such small animals is hardly possible and the techniques normally used for toxic residue analysis require for practical reasons sampling procedures such as mixed samples, which are insufficiently stratified to reveal small-scale distributions of the toxicant. Knowledge of the biology of soil animals suffers from the increasing neglect of autecological research. With the development of complex food webs, the indirect effects of chemical impact become increasingly important, particularly in the rhizosphere, which is a "hot spot" of biological activity,[31] where pesticides may become concentrated.[32] Jones[33] reports on the development of Aldicarb-metabolizing fungi in treated soils, which could indirectly affect soil mesofauna populations. Such localized microscale effects would be very difficult to be analyzed in a field situation.

In the current study, even after 3 or 4 years, there was little evidence of a convergence in the qualitative structure of the mesofaunal communities monitored. Each of the four sites has had its own specific history, which contributes to a certain degree of singularity in the faunal development. These results demonstrate the good suitability of qualitative parameters to indicate impact on the ecosystem even in the long run. The high structural complexity in space and time, however, may cause problems for the reproducibility and representativity of the results.

The amalgamation of species to trophic guilds assumes that soil animals have specific feeding habits and that feeding is an important regulating mechanism. In the literature, narrow specific food preferences are well documented for soil fauna, but also many supposedly specialized feeders are found to have unexpectedly catholic feeding habits.[34-36] However, for most soil animals detailed information on their feeding biology is lacking.[37] Because of this insufficient knowledge and because the allocation of taxa to specific trophic groups is at best crude, guild formation may be far from reality. However, this approach is still the best way to relate structural to functional ecosystem parameters. Problems resulting from an incomplete knowledge of the feeding biology of soil organisms and from an oversimplistic inclusion of the available information into conceptual models are reviewed by Walter et al.[38]

The microbial parameters measured in the current study were sensitive to Aldicarb application and showed different dynamics on the successional and grassland areas, suggesting the existence of site-specific microfloras (Vollmer in Reference 12, Schulz-Berendt in Reference 12). Close agreement between population trends for mesofauna guilds and microbiological data was observed for about 6 months, which is attributed mainly to a buildup within initial successional dynamics. Later, common trends for microbiological data and abundances are hard to establish.

It has been argued that measurement of functional ecosystem properties is the most appropriate means of assessing ecological impact, especially in an agricultural context when preservation of productivity and sustainability are the primary objectives to be achieved.[39] However, functional parameters are very integrative and may thus mask opposing effects which occur at a lower hierarchical level. This becomes very evident when species reactions are compared to reactions of groupings. Results illustrate that in order to get as reasonable an understanding of ecological impact as possible, it is necessary to assess effects at both detailed structural and integrative functional levels. As will be discussed below, there are severe limitations to accomplish this task. Keeping these in mind, computer simulation may be a useful tool for the extrapolation of data and in hypotheses building.[40]

A conceptual ecosystem model allows a critical evaluation of the representativity of the selected compartments and processes and of the significance and validity of the results.[41] Because of practical limitation to only some functionally important groups (e.g., it was not possible to include nematodes and protozoa in the study) and because of the missing structural differentiation of the microbial compartment, it is quite likely that agreement between population assessments and integrative measures of community function will at best be incomplete. Particularly in applied studies, a clear definition of the underlying conceptual model is necessary to estimate the probability of a "right judgment".

There remains an urgent need for research on the biology of soil animals and their relations to functional ecosystem parameters. On the one hand, soil animals may be sensitive and diverse bioindicators; on the other hand, ecological functions are needed to make an ecological assessment and to foresee the consequences of an impact. As long as there is an insufficient understanding of the field situation in terms of ecosystematic context and interactions therein, integration of soil zoological data with microbiological results seems only to be reasonable from microcosm studies with well-known experimental conditions. However, ecotoxicological studies ultimately aim at a protection of the environment which can only be assessed in the field. That is why, in spite of the complexity of the field situation and the insufficient ecological knowledge, ecotoxicological field studies are absolutely indispensable. These arguments lead to the conclusion that indication and judgment of an ecological impact as a prerequisite for a protection of the environment need complementary concepts with different experimental approaches.

ACKNOWLEDGMENTS

This paper would not have been possible without the generosity of my colleagues from the research group "Ecosystems and Soil Ecology", who allowed me to use their data: J. Müller (vegetation), H. Born (Enchytraeidae), and G. Vollmer and V. Schulz-Berendt (microflora). Also many thanks to an anonymous reviewer for very valuable suggestions.

REFERENCES

1. Baring, H.H., Die Milbenfauna eines Ackerbodens und ihre Beeinflussung durch Pflanzenschutzmittel. 2. Teil: Der Einfluß von Pflanzenschutzmitteln, *Z. Angew. Entomol.*, 41, 17, 1957.
2. Van de Bund, C.F., Changes in the soil fauna caused by the application of insecticides, *Boll. Zool. Agrar. Bachic.*, 7, 185, 1965.
3. Edwards, C., Soil pollutants and soil animals, *Sci. Am.*, 220, 88, 1969.
4. Edwards, C. and Thompson, A.R., Pesticides and the soil fauna, *Res. Rev.*, 45, 1, 1973.

5. Gregoire-Wibo, C., Influence de la repartition de l'Aldicarbe sur les microarthropodes édaphiques (Acariens et Collemboles), *Meded. Fac. Landbowwet. Rijksuniv. Gent*, 46, 629, 1981.

6. Van Straalen, N.M., Kraak, M.H.S., and Denneman, C.A.J., Soil microarthropods as indicators of soil acidification and forest decline in the Veluwe area, the Netherlands, *Pedobiologia*, 32, 47, 1988.

7. Karg, W., Bodenbiologische Untersuchungen über die Eignung von Milben, insbesondere von parasitiformen Raubmilben als Indikatoren, *Pedobiologia*, 8, 30, 1968.

8. Karg, W., Milben als Indikatoren zur Optimierung von Pflanzenschutzmaßnahmen in Apfelintensivanlagen, *Pedobiologia*, 18, 415, 1978.

9. Huhta, V., Ikonen, E., and Vilkamaa, P., Succession of invertebrate populations in artificial soil made of sewage sludge and crushed bark, *Ann. Zool. Fenn.*, 16, 223, 1979.

10. El Titi, A., Der Einfluß von Bodeninsektiziden und organischer Düngung auf Vertreter der Raubmilbenfamilie Rhodacaridae (Mesostigmata, Acarina), *Z. Pflanzenkr. Pflanzenschutz*, 93, 503, 1986.

11. Glockemann, B. and Larink, O., The influence of sewage sludge and heavy metal pollutants on mites, especially Gamasida in an agricultural soil, *Pedobiologia*, 33, 237, 1989.

12. Weidemann, G., Mathes, K., and Koehler, H., Auffindung von Indikatoren zur prospektiven Belastbarkeit von Ökosystemen, in Auffindung von Indikatoren zur prospektiven Bewertung der Belastbarkeit von Ökosystemen, Scheele, M. and Verfondern, M., Eds., *Jül-Spez*, 439, 7, 1988.

13. Thalmann, A., Zur Methodik der Bestimmung der Dehydrogenase-Aktivität im Boden mittels Triphenyltetrazoliumchlorid (TTC), *Landwirtsch. Forschung.*, 21, 248, 1968.

14. Beck, T., Über die Eignung von Modellversuchen bei der Messung der biologischen Aktivität von Böden, Bayer. *Landwirtsch. Jahrb.*, 50, 270, 1973.

15. Gerlach, A., Methodische Untersuchungen zur Bestimmung der Stickstoff-Netto-Mineralisation, *Scr. Geobot.*, 5, 1, 1973.

16. Dekker, A. and Houx, N.W.H., Simple determination of oxime carbamates in soil and environmental water by high pressure liquid chromatography, *J. Envir. Sci. Health*, B18, 379, 1983.

17. Marshall, E., The rise and decline of Temik, *Science*, 229, 1369, 1985.

18. Schmidt, G.H., *Pestizide und Umweltschutz*, Vieweg & Sohn, Braunschweig, Wiesbaden.

19. Karg, W., Die freilebenden Gamasina (Gamasides), Raubmilben, in *Die Tierwelt Deutschlands*, Dahl, F., and Peus, F., Eds., Fischer, Jena, 1971, 59.

20. Gisin, H., Collembolenfauna Europas, Museum d'histoire naturelle, Génève, 1960.

21. Fjellberg, A., Identification keys to Norwegian Collembola, *Norsk Entomologisk Forening*, 1980.

22. Nielsen, C.O. and Christensen, B., Studies on Enchytraeidae. VII. The Enchytraeidae. A critical revision and taxonomy of European species, *Nat. Jutl.*, 8–9, 1, 1959.

23. Bull, D.L., Metabolism of UC-21149 ⟨2-methyl-2-(methylthio)propionaldehyde O-(methyl-carbamoyl)oxime⟩ in cotton plants and soil in the field, *J. Econ. Entomol.*, 61, 1598, 1968.

24. Andrawes, N.R., Bagley, W.P., and Herrett, R.A., Fate and carry over properties of TEMIK aldicarb pesticides ⟨2-methyl-2-(methylthio)propionaldehyde O-(methylcarbamoyl-oxime⟩ in soil, *J. Agric. Food Chem.*, 19, 727, 1971.

25. Koehler, H. and Born, H., The influence of vegetation structure on the development of soil mesofauna, *Agric. Ecos. Environm.*, 27, 253, 1989.

26. Edwards, C.A. and Lofty, J.R., Nematicides and the soil fauna, *Proc. 6th Brit. Insect. Fungic. Conf.*, 1, 158, 1971.

27. Gregoire-Wibo, C., Incidences écologiques des traitements phytosanitaires en culture de betterave sucrière, essais expérimentaux en champ. II. Acariens, Polydesmes, Staphylins, Cryptophagides et Carabides, *Pedobiologia*, 25, 93, 1983.

28. Jones, A.S., Hornsby, A.G., Rao, P.S.C., and Anderson, M.P., Movement and degradation of Aldicarb residues in the saturated zone under citrus groves on the Florida ridge, *J. Contam. Hydrol.*, 1, 265, 1987.

29. Koehler, H., Zur Reproduzierbarkeit von Befunden zur Entwicklung der Bodenmikroarthropoden während der Anfangsphase von Sekundärsukzessionen, *Verh. Ges. Ökol.*, 19, 99, 1991.

30. Trepl, L., *Geschichte der Ökologie*, Athenäum, Frankfurt/M, 1987.

31. Clarholm, M., Interactions of bacteria, protozoa and plants leading to mineralization of soil nitrogen, *Soil Biol. Biochem.*, 17, 181, 1985.

32. Saad, F., Rückstandsdynamik des Pflanzenschutzmittelwirkstoffes Aldicarb in Erdbeerpflanzen und im Boden, Ph.D. thesis, TU Berlin, FB Landwirtschaft. Entwickl., 1972.

33. Jones, A.S., Metabolism of Aldicarb by five soil fungi, *J. Agric. Food Chem.*, 24, 115, 1976.

34. Sell, P., Caloglyphus sp. (Acarina: Acaridae), an effective nematophagous mite on root knot nematodes (Meloidogyne spp.), *Nematologica*, 34, 246, 1988.

35. Smrz, J. and Catska, V., Food selection of the field population of Tyrophagus putrescentiae (Schrank) (Araci, Acarida), *J. Appl. Entomol.*, 104, 329, 1987.

36. Walter, D.E., Hudgens, R.A., and Freckman, D.W., Consumption of nematodes by fungivorous mites, Tyrophagus spp. (Acarina: Astigmata: Acaridae), *Oecologia*, 70, 357, 1986.

37. Werner, M.R. and Dindal, D.L., Nutritional ecology of soil arthropods, in *Nutritional Ecology of Insects, Mites, Spiders, and Related Invertebrates*, Slansky, F. and Rodriguez, J.G., Eds., Wiley Interscience, New York, 1987, 815.

38. Walter, D.E., Kaplan, D.T., and Permar, T.A., Missing links: a review of methods used to estimate trophic links in soil food webs, *Agric. Ecos. Environ.*, 34, 399, 1991.

39. Crossley, D.A., Coleman, D.C., and Hendrix, P.W., The importance of the fauna in agricultural soils: research approaches and perspectives, *Agric. Ecos. Environ.*, 27, 47, 1989.

40. Mathes, K., Computersimulation: eine Methode zur Beurteilung von Umweltbelastungen, Ph.D. thesis, University of Bremen, FB2, 1987.

41. Mathes, K. and Weidemann, G., A baseline-ecosystem approach to the analysis of ecotoxicological effects, *Ecotoxicol. Environ. Saf.*, 20, 197, 1990.

undisturbed control plots. Here, some results about the alterations of an system properly, namely the nitrogen cycle in the soil, will be illustrated.

SOME EFFECTS OF THE INSECTICIDE ALDICARB ON THE NITROGEN CYCLE OF A RUDERAL ECOSYSTEM

Materials and Methods

The consequences of a single (systemic) insecticide application to the soil o 5 cm) of a ruderal ecosystem in regard to botanical, faunal, and micro-logical characteristics have been measured.[8] The experimental plots were pared by removing the topsoil (80 cm), picking out the larger stones, mixing, replacing the soil. In May 1985, after 1 month of settling, a plot of 150 m^2 contaminated with Aldicarb (Temik® 5G, 2.5 g/m^2 ai) and, together with a trol plot of equal size, raked to a depth of 5 cm and left for natural succession. both sites the first plants appeared after 3 weeks (*Chenopodium album* and apis arvensis*), and reached a coverage of 100% 6 weeks later.

In order to quantify the soil nitrogen balance during the vetetation period .5.85 to 7.10.85) the N_{min} content of the upper 80-cm soil depth was measured h a Kjeltec® analyzer system (Tecator) at the beginning and at the end of this iod. The N input (NH_4^+, NO_2^-, NO_3^-) by precipitation as well as the output m three lysimeters (surface 1.5 m^2, embedded to a depth of 80 cm in the und) of each plot were continuously sampled and analyzed by an automatic lyzer (Skalar). N net mineralization, changes of the N content of the microbial mass, and N_{min} content of the soil solution were determined by samples taken m the upper 0 to 4 and 4 to 8 cm of the soil immediately after Aldicarb lication and after 1, 2, 3, 4, 6, 8, 10, 12, and 20 weeks.

Six 100-cm^3 samples were taken from each plot, mixed, and stored in yethylene bags at 4°C. The N_{min} contents (NH_4^+, NO_2^-, NO_3^-) were measured extracting the soil and analyzing the extract with the automatic analyzer ntioned above. Net nitrogen mineralization was determined after Gerlach[9] by ubating soil samples in polyethylene bags at the sites for 3 weeks. After ubation, the N_{min} content was measured again. The difference in N_{min} content ore and after incubation yields the quantity of N net mineralization.

The N content of the microbial biomass was determined as N flush.[10] The uptake by the plants was estimated from the N content of the vegetation mass which was measured with the Kjeltec® analyzer system.

The described methods are discussed in detail by Schulz-Berendt.[11]

Results

The Aldicarb-induced changes in the nitrogen N balance are summarized in ble 1. Figure 2 shows the alterations to the nitrate content of the soil solution d to the leaching water during the first 54 weeks of succession. No standard

CHAPTER 32

On the Effects of Pesticides at the Ecosystem Level

K. Mathes

TABLE OF CONTENTS

0-87371-530-6/94/$0.00 + $.50

© 1994 by CRC Press, Inc.

ABSTRACT

The problem of anticipating the effects of pesticides upon terrestrial ecosystems is discussed theoretically and demonstrated by the effects of a single insecticide application ($2.5 \ g/m^2$ active ingredient (a.i.) Aldicarb) on the nitrogen-cycle of a ruderal ecosystem. The dominant plant of the contaminated site, the annual *Chenopodium album* (fat hen), showed an early senescence compared to the control site. As a consequence, nitrogen uptake was reduced while nitrogen mineralization proceeded. This resulted in increased nitrate-output by leaching. Consideration of some main properties emerging at the ecosystem-level of integration lead to the conclusion that the present procedure of admitting pesticides does not sufficiently take into account the uncertainties inherent in the prediction of ecological effects.

I. INTRODUCTION

One of the goals of ecotoxicological research is to produce proposals for the regulatory handling of pesticides. In doing so, a basic difficulty arises in translating the intentions of lawmakers into strategies and methods suited to the intended environmental protection. According to the German Law of Plant Protection, the Federal Biological Agency authorizes the marketing of a pesticide, if, among other conditions: ''by appropriate and correct use or as a consequence of such use the pesticide will not have (a) any harmful impact on the health of man and animals and on the ground water and (b) any other effects, especially on the 'balance of nature' (Naturhaushalt), which are not acceptable according to the actual state of scientific knowledge'' (§15(1) PflSchG,[1] translated by the author). These statements imply that ''harmful impact'' and ''balance of nature'' are specified concepts with an established scientific definition for which methods of quantification and prediction are available. However, fundamental problems exist in the realization of these concepts.

At present, in ecotoxicological hazard assessment, mainly the data of two sources are compared.[2] The assessed environmental fate of a chemical substance (exposure) is considered in relation to the response of some animal or plant species in laboratory tests (effect assessment). This currently accepted methodology makes a highly extended use of the assumption that a single experimental test situation can adequately produce information about various types of situations.[3] In a strict sense this implies identity between the experimental test and the class of situations it stands for. But in ecotoxicology we have the problem of dealing with a wide variety of biological objects and situations. The ''one for all'' philosophy implies

1. that by testing some species one obtains information concerning the possible reactions of many other species and
2. that the response of higher levels of biological organization than single species can be predicted from (eco)toxicological laboratory tests on a lower level of biological organization.

However, as the level of biological organization increases, nev[...] For example, nutrient cycling emerges as a phenomenon whi[...] the ecosystem level. The effects of xenobiotics on these new [...] testable at lower levels of biological organization. To calcul[...] ecosystem level from effects at the species level would require[...] edge of the toxic effects to organisms, but also the biology o[...] their interrelations need be known.[4,5] An appropriate knowle[...] trapolation procedures is not available in many cases.[5] In reg[...] mental process of bioelement recycling in the soil, as illustrate[...] causal relationships constituting the connections at the ecosy[...] from sufficiently known. In order to improve the information [...] ically induced changes of structure and function of terrestrial [...] German Ministry of Research has sponsored a number of resea[...] general conceptual and theoretical framework of these field exp[...] described elsewhere.[7] The effects of xenobiotics upon different [...] ecosystems were studied by comparing experimentally contami[...]

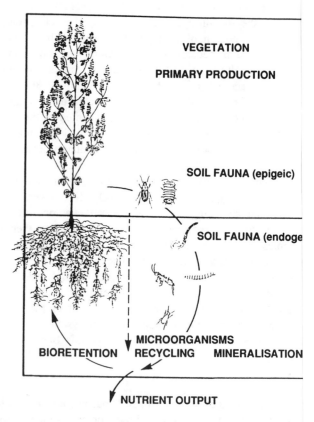

FIGURE 1. Bioelement recycling, a main ecosystem function wh[...] the interactions of soil biota.

CHAPTER 32

On the Effects of Pesticides at the Ecosystem Level

K. Mathes

TABLE OF CONTENTS

ABSTRACT

The problem of anticipating the effects of pesticides upon terrestrial eco-systems is discussed theoretically and demonstrated by the effects of a single insecticide application (2.5 g/m^2 active ingredient (a.i.) Aldicarb) on the nitrogen-cycle of a ruderal ecosystem. The dominant plant of the contaminated site, the annual *Chenopodium album* (fat hen), showed an early senescence compared to the control site. As a consequence, nitrogen uptake was reduced while nitrogen mineralization proceeded. This resulted in increased nitrate-output by leaching. Consideration of some main properties emerging at the ecosystem-level of integration lead to the conclusion that the present procedure of admitting pesticides does not sufficiently take into account the uncertainties inherent in the prediction of ecological effects.

I. INTRODUCTION

One of the goals of ecotoxicological research is to produce proposals for the regulatory handling of pesticides. In doing so, a basic difficulty arises in translating the intentions of lawmakers into strategies and methods suited to the intended environmental protection. According to the German Law of Plant Protection, the Federal Biological Agency authorizes the marketing of a pesticide, if, among other conditions: "by appropriate and correct use or as a consequence of such use the pesticide will not have (a) any harmful impact on the health of man and animals and on the ground water and (b) any other effects, especially on the 'balance of nature' (Naturhaushalt), which are not acceptable according to the actual state of scientific knowledge" (§15(1) PflSchG,[1] translated by the author). These statements imply that "harmful impact" and "balance of nature" are specified concepts with an established scientific definition for which methods of quantification and prediction are available. However, fundamental problems exist in the realization of these concepts.

At present, in ecotoxicological hazard assessment, mainly the data of two sources are compared.[2] The assessed environmental fate of a chemical substance (exposure) is considered in relation to the response of some animal or plant species in laboratory tests (effect assessment). This currently accepted methodology makes a highly extended use of the assumption that a single experimental test situation can adequately produce information about various types of situations.[3] In a strict sense this implies identity between the experimental test and the class of situations it stands for. But in ecotoxicology we have the problem of dealing with a wide variety of biological objects and situations. The "one for all" philosophy implies

1. that by testing some species one obtains information concerning the possible reactions of many other species and
2. that the response of higher levels of biological organization than single species can be predicted from (eco)toxicological laboratory tests on a lower level of biological organization.

However, as the level of biological organization increases, new properties arise. For example, nutrient cycling emerges as a phenomenon which is specific for the ecosystem level. The effects of xenobiotics on these new properties are not testable at lower levels of biological organization. To calculate effects at the ecosystem level from effects at the species level would require not only knowledge of the toxic effects to organisms, but also the biology of the species and their interrelations need be known.[4,5] An appropriate knowledge base for extrapolation procedures is not available in many cases.[5] In regard to the fundamental process of bioelement recycling in the soil, as illustrated in Figure 1, the causal relationships constituting the connections at the ecosystem level are far from sufficiently known. In order to improve the information concerning chemically induced changes of structure and function of terrestrial ecosystems, the German Ministry of Research has sponsored a number of research projects.[6] The general conceptual and theoretical framework of these field experiments has been described elsewhere.[7] The effects of xenobiotics upon different types of terrestrial ecosystems were studied by comparing experimentally contaminated ecosystems

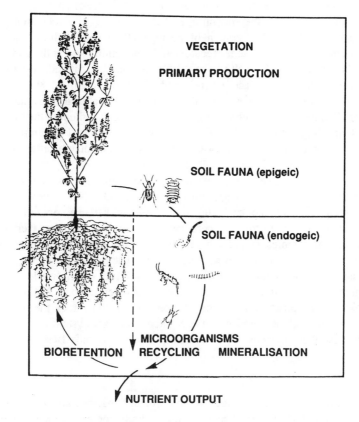

FIGURE 1. Bioelement recycling, a main ecosystem function which results from the interactions of soil biota.

with undisturbed control plots. Here, some results about the alterations of an ecosystem properly, namely the nitrogen cycle in the soil, will be illustrated.

II. SOME EFFECTS OF THE INSECTICIDE ALDICARB ON THE NITROGEN CYCLE OF A RUDERAL ECOSYSTEM

A. Materials and Methods

The consequences of a single (systemic) insecticide application to the soil (0 to 5 cm) of a ruderal ecosystem in regard to botanical, faunal, and microbiological characteristics have been measured.[8] The experimental plots were prepared by removing the topsoil (80 cm), picking out the larger stones, mixing, and replacing the soil. In May 1985, after 1 month of settling, a plot of 150 m² was contaminated with Aldicarb (Temik® 5G, 2.5 g/m² ai) and, together with a control plot of equal size, raked to a depth of 5 cm and left for natural succession. At both sites the first plants appeared after 3 weeks (*Chenopodium album* and *Sinapis arvensis*), and reached a coverage of 100% 6 weeks later.

In order to quantify the soil nitrogen balance during the vetetation period (22.5.85 to 7.10.85) the N_{min} content of the upper 80-cm soil depth was measured with a Kjeltec® analyzer system (Tecator) at the beginning and at the end of this period. The N input (NH_4^+, NO_2^-, NO_3^-) by precipitation as well as the output from three lysimeters (surface 1.5 m², embedded to a depth of 80 cm in the ground) of each plot were continuously sampled and analyzed by an automatic analyzer (Skalar). N net mineralization, changes of the N content of the microbial biomass, and N_{min} content of the soil solution were determined by samples taken from the upper 0 to 4 and 4 to 8 cm of the soil immediately after Aldicarb application and after 1, 2, 3, 4, 6, 8, 10, 12, and 20 weeks.

Six 100-cm³ samples were taken from each plot, mixed, and stored in polyethylene bags at 4°C. The N_{min} contents (NH_4^+, NO_2^-, NO_3^-) were measured by extracting the soil and analyzing the extract with the automatic analyzer mentioned above. Net nitrogen mineralization was determined after Gerlach[9] by incubating soil samples in polyethylene bags at the sites for 3 weeks. After incubation, the N_{min} content was measured again. The difference in N_{min} content before and after incubation yields the quantity of N net mineralization.

The N content of the microbial biomass was determined as N flush.[10] The N uptake by the plants was estimated from the N content of the vegetation biomass which was measured with the Kjeltec® analyzer system.

The described methods are discussed in detail by Schulz-Berendt.[11]

B. Results

The Aldicarb-induced changes in the nitrogen N balance are summarized in Table 1. Figure 2 shows the alterations to the nitrate content of the soil solution and to the leaching water during the first 54 weeks of succession. No standard